Pd-based Membranes

Pd-based Membranes

Overview and Perspectives

Special Issue Editors

Thijs Peters

Alessio Caravella

MDPI • Basel • Beijing • Wuhan • Barcelona • Belgrade

Special Issue Editors
Thijs Peters
Sustainable Energy Technology
SINTEF Industry
Norway

Alessio Caravella
University of Calabria (UNICAL)
Italy

Editorial Office
MDPI
St. Alban-Anlage 66
4052 Basel, Switzerland

This is a reprint of articles from the Special Issue published online in the open access journal *Membranes* (ISSN 2077-0375) from 2018 to 2019 (available at: https://www.mdpi.com/journal/membranes/special_issues/Pd_based_membranes)

For citation purposes, cite each article independently as indicated on the article page online and as indicated below:

LastName, A.A.; LastName, B.B.; LastName, C.C. Article Title. *Journal Name* **Year**, *Article Number*, Page Range.

ISBN 978-3-03897-702-5 (Pbk)
ISBN 978-3-03897-703-2 (PDF)

Contents

About the Special Issue Editors

Thijs Peters is a senior research scientist and project manager at the Department of Sustainable Energy Technology at SINTEF Industry in Oslo. He graduated with a Ph.D. from the Technical University of Eindhoven in 2005 within the field of process development and started at SINTEF shortly thereafter. His research interests span from process chemistry, membrane technology, hydrogen production, CO2 capture, and energy efficiency, to gas separation technologies. He has 54 publications in international peer-reviewed journals (h-index = 21) and has contributed to 10 book chapters and >150 conferences.

Alessio Caravella is currently Associate Professor chairing the courses of Thermodynamics and Chemical Reactors Design at the Dept. of Environmental and Chemical Engineering (DIATIC) of the University of Calabria (Italy), as well as Associate Researcher at the Institute on Membrane Technology (ITM-CNR, Italy). His main research activities focus on the experimental analysis and simulation of hierarchical catalytic structures and membrane-assisted reaction and separation devices, with a particular regard to the application of Pd-based membranes devices to hydrogen purification. Recently, he is also investigating the adsorption properties of zeolites and coordination polymer particles for hydrogen separation from light gases.

Preface to "Pd-based Membranes"

This Special Issue of *Membranes* focuses on several new aspects of Pd-based membranes for hydrogen separation and their main applications. The papers composing it should be read as single chapters of a whole story, starting from membrane fabrication up to the main applications, passing through materials investigations and technology development. The next section provides an overview of the highlights of each "chapter" reported in a bottom-up order starting with Pd membrane manufacturing. We hope that readers enjoy this Special Issue and gain inspiration from it for their own work. In the end, technological changes are the fruit of ideas planted as seeds in researchers' minds: the more individual minds are connected to each other, the higher the probability of creating originality. In this sense, this Special Issue represents a small attempt to increase the connectivity among interested minds, as we provide our contributions to technological innovation.

Thijs Peters, Alessio Caravella
Special Issue Editors

membranes

MDPI

Editorial

Pd-Based Membranes: Overview and Perspectives

Thijs Peters [1,*] and Alessio Caravella [2]

1 SINTEF Industry, 0314 Oslo, Norway

2 Department of Environmental and Chemical Engineering (DIATIC), University of Calabria, Via P. Bucci, Cubo 44A, 87036 Rende (CS), Italy; alessio.caravella@unical.it

* Correspondence: thijs.peters@sintef.no; Tel.: +47-982-43-941

Received: 30 January 2019; Accepted: 30 January 2019; Published: 1 February 2019

Abstract: Palladium (Pd)-based membranes have received a lot of attention from both academia and industry thanks to their ability to selectively separate hydrogen from gas streams. Integration of such membranes with appropriate catalysts in membrane reactors allows for hydrogen production with CO_2 capture that can be applied in smaller bioenergy or combined heat and power (CHP) plants, as well as in large-scale power plants. Pd-based membranes are, therefore, regarded as a Key Enabling Technology (KET) to facilitate the transition towards a knowledge-based, low carbon and resource-efficient economy. This Special Issue of the journal *Membranes* on "Pd-based Membranes: Overview and Perspectives" contains nine peer-reviewed articles. Topics include manufacturing techniques, understanding of material phenomena, module and reactor design, novel applications, and demonstration efforts and industrial exploitation.

Keywords: palladium-based membrane; hydrogen; manufacturing; demonstration

1. Introduction

This Special Issue of *Membranes* focuses on several new aspects of Pd-based membranes for hydrogen separation and their main applications. The papers comprising the Special Issue should be read as single chapters of a whole story, starting from membrane fabrication, passing through materials investigations and technology development, and concluding with the main applications. The next section provides an overview of highlights from each "chapter" reported in order, starting with Pd-membrane manufacturing.

2. Highlights of This Special Issue

2.1. Pd-Membrane Manufacturing

Thin state-of-the-art Pd-based membranes in general are being constituted of a thin selective Pd or Pd-alloy layer applied onto a porous support providing for mechanical strength. Dense metal membranes display extremely high levels of selectivity, however, a lower thickness limit seemingly exists for which a dense layer can be obtained [1]. This thickness limit increases with surface roughness and pore size in the support's top layer [2]. Clearly, this relationship makes strong demands on the pore size distribution of the supports [3,4]. It is, therefore, usual to carry out some form of pre-treatment or surface modification of the support to improve the final quality of the membrane. An overview of such strategies, and available commercial support materials, is accurately described by Alique et al. in this Special Issue [4]. Currently, several techniques for the determination of the pore size distribution of porous supports exist, but there is still a lack of efficient methods for determination of the size of the pore opening [5], as well as of the defect distribution of dense Pd-based composite membranes. In this Special issue of *Membranes*, a novel "modified" liquid–liquid displacement porometry (MLLDP) to quantify the pore opening size is introduced by Zheng et al. [5]. This method can operate under reasonably low pressures for a wide spectrum of pore sizes, due to the relatively lower liquid–liquid interfacial tensions compared to gas–liquid surface tensions. In addition, the "entraining phenomenon" can be eliminated in the MLLDP method. As with other techniques, the accuracy of MLLDP, however, decreases with increasing pore size. In the article, the applicability of the MLLDP method for the measurement of the defect size distribution for Pd composite membranes is also shown. This may be an interesting technique to assess pinhole formation during long-term operation.

Several technologies can be used to apply a thin layer of Pd or Pd-based alloys, onto a porous support. For the most commonly applied technique, electroless plating, a thorough review of recent developments is presented by Alique et al. [4]. Strategies to improve the deposition itself, e.g., to increase the film homogeneity and to reduce the carbon deposits, are described in this article. Moreover, membrane repair and protection strategies are introduced. In addition, Wunsch et al. present an alternative cost-effective technique for metal-layer deposition by suspension plasma spraying, using supports composed of porous sinter metallic supports made of Crofer-22 APU deposited with a YSZ diffusion-barrier layer [6]. Advantages lie in the short time requirements for deposition, substrates do not have to be activated as in the case of electroless plating, and no metal-loaded liquid waste is produced. Initial results were not satisfactory, however, since deposited layers had remaining open porosity, but work is ongoing [6].

2.2. Material Investigations

For metal membranes to be massively integrated into industrial processes to separate hydrogen from gas mixture, it is crucial to assess their performance and stability under actual operating conditions. In particular, as these types of membranes are thought to operate at relatively high temperatures (300–600 °C) and high pressure (2–50 bar and even more), while also in transient conditions in which they can be subjected to rapid changes, it is essential to understand the behavior of the materials they are made of under such conditions.

In this Special Issue, three papers regarding novel and non-conventional studies of materials are offered. In particular, Vicinanza et al. [7] studied the heat treatment of Pd-based membranes, separating the single contributions of both membrane surfaces (i.e., on the feed and permeate side). In their work, they consider three different membrane thicknesses, from whose analysis the effects of adsorption and desorption are disentangled, quantifying the surface phenomena influence, and also in terms of the apparent Arrhenius parameters for permeation before and after the heat treatment [7].

Complementary to the work of Vicinanza et al. [7], Løvvik et al. [8] carried out an interesting and specific work on the influence of grain boundary segregation of bulk in Pd-Ag-Cu membranes, an area to which the literature has generally paid relatively poor attention. Specifically, this study is based on first-principles electronic structure calculations performed on realistic atomic-scale models of binary Pd-Cu and ternary Pd-Cu-Ag alloys. In this way, a systematic approach to designing metal alloys is introduced, which opens up the possibility of more precisely predicting the behaviour of metal lattices a priori, thus reducing the number of experimental tests required and the costs related to the development of new membrane alloy materials.

In the third paper in this section, Bellini et al. [9] provide an original review on thermodynamic aspects related to hydrogen-metal systems in non-ideal conditions (i.e., pressure-dependent diffusivity and solubility [10–13]). Analysing information drawn from several studies in the open literature they show a systematic thermodynamic approach based on the chemical potential of the Pd-H system to deal with modeling of hydrogen solubility in the lattice. In particular, an explicit expression for the activity of H atoms in the lattice is obtained, allowing membrane behavior to be modeled under conditions of interest for real industrial applications.

2.3. Module and Reactor Design

This Special Issue of *Membranes* would not be complete if it did not provide an insight into novel module and reactor design. Here, micro-membrane reactors that enhance heat management, reduce gas phase diffusion limitations, and increase the membrane area to reactor volume ratio compared to traditional tubular reactors are introduced [14,15]. The papers dealing with this topic are those of Wunsch et al., which report a number of aspects related to micro-membrane reactors, ranging from reactor configuration development, feasible and low-cost fabrication techniques of micro-membranes, and optimal coupling and integration of reactive and separating processes in single compact modules [6,16]. Specifically, Dittmeyer and colleagues at the Institute for Micro Process Engineering at the Karlsruhe Institute of Technology (KIT) have developed micro-membrane reactors composed of stacks of sub-modules including multiple reactive and permeative stages. Even though the idea of using staged membrane reactors is not new [17], the technological approach at KIT is to develop compact systems so as to minimize the drawbacks of larger-scale devices, such as heat removal, concentration polarization and a relatively low membrane surface area per catalyst volume (ca. 10^3–10^6 m^{-1}). The main applications of these micro devices, for now, are reforming of methane and dehydrogenation of liquid organic hydrogen carriers (LOHCs) [6,16], which we also refer to in Section 2.4.

2.4. *Applications of Pd-based Membranes*

The application of Pd-based membrane technology is currently mainly focused on producing ultrapure hydrogen from fossil sources. As an alternative to hydrogen purification through partial oxidation (PROX) and pressure swing adsorption (PSA), Pd-based membranes have received much attention in the last 30 years because they combine the reforming reaction for hydrogen generation and its separation/purification. The majority of these studies investigate Pd-based membrane reactors in process schemes involving reforming of methane. For example, a 40 Nm^3/h-class membrane reformer (MRF) system for H_2 production has been developed by Tokyo Gas, and its long-term durability and reliability have been demonstrated over 8000 h [18,19]. Along the same lines, Pd-based membrane integration is investigated in the fuel processor of distributed combined heating and power (CHP) plants employing fuel cell technology [20], leading to a drastic fuel processing plant simplification.

Many other alternative liquid hydrocarbons and oxygenates may as well be used for the production of hydrogen at a smaller scale in reforming or gasification processes, e.g., methanol, ethanol, glycerol or diesel, originating either from biomass or fossil sources. Among the various renewable fuels, methanol is an interesting hydrogen source because at room temperature it is liquid, and therefore easy to handle and to store. Furthermore, it shows a relatively high H/C ratio and low reforming temperature, ranging from 200 to 300 °C. In their contribution, Iulianelli et al., describe the progress in the last decades with respect to modeling studies on methanol steam reforming in membrane reactors [21].

Pd-based membrane technology is also considered in gas-to-liquid (GTL) processes and chemical synthesis, such as alkane dehydrogenation (DH) reactions. These applications, and their technical feasibility verified at the pilot level, are presented by Palo et al. [22]. The results achieved showed that membrane reactors can be effectively used in all the mentioned applications. It should be noted, however, that in most of the proposed solutions, the concept of the membrane reactor is based on the application of a sequence of reaction–separation–reaction units rather than on the application of a reactor in which the catalyst and membrane are present in the same process unit, according to the concept of process intensification. This is driven by particular cases where there is a mismatch between optimal operating conditions for catalyst and membrane operation, thereby also impacting on the operation and maintenance of the reaction system. One specific case of this is, for example, the dehydrogenation of propane, where coke formation on the surface of Pd-based membranes prevents continuous operation of an integrated membrane DH process at the conventionally applied temperatures required for the catalysts, i.e., 500–600 °C [23].

Pd-based membranes have also been applied in hydrogen and chemical heat storage systems, in small and medium scale, from renewable energy. Conventional storage solutions store hydrogen physically, by compression or liquefaction, to increase the low volumetric energy density compared to atmospheric conditions. Liquid organic hydrogen carriers (LOHCs), on the other hand, propagate the reversible chemical binding of hydrogen to an organic liquid. The LOHC can store 6.2 wt% hydrogen when fully loaded, which corresponds to a storage density of 17.5 L_{LOHC}/kg_{H2}. In general though, dehydrogenation is technically more difficult to implement than hydrogenation. As a solution, Pd-based membrane reactors have already been applied to facilitate the dehydrogenation of LOHC methylcyclohexane with promising results. For example, Wunsch presents an intensified LOHC-dehydrogenation, applying a multi-stage microreactor and Pd-Based membrane process design, in this Special Issue of *Membranes* [6,16]. Simulations were carried out, which showed that the described approach can drastically intensify the whole dehydrogenation process, in addition to the in situ purification of the hydrogen.

3. Final Remarks

Overall, the editors are convinced that metal membranes, and Pd-based membranes in particular, have a lot more to contribute than what has already been demonstrated world-wide. We hope that readers enjoy this Special Issue and gain inspiration from it for their own work. In the end, technological changes are the fruit of ideas planted as seeds in researchers' minds: the more that individual minds are connected to each other, the higher the probability of creating originality. In this sense, this Special Issue represents a small attempt to increase the connectivity among interested minds, as we provide our contributions to technological innovation.

Funding: The support from the European Union and the Research Council of Norway (RCN) through the RCN-CLIMIT (Project No: 215666) program, and the FCH-JU AutoRE project (Contract no.: 671396) is gratefully acknowledged. A. Caravella has received funding through the "Programma Per Giovani Ricercatori *Rita Levi Montalcini*" granted by the Italian "Ministero dell'Istruzione, dell'Università e della Ricerca, MIUR" (Grant No: PGR12BV33A), which is gratefully acknowledged.

Acknowledgments: The editors acknowledge all the contributors to this Special Issue and thank them for generously taking the time and effort to prepare a manuscript.

Conflicts of Interest: The authors declare no conflicts of interest. The funders had no role in the design of the study; in the collection, analyses, or interpretation of data; in the writing of the manuscript, and in the decision to publish the results.

References

1. Bredesen, R.; Jordal, K.; Bolland, A. High-temperature membranes in power generation with CO_2 capture. *Chem. Eng. Process.* **2004**, *43*, 1129–1158. [CrossRef]
2. Mardilovich, I.P.; Engwall, E.; Ma, Y.H. Dependence of hydrogen flux on the pore size and plating surface topology of asymmetric Pd-porous stainless steel membranes. *Desalination* **2002**, *144*, 85–89. [CrossRef]
3. Sun, G.B.; Hidajat, K.; Kawi, S. Ultra thin Pd membrane on alpha-Al_2O_3 hollow fiber by electroless plating: High permeance and selectivity. *J. Membr. Sci.* **2006**, *284*, 110–119. [CrossRef]
4. Alique, D.; Martinez-Diaz, D.; Sanz, R.; Calles, A.J. Review of Supported Pd-Based Membranes Preparation by Electroless Plating for Ultra-Pure Hydrogen Production. *Membranes* **2018**, *8*, 5. [CrossRef]
5. Zheng, L.; Li, H.; Yu, H.; Kang, G.; Xu, T.; Yu, J.; Li, X.; Xu, H. "Modified" Liquid-Liquid Displacement Porometry and Its Applications in Pd-Based Composite Membranes. *Membranes* **2018**, *8*, 29. [CrossRef] [PubMed]
6. Wunsch, A.; Kant, P.; Mohr, M.; Haas-Santo, K.; Pfeifer, P.; Dittmeyer, R. Recent Developments in Compact Membrane Reactors with Hydrogen Separation. *Membranes* **2018**, *8*, 107. [CrossRef]
7. Vicinanza, N.; Svenum, I.H.; Peters, T.; Bredesen, R.; Venvik, H. New Insight to the Effects of Heat Treatment in Air on the Permeation Properties of Thin Pd77%Ag23% Membranes. *Membranes* **2018**, *8*, 92. [CrossRef]
8. Løvvik, O.M.; Zhao, D.; Li, Y.; Bredesen, R.; Peters, T. Grain Boundary Segregation in Pd-Cu-Ag Alloys for High Permeability Hydrogen Separation Membranes. *Membranes* **2018**, *8*, 81. [CrossRef]
9. Bellini, S.; Sun, Y.; Gallucci, F.; Caravella, A. Thermodynamic Aspects in Non-Ideal Metal Membranes for Hydrogen Purification. *Membranes* **2018**, *8*, 82. [CrossRef]

10. Peters, T.A.; Stange, M.; Bredesen, R. On the high pressure performance of thin supported Pd-23%Ag membranes—Evidence of ultrahigh hydrogen flux after air treatment. *J. Membr. Sci.* **2011**, *378*, 28–34. [CrossRef]
11. Hara, S.; Ishitsuka, M.; Suda, H.; Mukaida, M.; Haraya, K. Pressure-Dependent Hydrogen Permeability Extended for Metal Membranes Not Obeying the Square-Root Law. *J. Phys. Chem. B* **2009**, *113*, 9795–9801. [CrossRef] [PubMed]
12. Flanagan, T.B.; Wang, D. Exponents for the pressure dependence of hydrogen permeation through Pd and Pd-Ag alloy membranes. *J. Phys. Chem. C* **2010**, *114*, 14482–14488. [CrossRef]
13. Caravella, A.; Hara, S.; Drioli, E.; Barbieri, G. Sieverts law pressure exponent for hydrogen permeation through Pd-based membranes: Coupled influence of non-ideal diffusion and multicomponent external mass transfer. *Int. J. Hydrogen Energy* **2013**, *38*, 16229–16244. [CrossRef]
14. Dittmeyer, R.; Boeltken, T.; Piermartini, P.; Selinsek, M.; Loewert, M.; Dallmann, F.; Kreuder, H.; Cholewa, M.; Wunsch, A.; Belimov, M.; et al. Micro and micro membrane reactors for advanced applications in chemical energy conversion. *Curr. Opin. Chem. Eng.* **2017**, *17*, 108–125. [CrossRef]
15. Bredesen, R.; Peters, T.A.; Boeltken, T.; Dittmeyer, R. Pd-Based Membranes in Hydrogen Production for Fuel cells. In *Process Intensification for Sustainable Energy Conversion*; John Wiley & Sons: Hoboken, NJ, USA, 2015; p. 209.
16. Wunsch, A.; Mohr, M.; Pfeifer, P. Intensified LOHC-Dehydrogenation Using Multi-Stage Microstructures and Pd-Based Membranes. *Membranes* **2018**, *8*, 112. [CrossRef] [PubMed]
17. Caravella, A.; Di Maio, F.P.; Di Renzo, A. Optimization of membrane area and catalyst distribution in a permeative-stage membrane reactor for methane steam reforming. *J. Membr. Sci.* **2008**, *321*, 209–221. [CrossRef]
18. Yakabe, H.; Kurokawa, H.; Shirasaki, Y.; Yasuda, I. Operation of a palladium membrane reformer system for hydrogen production: The case of Tokyo Gas Palladium Membrane Technology for Hydrogen Production. In *Carbon Capture and Other Applications*; Woodhead Publishing: Cambridge, UK, 2015; pp. 303–318.
19. Kurokawa, H.; Yakabe, H.; Yasuda, I.; Peters, T.; Bredesen, R. Inhibition effect of CO on hydrogen permeability of Pd-Ag membrane applied in a microchannel module configuration. *Int. J. Hydrogen Energy* **2014**, *39*, 17201–17209. [CrossRef]
20. Loreti, G.; Facci, A.L.; Peters, T.; Ubertini, S. Numerical modeling of an automotive derivative polymer electrolyte membrane fuel cell cogeneration system with selective membranes. *Int. J. Hydrogen Energy* **2018**, in press. [CrossRef]
21. Iulianelli, A.; Ghasemzadeh, K.; Basile, A. Progress in Methanol Steam Reforming Modelling via Membrane Reactors Technology. *Membranes* **2018**, *8*, 65. [CrossRef]
22. Palo, E.; Salladini, A.; Morico, B.; Palma, V.; Ricca, A.; Iaquaniello, G. Application of Pd-Based Membrane Reactors: An Industrial Perspective. *Membranes* **2018**, *8*, 101. [CrossRef]
23. Peters, T.A.; Liron, O.; Tschentscher, R.; Sheintuch, M.; Bredesen, R. Investigation of Pd-based membranes in propane dehydrogenation (PDH) processes. *Chem. Eng. J.* **2016**, *305*, 191–200. [CrossRef]

membranes

MDPI

Article

Intensified LOHC-Dehydrogenation Using Multi-Stage Microstructures and Pd-Based Membranes

Alexander Wunsch, Marijan Mohr and Peter Pfeifer *

Institute for Micro Process Engineering, Karlsruhe Institute for Technology, 76344 Eggenstein-Leopoldshafen, Germany; alexander.wunsch@kit.edu (A.W.); Marijan.Mohr@gmx.de (M.M.)

* Correspondence: peter.pfeifer@kit.edu; Tel.: +49-721-608-24767

Received: 28 September 2018; Accepted: 14 November 2018; Published: 19 November 2018

Abstract: Liquid organic hydrogen carriers (LOHC) are able to store hydrogen stably and safely in liquid form. The carrier can be loaded or unloaded with hydrogen via catalytic reactions. However, the release reaction brings certain challenges. In addition to an enormous heat requirement, the released hydrogen is contaminated by traces of evaporated LOHC and by-products. Micro process engineering offers a promising approach to meet these challenges. In this paper, a micro-structured multi-stage reactor concept with an intermediate separation of hydrogen is presented for the application of perhydro-dibenzyltoluene dehydrogenation. Each reactor stage consists of a micro-structured radial flow reactor designed for multi-phase flow of LOHC and released hydrogen. The hydrogen is separated from the reactors' gas phase effluent via PdAg-membranes, which are integrated into a micro-structured environment. Separate experiments were carried out to describe the kinetics of the reaction and the separation ability of the membrane. A model was developed, which was fed with these data to demonstrate the influence of intermediate separation on the efficiency of LOHC dehydrogenation.

Keywords: LOHC; dehydrogenation; multi-stage; PdAg-membrane; micro reactor; hydrogen purification

---

1. Introduction

In the context of the energy transition, various technologies are investigated to store fluctuating renewable energy over periods of varying lengths. In contrast to batteries, electrical energy can be converted into chemical energy in the form of hydrogen by electrolysis. This hydrogen may serve as a particularly clean and climate-neutral energy source for mobility and stationary applications in the future. With approximately 33 kWh/kg, hydrogen has the highest mass related energy density of all fuels. However, storage is difficult because the substance is a gas under atmospheric conditions and has a low volumetric energy density. Common solutions such as compressed hydrogen at 350 or 700 bar (0.8 or 1.3 kWh_{th}/L_{H2}) and liquid hydrogen (2.4 kWh_{th}/L_{H2}) only provide partial benefits due to the high-risk potential and the difficult handling. An alternative technology is the storage of hydrogen in a so-called Liquid Organic Hydrogen Carrier (LOHC). The LOHC can be loaded with hydrogen (LOHC+) or unloaded (LOHC-) by reversible hydrogenation. The LOHC serves as a "deposit bottle" [1–4]. The continuous further development of the hydrogen network calls for a technology that pursues a decentralized approach. For example, micro-structured dehydrogenators with palladium membranes can be used in hydrogen filling stations, trains, or tankers to provide high-quality hydrogen from compact dehydrogenation units.

The perhydro-dibenzyltoluene (18H-DBT, LOHC+)/dibenzyltoluene (0H-DBT, LOHC-) system proves to be a promising LOHC since the aromatic compound can absorb up to nine molecules of hydrogen [5]. DBT is commercially used as a heat transfer fluid (e.g., Marlortherm SH from

Sasol) with a great availability. In addition to a storage density of 57 $g_{H2}/L_{18H\text{-}DBT}$ (equivalent to 1.9kWh$_{H2}/L_{18H\text{-}DBT}$), the isomeric mixture is easy to store under atmospheric conditions due to its high stability. Furthermore, the material system is flame-resistant, neither explosive nor toxic, and is, therefore, not considered a hazardous good. This results in attractive opportunities to use the existing infrastructure via tank trucks for the distribution of the LOHC [6–8].

The loading/unloading of the LOHC can only take place under certain conditions (pressure, temperature) and in the presence of a catalyst. Thermodynamically favorable conditions for the release are temperatures between 280–330 °C and pressures below 5 bar. Under these conditions, the LOHC is dehydrogenated in the liquid state. During the reaction, a multiphase flow forms due to the released hydrogen. Due to the stoichiometry of 9 $mol_{H2}/mol_{18H\text{-}DBT}$, an enormous gas quantity forms even at low conversion rates in comparison to the liquid amount. As a result, part of the LOHC evaporates due to its non-negligible vapor pressure and is detectable in traces in the product gas despite subsequent condensation. Furthermore, low-boiling by-products are also found in the product gas. In order to supply hydrogen in high purity, e.g., for the operation of fuel cells, a purification step is necessary. Good heat management is also crucial for high catalyst and reactor related hydrogen productivity. Approximately 71 kJ/mol_{H2} heat is required for the release, which leads to a huge heat demand when the carrier is completely dehydrogenated (639 kJ/$mol_{18H\text{-}DBT}$) [5].

Micro-process engineering offers two promising approaches—the use of a micro-structured reactor allows an almost isothermal reaction environment, which prevents "cold spots" by limited heat input. Furthermore, it has already been shown that Pd-based membranes benefit from a micro-structured environment to separate hydrogen with high efficiency and purity from a gas mixture [9–16]. The combination of membrane and reaction in microreactors has been demonstrated as highly beneficial over conventional systems in these studies. Nevertheless, in the current reaction system, a fundamental difference exists, which is related to two major obstacles in the combination of membrane and micro-reactor. This includes the occurrence of liquid, which can prevent reasonable hydrogen permeation through the membrane, and the difficult phase contact between the liquid and the catalyst. A combination of the two process steps may be difficult but not inconceivable in previously proposed micro-reactor arrangements. In our opinion, the easiest approach to reasonable process intensification may be possible by a multi-stage dehydrogenation process with intermediate separation of hydrogen by Pd-based membranes (see Figure 1) with also an intelligent microchannel design fostering phase contacting presented in our current study. Overall, the multistage concept is attractive in two aspects. The separation removes most of the gas for the consecutive reactor stage in addition to its original function, which is the purification of hydrogen. Hydrogen removal minimizes the negative influence of the gas phase on the residence time of the liquid in the reactor part. Typical solutions to foster the conversion like an increase of the reaction temperature and higher residence times can cause catalyst deactivation by coke formation on the catalysts' surface area. With the multi-stage approach, it may be possible to avoid the deactivation since, with an increasing number of reactor stages, less LOHC per stage must be converted and dry-out of the catalyst surface is less pronounced. A temperature stepping of the various reactors may further overcome the kinetic inhibition by increased formation of unloaded LOHC. However, the separation only works under a concentration gradient of hydrogen between the retentate and permeate side and, thus, the overall system pressure must be increased compared to the conventional process. This leads to decreasing equilibrium conversion.

An integrated system, not explored in this study, represents a virtually infinite number of separation stages, which, at first sight, is much more advantageous. However, the previously mentioned obstacle that the LOHC could at least partially wet the membrane surface area may lead to a drop in the separation efficiency. Furthermore, it is also unknown how strongly gas and liquid interact with each other in the microreactor and how much the residence times differ. Therefore, the multi-stage concept is ideally suited for investigating these phenomena in more detail.

According to our calculations with already applied membrane modules [16], considering the relatively low concentrations of other components in the gas phase of a separation at low reaction pressures (4–5 bar (a)) is feasible and the disadvantage of the shifted reaction equilibrium is, therefore, acceptable. In this work, the micro-structured multi-stage reactor concept with intermediate hydrogen separation (schematically shown in Figure 1) is modelled based on experimental data of the reactor and membrane separator.

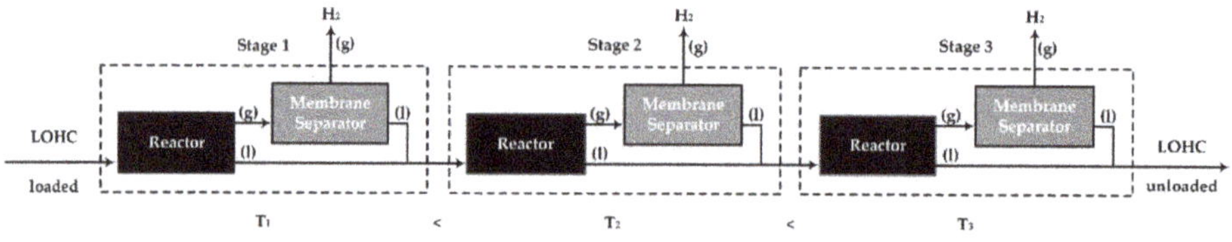

Figure 1. Schematic of the multi-stage approach with intermediate separation of hydrogen for the dehydrogenation of liquid organic hydrogen carriers. Most of the produced hydrogen is removed from the system after each reactor stage.

2. Materials and Methods

2.1. Radial Flow Reactor

A micro-structured radial flow reactor was developed for the challenging multiphase reaction. A CAD scheme of the microstructure can be seen in Figure 2a. The reactants enter at the center of the reaction chamber, flow out radially, and are collected in a ring. The microstructure is divided into eight equally sized areas. The separating fins possess a curvature to avoid preferred fluid movement. Within each area, hexagonal arranged pins with a distance of 1.2 mm to each other are arranged for better heat flux and catalyst bed stabilization. In the cavity of this structure, catalyst particles of a size 200 to 300 μm are distributed compactly to form a catalyst bed. Along the reactor length, in this case the radius, the flow cross-section is continuously increasing. The reason for this is a reduction in residence time due to the extreme formation of gas, which can be partly or completely compensated by this shaping depending on the degree of dehydrogenation (DoDH). Figure 2b shows the reactor with an integrated microstructure. The loaded LOHC is fed from below and then enters the microstructure. The radially removed product mixture is collected and transported to one outlet on the side. A steel plate above the microstructure can be replaced by a window, which makes an optical inspection of the processes in the catalyst bed possible. The reactor is electrically heated via heating cartridges and sealed between the individual components by flat graphite gaskets.

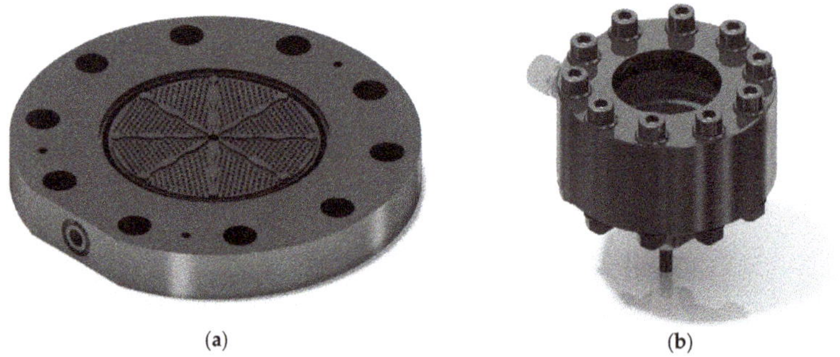

(**a**) (**b**)

Figure 2. CAD schematic of the radial flow reactor: (**a**) The microstructure in detail including the space between the pins, which is filled with catalyst particles. The reactants are entering in the center and the reaction mixture flows from the inside to the outside and is removed in the form of a ring. (**b**) View of the reactor. A look inside is possible by using a glass plate pressed onto the structure.

In the reactor model, it is assumed that the flow is approximately similar to that of a Plug Flow Reactor (PFR). Thus, a 1-dimensional problem, referred to the radius as coordinate, is assumed. A Tanks-in-Series model, which represents fluid-connected cylindrical rings as segments of the reactor along the radius coordinate with at least 50 tanks, is applied for the description of the PFR. A program flow chart can be seen in Figure 3. After defining the input parameters and setting the start values, the program calculates the CSTR (Continuous stirred tank reactor) cascade. In each tank (CSTR), the molar flows of the gas phase $\dot{N}_{H2}$ and liquid phase $\dot{N}^L$ as well as the molar fractions x_i are determined. In a mixer, gas and liquid phase are summed up to molar fractions of z_i and a flashbox is applied to calculate a new gas-liquid distribution of all species for the next tank based on an ideal gas law or with the modified UNIFAC model. Subsequently, it is checked whether the liquid phase is already saturated with Hydrogen. If not especially at the inlet of the reactor, the generated hydrogen is assumed to absorb until the saturation is exceeded and the free gas phase is formed. Physisorption of hydrogen in the LOHC was taken into account by Henry's law. In addition to the reactor geometry, necessary substance data were also implemented [17–20]. The modelling was performed with the software Matlab®and the equations were solved with the solver "fsolve".

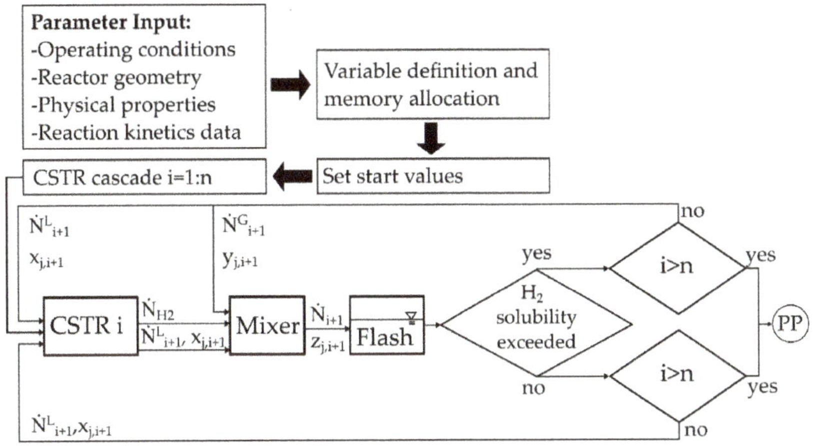

Figure 3. Program flow chart of the model for the radial flow reactor. In this Tanks-in-series model, the phase equilibrium and physical absorption is calculated at every single tank.

The formation of hydrogen takes place under consecutive forming of double bonds at the catalyst surface. Due to the unfavorable energetic conditions of many possible intermediates, only the hydrogenation stages with fully hydrogenated/dehydrogenated C6-rings can experimentally be determined as species (12H-DBT, 6H-DBT) [21]. In the DBT reaction network, there are, thus, four species considered, each with a large number of isomers. Due to the complexity of the reaction kinetics, adsorption and desorption phenomena were neglected. Since a catalyst in egg-shell configuration is used, only an external mass transport limitation can exist. Furthermore, due to the limited availability of isomer data, it was further assumed that the isomers of the respective LOHC species are chemically and physically identical. This results in three equilibrium reactions shown below.

$$18H - DBT \underset{k_4}{\overset{k_1}{\rightleftarrows}} 12H - DBT + 3H_2 \tag{1}$$

$$12H - DBT \underset{k_5}{\overset{k_2}{\rightleftarrows}} 6H - DBT + 3H_2 \tag{2}$$

$$6H - DBT \underset{k_6}{\overset{k_3}{\rightleftarrows}} 0H - DBT + 3H_2 \tag{3}$$

The experiments were carried out under conditions where hydrogenation can be neglected compared to dehydrogenation, i.e., at low conversion and far away from thermodynamic equilibrium.

$$k_1 \gg k_4, \;\; k_2 \gg k_5, \;\; k_3 \gg k_6 \tag{4}$$

For the experiments, an almost complete hydrogenated perhydro-dibenzyltoluene (hydrogenation degree 96.1%, provided by Hydrogenious Technologies GmbH) was used as starting material. This means that there is nearly exclusively 18H-DBT in the feedstock, which is why only the first reaction was considered in the experiments (see Equation (1)). It was also taken care in the experiments that 6H-DBT and 0H-DBT were below the quantification limit. Nevertheless, the determined reaction constant k_1 was used as the reaction constant of the other reactions in the model, i.e., the further simulation.

$$k_1 = k_2 = k_3 = k \tag{5}$$

To keep the conversion low, the catalyst bed was diluted with inert material to 25 wt%.

In summary of the kinetic assumptions including further a reaction order of $n = 1$, the following simple rate expression results for all reaction steps.

$$r_i = k \cdot c_i^n \text{ with } c_i = \frac{x_{i,out} \cdot \rho_{LOHC,mix}^{liquid}}{\widetilde{M}_{LOHC,mix}} \tag{6}$$

The reaction constant is described by the modified Arrhenius equation by applying a reference temperature.

$$k = k_{T,ref} \cdot \exp\left(\frac{-E_{A,R}}{R} \cdot \left(\frac{1}{T} - \frac{1}{T_{ref}}\right)\right) \tag{7}$$

A total number of six reaction temperatures were experimentally investigated with each having four different residence times, so that a total of 24 measured data points were generated. The degree of dehydrogenation was determined by NMR spectroscopy [21]. A micro-ring gear pump was used to adjust the modified mass-related residence time, according to the following equation by applying the initial LOHC feed.

$$\tau_{mod} = \frac{m_{cat}}{\dot{m}_{LOHC}} \tag{8}$$

2.2. Membrane Apparatus

A membrane module with 17 microchannels (width × depth × length: 500 µm × 300 µm × 4 cm), which was described in detail in References [9,16], was used as separation device (see Figure 4). The used PdAg membrane was produced by SINTEF/Oslo via magnetron sputtering and has a thickness of approximately 10 µm. The membranes are produced by deposition on the perfect surface of 6 inch silicon wafers from an alloyed Pd77Ag23 target. Subsequently, the film was removed from the wafer, which allowed for the integration in a module [22].

In general, PdAg is more suitable for the LOHC application than pure palladium due to lower operating temperatures (300–350 °C) and higher hydrogen permeation flux, which fit to the dehydrogenation operating conditions. Microsieves from both sides were used to mechanically stabilize the membrane. Based on the number of holes in the sieve, an effective membrane area of 1.5 cm^2 was calculated. Good stability of these membranes has been reported several times in various studies. However, long-term operation of these membranes under the load with DBT has not been experimentally verified yet. This part will follow in further studies.

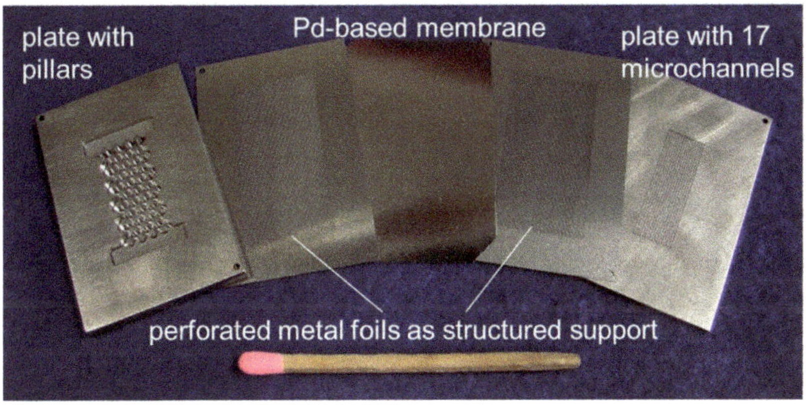

Figure 4. Picture of the used microstructured membrane separation module [16]. Due to low operating temperatures (300–350 °C), a PdAg membrane provided by SINTEF/Oslo was selected.

The separation of hydrogen via Pd-based membranes is a multi-step mechanism. If the limiting step is bulk diffusion in the Pd lattice, the flux can be described by combining Fick's and Sieverts' law with a square root dependence on the hydrogen pressure.

$$\dot{F}_{H_2} = \Pi \cdot \left(p_{H_2,Ret}^{0.5} - p_{H_2,Perm}^{0.5}\right) \tag{9}$$

Thus, the Flux is dependent on the permeance and the square root of the partial pressure gradient of hydrogen between the permeate side and the retentate side. The permeance Π is defined as the quotient of temperature-dependent and material specific permeability Q and membrane thickness s. The temperature-dependency follows the Arrhenius expression below.

$$\Pi = \frac{Q}{s} = \frac{Q_0 \cdot \exp\left(-\frac{E_{A,M}}{RT}\right)}{s} \tag{10}$$

The model reported in Reference [9] was used for fitting the experimental data and was used to determine the activation energy and the pre-exponential factor. Experiments were conducted at ambient hydrogen pressure on the permeate side (no sweep gas). To determine the permeance of the PdAg-membrane, a temperature and pressure variation at a constant hydrogen flow of 250 mL/min was carried out.

2.3. Multi-Stage Reactor Concept with Intermediate Hydrogen Separation

For describing the sequence of devices, both mathematical models were linked according to the connected fluid streams. Figure 1 shows an example of a three-stage process. The gas and liquid phases are separated at the reactor outlet. The liquid product flow of stage n is fed directly into the following reactor stage $n + 1$. The gas phase, on the other hand, only enters the membrane separation module where it is separated from LOHC species in the gas phase. As a consequence, LOHC in the gas phase condenses on the membrane during the separation and the associated partial pressure reduction of the hydrogen. The condensed flow is also fed to the next reactor stage. To consider condensation for volume contraction and partial pressure change, the membrane was divided into n sections and a flash calculation (as described for the reactor simulation—ideal or by modified UNIFAC model) was integrated, which is shown schematically in Figure 5.

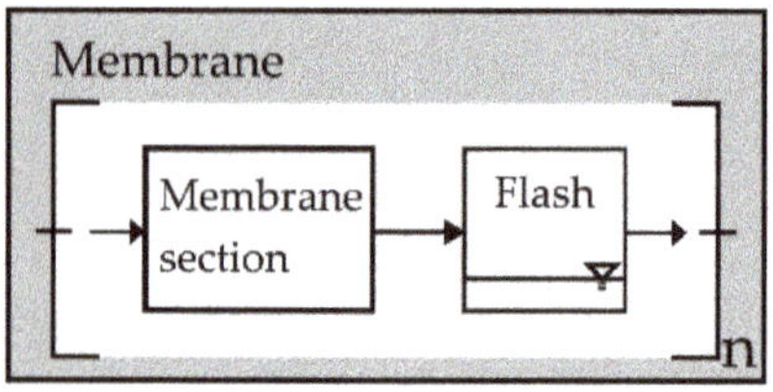

Figure 5. Modeling of the separation stage. The device was split into n parts to integrate a flash that takes into account the condensation of the LOHC.

3. Results and Discussion

In this section, the experimental results are first discussed and then the multi-staged reactor concept with intermediate separation of hydrogen is evaluated.

3.1. Determination of the Reaction Kinetics

Figure 6 shows the concentration curves as a function of the modified residence time. An almost linear dependence can be seen for all temperatures. This is in agreement with expectations and shows that the residence time of the liquid at low conversions is close to the hydrodynamic residence time. The solid line indicates the degree of dehydrogenation of the feed stream due to incompleteness of hydrogenation. Dehydrogenation degrees between 1% and 9% were achieved in the experiments, which are low enough to be analyzed in a differential approach.

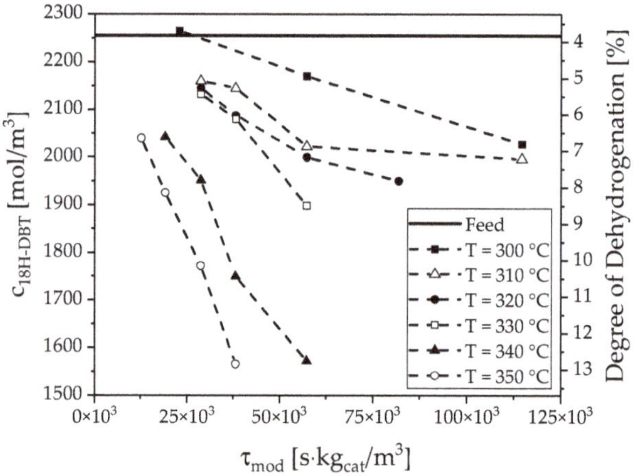

Figure 6. Concentration of 18H-DBT as a function of residence time at six different temperatures (300–350 °C) and 4 bar(a).

To determine the temperature dependency of the reaction, the logarithmic reaction rate was plotted as a function of temperature (see Figure 7). The linear fit describes the measuring points with satisfying quality, considering the complexity of the multiphase reaction system.

The following parameters could be determined from the fit.

$$E_{A,R} = 156.8 \pm 28.5 \ \text{kJ/mol} \tag{11}$$

$$k_{T,ref} = 2.637 \times 10^{-6} \pm 0.307 \times 10^{-6} \ \text{m}^3/(\text{kg}_{\text{Cat}}\text{s}) \tag{12}$$

The model parameter can be compared to a study from the researchers at Erlangen [10]. They operated a batch reactor to determine the kinetics of H18-DBT on the Pt catalyst similar to ours. Their parameters were obtained from experiments with higher conversions. This allows determining a reaction order, which was calculated in the range of 2. However, this result is then already influenced by the back reactions as well as possible transfer-hydrogenations occurring between the species. Thus, their reported activation energy and the pre-exponential factor consequently differ. A value of $E_{A,R}$ of roughly 120 kJ/mol is calculated. Nevertheless, while higher conversions would be required to investigate the actual wetting of the catalyst (see Section 3.3), back-reactions and transfer-hydrogenations are influencing the results. More experiments will follow where a second or third stage entry concentration will be fed to the reactor to detail the kinetics further.

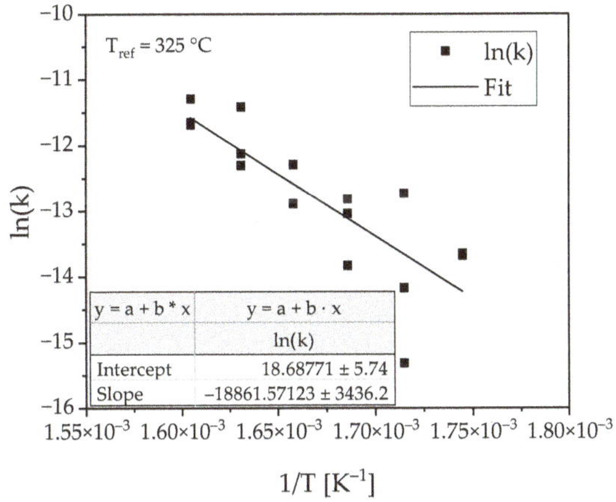

Figure 7. Arrhenius plot with average reaction constants and linear fit.

3.2. Membrane Characterization

The measured flux in the membrane device is plotted as a function of the difference between the square roots of the hydrogen partial pressures (see Figure 8). The measuring points at constant temperature follow the expected linear dependence. The Sieverts' law can, therefore, describe permeation. The slope of each trend line represents the permeance.

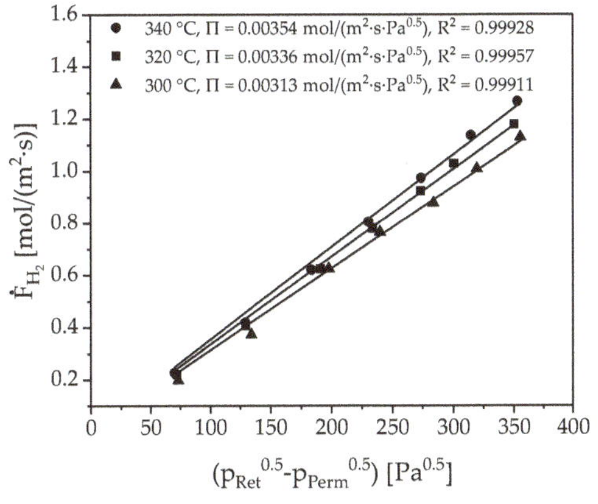

Figure 8. Sieverts plot of the permeation experiments. The permeance was determined for three temperatures by varying retentate pressure at an ambient hydrogen pressure at the permeate side.

If the logarithmus naturalis of permeance is plotted as a function of the reciprocal temperature, as seen in Figure 9, the activation energy and the pre-exponential factor can also be determined. The fit describes the measurement data with satisfying quality.

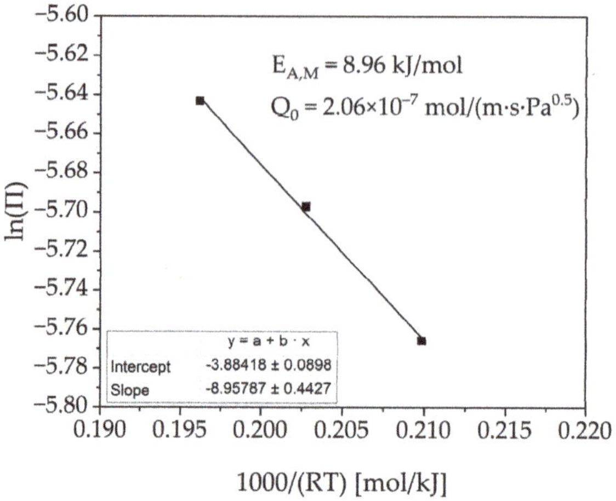

Figure 9. Fitted Arrhenius-relation of the measured permeance at three different temperatures.

The following experimentally determined values were obtained, which seem to align to our previous studies [9] including the literature review.

$$E_{A,M} = 8.96 \pm 0.44 \ \text{kJ/mol} \tag{13}$$

$$Q_0 = 2.06 \times 10^{-7} \pm 0.05 \times 10^{-7} \ \text{mol}/\left(\text{m} \cdot \text{s} \cdot \text{Pa}^{0.5}\right) \tag{14}$$

3.3. Evaluation of the Multi-Stage Reactor Approach with Intermediate Hydrogen Separation

Based on the experimentally determined correlations, the results of the simulations with the multi-stage reactor concept with intermediate separation of hydrogen are presented in the following. To describe the unknown influence of the gas phase on the residence time of the liquid and the wetting of the catalyst a functional correlation between the effectively wetted catalyst mass and the real residence time of the liquid phase was introduced via the correction factor α. For this correction factor, a function was chosen, which correlates α to the void fraction of the liquid phase and the quotient of liquid and gas residence time as an exponent of the void fraction.

$$m_{Cat}^{eff} = \alpha \cdot m_{Cat} \tag{15}$$

$$\alpha = \ln\left(1 + \varepsilon_L^b \cdot (e - 1)\right) \tag{16}$$

$$b = \frac{\tau_{gas}}{\tau_{liq}} \tag{17}$$

As conversion increases, the liquid phase volume fraction ε_L decreases and, finally, the effectively wetted catalyst mass that can serve the active surface for dehydrogenation of the liquid species decreases (see Figure 10). The chosen function allows us to describe the limiting cases. If gas and liquid velocity are identical, the void fraction and α have an almost linear dependence $\alpha \approx \varepsilon_L$ since $\tau_{gas} \approx \tau_{liq}$ or $b = 1$. According to our first observations in the packed bed microreactor system, it seems that part of the evolving gas phase can escape underneath the glass plate at much higher velocity and ε_L can be greater than the expected value, which means that $\tau_{gas} \rightarrow 0$ i.e., high superficial velocity of the gas. This results in a much better case of $\alpha = 1$. All intermediate conditions where $\tau_{gas} < \tau_{liq}$ can be controlled by using parameter b, the residence time proportion (IRTP) is shown in the figure below; calculation according Equation (17).

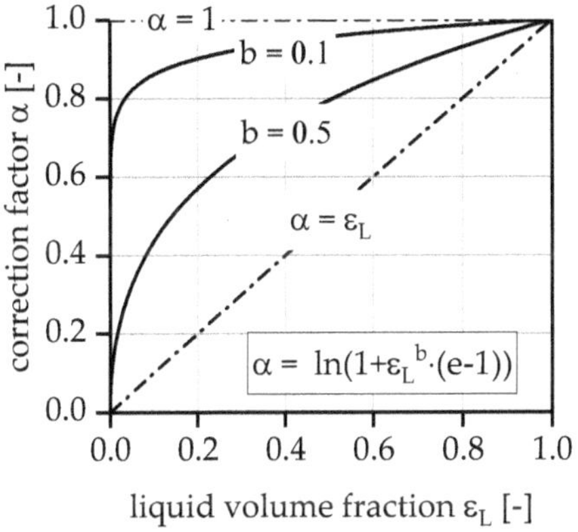

Figure 10. Constructed relationship between correction factor α and liquid void fraction ε_L and the residence time proportion b.

Based on these assumptions and the experimentally determined data, simulations were performed based on the geometry of the investigated lab scale single-foil reactor system. The results can be seen in Figure 11 for a fixed reaction temperature and feed flow. First, it was investigated how the number of separation stages affects the overall degree of dehydrogenation (DoDH) with varying distribution of gas and liquid phases at a constant total catalyst mass in the reactor arrangement (see Figure 11a). That means that a single reactor is compared to two, three, four, and five reactors with half, third, fourth, or fifth mass in each reactor, respectively. It can be seen that the DoDH can be increased slightly by the increasing number of intermediate separations (b = 0.1) if the residence time of the gas phase is short while good catalyst wetting is the consequence. Under these conditions, the membrane application is less valuable, i.e., does not provide reasonable advantage with regard to costs and system size. This could probably be improved by lowering the total catalyst mass i.e., when a lower total conversion is obtained. Nevertheless, the longer the gas remains or the less catalyst is wetted due to similar residence time of gas and liquid, the more the process can be intensified (b = 0.5) via the stepwise hydrogen separation. The DoDH can be practically doubled with a five-stage process. If the DoDH is monitored over this five-stage process (see Figure 11b), it becomes clear that the DoDH increases strongly independent of the residence time distribution between the gas and the liquid phase in both cases. With more efficient dehydrogenation (b = 0.1), however, the DoDH per stage (DoDH/stage) decreases after the first stage while it remains almost constant at b = 0.5. This can be explained by the inhibition of the reaction by the resulting completely dehydrated dibenzyltoluene.

One major issue that needs to be resolved in the overall model is the possible reaction of gaseous LOHC species on the dry region of the catalyst. This must be investigated in the future by separate gas phase reaction experiments and inclusion in the overall model. Nevertheless, we believe that the contribution will be small compared due to the low partial pressure of LOHC species in the gas phase and compared to the surface coverage with liquid species. Thus, process intensification by the multi-stage system will definitely remain dominant.

We further plan to extend the kinetics with the back reaction in the future. This is relevant when the overall conversion increases towards thermodynamic equilibrium due to transfer hydrogenation happening between the different hydrogenated intermediates of dibenzyltoluene. Under those conditions, the reaction rate can drop quite considerably. Such effect would further argue in the direction of even higher process intensification by the suggested multi-stage concept, i.e., when technical and economical relevant conversion is desired. Therefore, differential conversion with the first stage, which is a partially dehydrogenated product, are required to build up an even more complex kinetic model. Lastly, it is planned to perform optimization simulations to larger scale dehydrogenation systems based on the more detailed modeling.

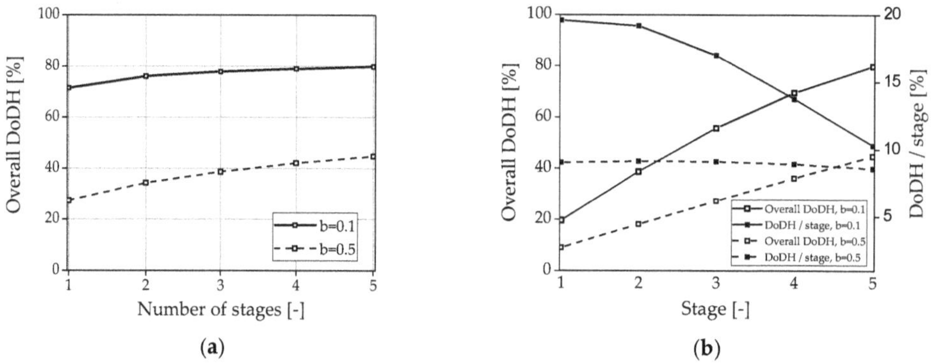

Figure 11. Simulation results (n_{CSTR} = 100) of the multi-staged approach with intermediate hydrogen separation carried out at 335 °C and 20 g/h feed: (**a**) overall DoDH with varying stage numbers at the same overall catalyst mass for two different parameters *b*. (**b**) Overall DoDH and DoDH per stage for a five-stage reactor concept.

4. Conclusions

In this work, a microstructured multi-stage reactor concept with intermediate separation of hydrogen for the purpose of dehydrogenation of perhydro-dibenzyltoluene was investigated. First, the kinetics for the reaction and the separation of the hydrogen were determined. The experimental data were used to feed a developed model describing the multi-stage approach. Simulations were carried out, which show that the described approach can drastically intensify the whole dehydrogenation process in addition to the purification of the hydrogen especially under conditions where gas has no superficial velocity. PdAg membranes are particularly suitable for use in this context due to their relatively high permeance at a low operation temperature (300–350 °C). Back reaction and gas phase reactions will be included in the model in future work to describe the promising effects toward process intensification of the intermediate hydrogen separation. Moreover, long-term testing of a lab model will further deliver data on stability, which will then be used to scale a plant for larger throughput and to perform life cycle analysis (costing and environmental suitability) of the proposed dehydrogenation system.

Author Contributions: Conceptualization, A.W., M.M. and P.P.; Methodology, A.W., M.M. and P.P.; Software, M.M.; Validation, A.W., M.M. and P.P.; Formal analysis, A.W., M.M. and P.P.; Investigation, A.W. and M.M.; Resources, A.W. and M.M.; Data curation, M.M.; Writing—original draft preparation, A.W.; Writing—review and editing, A.W. and P.P.; Visualization, A.W. and M.M.; Supervision, P.P.; Project administration, P.P.; Funding acquisition, P.P.

Funding: The authors gratefully acknowledge funding by the German Federal Ministry of Education and Research (BMBF) within the Kopernikus Project P2X: Flexible use of renewable resources—exploration, validation, and implementation of 'Power-to-X' concepts.

Acknowledgments: The author thanks Franziska Auer and Michael Geißelbrecht from FAU-CRT Erlangen for NMR measurements, vapor pressure data, and good cooperation within the Kopernikus Project RC-B1. The author also thanks SINTEF/Norway for fabricating and providing PdAg-membranes.

Conflicts of Interest: The authors declare no conflict of interest.

References

1. Teichmann, D.; Arlt, W.; Wasserscheid, P.; Freymann, R. A future energy supply based on Liquid Organic Hydrogen Carriers (LOHC). *Energy Environ. Sci.* **2011**, *4*, 2767–2773. [CrossRef]
2. Teichmann, D.; Arlt, W.; Wasserscheid, P. Liquid Organic Hydrogen Carriers as an efficient vector for the transport and storage of renewable energy. *Int. J. Hydrog. Energy* **2012**, *37*, 18118–18132. [CrossRef]
3. Preuster, P.; Papp, C.; Wasserscheid, P. Liquid Organic Hydrogen Carriers (LOHCs): Toward a Hydrogen-free Hydrogen Economy. *Acc. Chem. Res.* **2017**, *50*, 74–85. [CrossRef] [PubMed]
4. Preuster, P.; Alekseev, A.; Wasserscheid, P. Hydrogen storage technologies for future energy systems. *Annu. Rev. Chem. Biomol. Eng.* **2017**, *8*, 445–471. [CrossRef] [PubMed]
5. Bruckner, N.; Obesser, K.; Bosmann, A.; Teichmann, D.; Arlt, W.; Dungs, J.; Wasserscheid, P. Evaluation of industrially applied heat-transfer fluids as Liquid Organic Hydrogen Carrier systems. *ChemSusChem* **2014**, *7*, 229–235. [CrossRef] [PubMed]
6. Markiewicz, M.; Zhang, Y.Q.; Bosmann, A.; Bruckner, N.; Thoming, J.; Wasserscheid, P.; Stolte, S. Environmental and health impact assessment of Liquid Organic Hydrogen Carrier (LOHC) systems—Challenges and preliminary results. *Energy Environ. Sci.* **2015**, *8*, 1035–1045. [CrossRef]
7. IFA-Institut-für-Arbeitsschutz-der-Deutschen-Gesetzlichen-Unfallversicherung. *Material Safety Data Sheet (MSDS)—Dibenzyltoluene, Isomers: GESTIS-Stoffdatenbank*; IFA: Berlin, Germany, 2018.
8. Hydrogenious-Technologies-GmbH. *Material Safety Data Sheet (MSDS)—Verordnung (EG) Nr. 1907/2006 Version 1.1*; Hydrogenious-Technologies-GmbH: Erlangen, Germany, 2017.
9. Boeltken, T.; Belimov, M.; Pfeifer, P.; Peters, T.A.; Bredesen, R.; Dittmeyer, R. Fabrication and testing of a planar microstructured concept module with integrated palladium membranes. *Chem. Eng. Process.* **2013**, *67*, 136–147. [CrossRef]
10. Preuster, P. Entwicklung eines Reaktors zur Dehydrierung Chemischer Wasserstoffträger als Bestandteil eines Dezentralen, Stationären Energiespeichers Dr.-Ing. Ph.D. Thesis, Friedrich-Alexander Universität Erlangen-Nürnberg, Erlangen, Germany, 2017.
11. Boeltken, T.; Wunsch, A.; Gietzelt, T.; Pfeifer, P.; Dittmeyer, R. Ultra-compact microstructured methane steam reformer with integrated Palladium membrane for on-site production of pure hydrogen: Experimental demonstration. *Int. J. Hydrog. Energy* **2014**, *39*, 18058–18068. [CrossRef]
12. Kreuder, H.; Muller, C.; Meier, J.; Gerhards, U.; Dittmeyer, R.; Pfeifer, P. Catalyst development for the dehydrogenation of MCH in a microstructured membrane reactor-For heat storage by a Liquid Organic Reaction Cycle. *Catal. Today* **2015**, *242*, 211–220. [CrossRef]
13. Kreuder, H.; Boeltken, T.; Cholewa, M.; Meier, J.; Pfeifer, P.; Dittmeyer, R. Heat storage by the dehydrogenation of methylcyclohexane—Experimental studies for the design of a microstructured membrane reactor. *Int. J. Hydrog. Energy* **2016**, *41*, 12082–12092. [CrossRef]
14. Dittmeyer, R.; Boeltken, T.; Piermartini, P.; Selinsek, M.; Loewert, M.; Dallmann, F.; Kreuder, H.; Cholewa, M.; Wunsch, A.; Belimov, M.; et al. Micro and micro membrane reactors for advanced applications in chemical energy conversion. *Curr. Opin. Chem. Eng.* **2017**, *17*, 108–125. [CrossRef]
15. Cholewa, M.; Durrschnabel, R.; Boukis, N.; Pfeifer, P. High pressure membrane separator for hydrogen purification of gas from hydrothermal treatment of biomass. *Int. J. Hydrog. Energy* **2018**, *43*, 13294–13304. [CrossRef]
16. Cholewa, M.; Zehner, B.; Kreuder, H.; Pfeifer, P. Optimization of membrane area to catalyst mass in a microstructured membrane reactor for dehydrogenation of methylcyclohexane. *Chem. Eng. Process.* **2018**, *125*, 325–333. [CrossRef]
17. Aslam, R.; Muller, K.; Ant, W. Experimental study of solubility of water in Liquid Organic Hydrogen Carriers. *J. Chem. Eng. Data* **2015**, *60*, 1997–2002. [CrossRef]
18. Aslam, R.; Muller, K.; Muller, M.; Koch, M.; Wasserscheid, P.; Arlt, W. Measurement of hydrogen solubility in potential Liquid Organic Hydrogen Carriers. *J. Chem. Eng. Data* **2016**, *61*, 643–649. [CrossRef]
19. Aslam, R.; Muller, K. Adsorption isotherm of dibenzyl toluene and its partially hydrogenated forms over phenyl hexyl silica. *Mod. Chem. Appl.* **2017**, *5*. [CrossRef]
20. Muller, K.; Stark, K.; Emel'yanenko, V.N.; Varfolomeev, M.A.; Zaitsau, D.H.; Shoifet, E.; Schick, C.; Verevkin, S.P.; Arlt, W. Liquid Organic Hydrogen Carriers: Thermophysical and thermochemical studies of benzyl- and dibenzyl-toluene derivatives. *Ind. Eng. Chem. Res.* **2015**, *54*, 7967–7976. [CrossRef]

21. Do, G.; Preuster, P.; Aslam, R.; Bosmann, A.; Muller, K.; Arlt, W.; Wasserscheid, P. Hydrogenation of the liquid organic hydrogen carrier compound dibenzyltoluene—Reaction pathway determination by H-1 NMR spectroscopy. *React. Chem. Eng.* **2016**, *1*, 313–320. [CrossRef]
22. Peters, T.A.; Stange, M.; Bredesen, R. 2—Fabrication of palladium-based membranes by magnetron sputtering. In *Palladium Membrane Technology for Hydrogen Production, Carbon Capture and Other Applications*; Doukelis, A., Panopoulos, K., Koumanakos, A., Kakaras, E., Eds.; Woodhead Publishing: Cambridge, UK, 2015; pp. 25–41.

membranes

MDPI

Article

Recent Developments in Compact Membrane Reactors with Hydrogen Separation

Alexander Wunsch, Paul Kant, Marijan Mohr, Katja Haas-Santo, Peter Pfeifer and Roland Dittmeyer *

Institute for Micro Process Engineering, Karlsruhe Institute of Technology, 76344 Eggenstein-Leopoldshafen, Germany; alexander.wunsch@kit.edu (A.W.); paul.kant@kit.edu (P.K.); marijan.mohr@gmx.de (M.M.); katja.haas-santo@kit.edu (K.H.-S.); peter.pfeifer@kit.edu (P.P.)

* Correspondence: roland.dittmeyer@kit.edu; Tel.: +49-721-6082-3114

Received: 1 October 2018; Accepted: 9 November 2018; Published: 14 November 2018

Abstract: Hydrogen production and storage in small and medium scale, and chemical heat storage from renewable energy, are of great interest nowadays. Micro-membrane reactors for reforming of methane, as well as for the dehydrogenation of liquid organic hydrogen carriers (LOHCs), have been developed. The systems consist of stacked plates with integrated palladium (Pd) membranes. As an alternative to rolled and electroless plated (Pd) membranes, the development of a cost-effective method for the fabrication of Pd membranes by suspension plasma spraying is presented.

Keywords: membrane reactor; hydrogen; palladium; microstructured; LOHC; suspension plasma spraying

---

1. Introduction

Hydrogen is not only a valuable resource for fuel cells and an efficient solution for sustainable mobility. It is used in large quantities for the production of ammonia, alcohols, and fertilizers, and for cracking of heavy crude oil factions and hydrodesulphurization of fuels. Smaller quantities of hydrogen are needed in the electronic and metallurgical industry. Large scale users produce their hydrogen demand on site, mainly via methane steam reforming and purification via pressure swing absorption. Small scale users cannot produce their hydrogen demand economically on site, since the pressure swing absorption does not scale down economically. Therefore, hydrogen is transported to these consumers by truck.

The transportation of hydrogen by truck, usually in compressed or liquefied state, is expensive and energy intensive, due to the low density even of liquefied hydrogen (70 kg/m^3), and safety aspects on the road. In addition, the transportation, including compression or liquefaction, adds heavily to the carbon dioxide (CO_2) balance.

A technical innovation must, therefore, be sought in order to supply small consumers of hydrogen—which, after all, require hydrogen consumption of about 5% of the total and, thus, almost 3 million tons per year—through decentralized production. This is not only economically favorable, but also allows the use of bio-gases, instead of natural gas, as a largely CO_2-neutral hydrogen production which is not burdened by complex transport routes. Compact membrane reformers are an attractive option to produce hydrogen in small to medium size quantities in decentralized locations, on site, for low pressure applications by steam reforming of methane or natural gas. Using microstructured devices, very compact units can be realized. These compact reformers are built of thin metal sheets integrating different functions, such as reforming, separation of hydrogen from the gas by a membrane, and heat integration by internal combustion. The very large ratio of channel surface area to reactor volume leads to good heat and mass transfer properties. The main challenges are the palladium membrane and the integration of the catalyst in the microstructured channels. Advances in the last years in the integration of thinner palladium membranes, to achieve higher efficiency, have been made. Mainly thin supported-metal foils have been employed, also, electroless plating (ELP) is frequently used for laboratory-scale membrane reactors. However, these preparation techniques are expensive, and mostly not suitable for higher quantities, as they are cost-intensive, and manufacturing is time consuming [1].

Besides hydrogen production, the storage of the gas is expensive, due to low volumetric density. In the context of building up the hydrogen infrastructure, a technology is needed that makes storage and distribution lucrative. The use of organic liquids with the ability to bind hydrogen reversibly has great potential in this area. Under certain conditions, and in the presence of a catalyst, the liquid can be hydrogenated and, if required, dehydrogenated again. In loaded or unloaded state, the liquid organic hydrogen carriers (LOHCs) can be transported or stored practically without losses. However, an enormous amount of heat is required for dehydrogenation. In addition, small amounts of unwanted compounds that contaminate the hydrogen are also formed by reaction. Microstructured devices can help out—due to the small dimensions, an almost isothermal reaction management is possible. The use of a Pd-based membrane, enclosed by microstructures, can purify the hydrogen with high efficiency.

In the present article, the concepts for highly integrated compact membrane reformers, the fabrication of thin Pd membranes by suspension plasma spraying, and the application of a membrane reactor concept for the dehydrogenation of liquid organic hydrocarbons, are presented.

2. Microstructured Membrane Reactors—μEnhancer 2.0

Several studies and reviews on compact small-scale hydrogen production have been published in the recent years [2–4]. Most groups used Pd or Pd alloys as unsupported thin films, or deposited membranes, for in situ hydrogen removal. Rahimpour et al. reviewed applications and preparation methods of palladium membranes. The main preparation methods for Pd-based membranes mentioned by Rahimpour et al. include electroless plating, metal organic chemical vapor deposition (MOCVD), and physical vapor deposition (PVD) plasma sputtering [5], while suspension plasma spraying (SPS), an also promising technique [6], is not mentioned. Fernandez et al. presented a paper on the research at Tecnalia, and TU/e using a fluidized bed membrane reactor concept with methane or biogas as feedstock [7], underlining the feasibility of compact small-scale hydrogen production. In the next chapter, microstructured reactor concepts will be described.

Designs

All reformer designs developed at the Institute for Micro Process Engineering (IMVT) are modular. Modules with specific functions and production capacity are stacked in a reactor allowing easy adjusting of production capacity.

Based on the first designs of a planar microstructured membrane reactor [8,9], two designs for the second generation of reaction modules were developed at IMVT (μEnhancer 2.0). The first of the two newly developed designs is made up of two modules, one combustion module and one reformer module with integrated hydrogen purification. The second design combines combustion, reforming, and hydrogen purification in one module.

All microstructured modules are built by stacking microstructured stainless steel sheets; the functionalities as gas distribution, reaction channels with catalyst for combustion and reforming, and hydrogen separation via Pd membranes, are realized by the design of the plates and stacks. Laser welding of the stack ensures gas tightness. The hydrogen separation membrane can be incorporated either by laser welding of thin Pd foils, or by coating of porous metallic composite supports with Pd or Pd alloys.

The main part of the reactors is the reforming of methane. In the μEnhancer 2.0 designs, reforming is carried out in two steps: one pre-reforming step, in which a sufficiently high hydrogen partial pressure for separation is achieved; and one reforming step, in which hydrogen is removed from the reaction channels, simultaneously to the reaction, to push the reaction to higher conversion rates. The heat of reaction for the reforming is supplied via combustion of either additional fuel or the retentate of the reformer layer/module. A comprehensible scheme of the combination of the three functionalities (combustion, reforming, and hydrogen separation) in one module is shown in Figure 1. Detailed schemes of the modules of both newly developed μEnhancer 2.0 designs are shown in Figure 2.

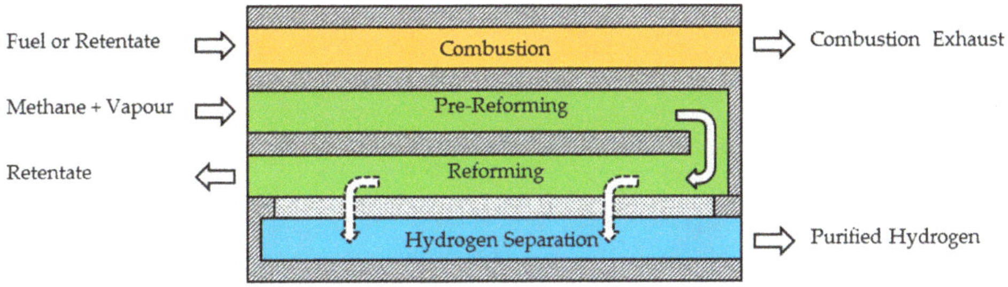

Figure 1. Rough scheme of the combination of the three functionalities (combustion for heat supply, reforming and hydrogen separation) of a fully integrated methane steam reformer in one module.

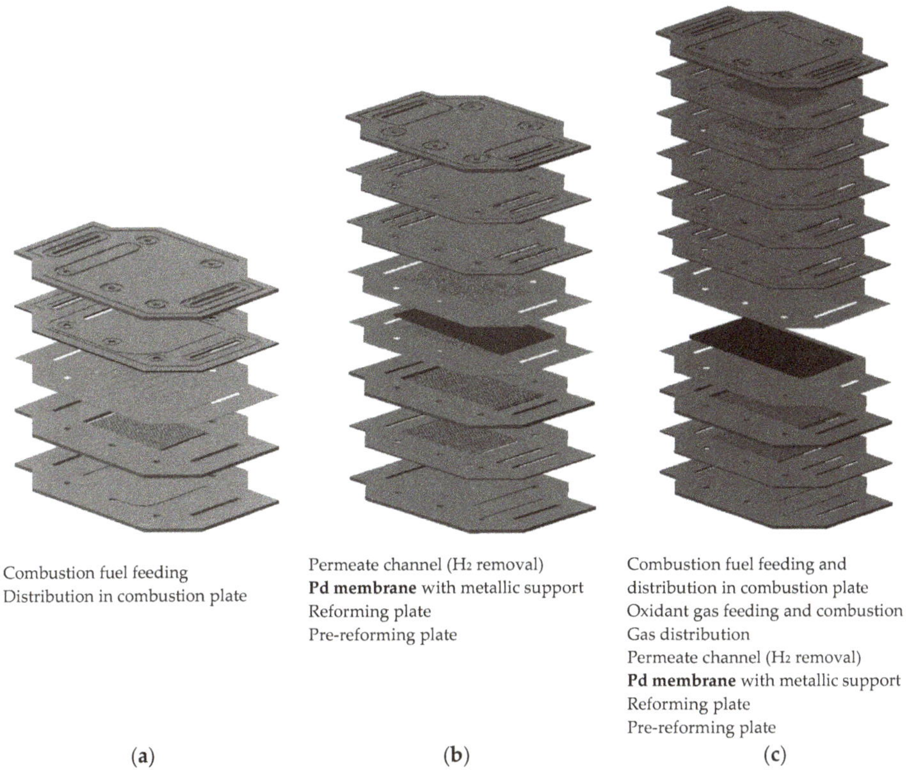

Figure 2. Stack designs for combustion, reformer, and integrated module for the reforming of methane with integrated hydrogen separation: (**a**) combustion; (**b**) reforming; (**c**) integrated.

The hydrogen separation part of the stack, shown in Figure 2, is the integral part of both designs. The combustion modules in both designs include a distribution layer for the air needed for combustion, in order to provide uniform heating over the whole area.

In the first design with separated combustion and reforming modules, the process flexibility is quite high, as the temperature can be changed by regulation of the combustion feed gases.

In the second design, a layer for combustion fueled by retentate or new fuel is integrated. This layer is directly adjacent to the reforming, as well as the hydrogen separation by the Pd membrane. This integrated design saves about 15% material and 25% space (height) in comparison to the first design. For comparison of number of plates and stack height, see Table 1.

Table 1. Height and number of plates of single and combined modules.

	Combustion	Reforming	Stacked (Comb. + Ref.)	Integrated
Plates	5	8	13	11
Height	7.2 mm	9.4 mm	16.6 mm	12.4 mm

The challenges in fabrication techniques are high, and the operational flexibility is reduced, and all process conditions for the reactions have to be matched. However, the compactness of this system is very high, and so are the hydrogen yield and productivity for a well-balanced system. An alternative serial arrangement of the functional modules is also possible: first, a microchannel reformer followed by a combined water gas shift membrane reactor, and even these functions can be separated into a microchannel water–gas shift reactor followed by a gas separation unit with a palladium membrane. This modular approach, enabled by micro process engineering, offers important process flexibility, according to the needs present in the distinct setting.

The new ultra-compact reactor system for on-site production, in which the described modules are combined/connected to supply tubes, etc., is shown in Figure 3. These reactor systems can be easily integrated in decentralized processes with only small hydrogen demand.

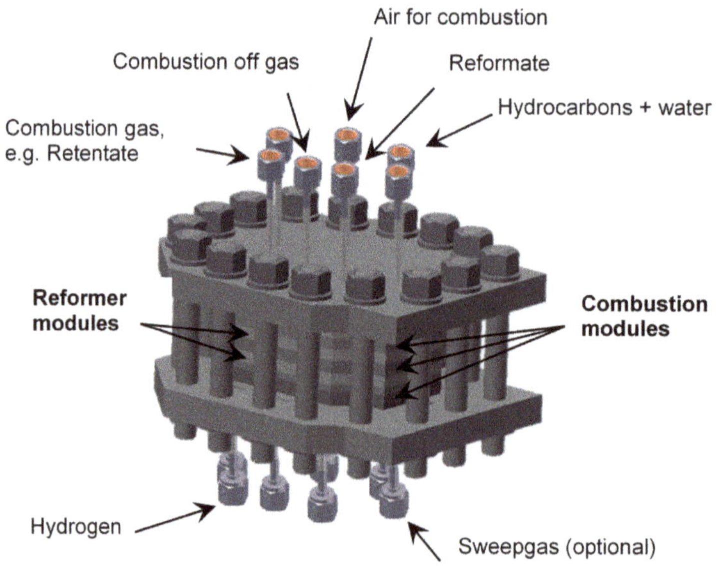

Figure 3. Schematic representation of the modular membrane reactor system with reformer and combustion modules, with inlets and outlets for reaction gases.

However, for a larger amount of facilities, it is of economic interest to reduce the production costs for the palladium membranes. An approach to substitute the commonly via electroless plating or cold rolling-produced Pd membranes, by ones fabricated by SPS, is presented in the next chapter.

3. Palladium Based Composite Membranes via Suspension Plasma Spraying

3.1. State of the Art: Membrane Materials

The idea of using palladium as active membrane material for high temperature hydrogen separation goes back more than 100 years. Already in 1916, Snelling patented an "Apparatus for Separating Gases", which is basically an electroplated palladium membrane on a tubular ceramic support [10]. Meanwhile, a vast number of materials has been proposed as an alternative to the expensive and sensitive palladium, among which are palladium alloys and cermets, group V metals, and alloys containing these metals. Extensive overviews on membrane materials can, for instance, be found in [5,11–13]. Palladium is very sensitive towards poisoning, for example, by sulfur species or carbon monoxide, and unstable when submitted to temperature cycles in hydrogen atmospheres crossing the α- to β-phase transition temperature around 293 °C [5,11–15]. However, some palladium-based alloys overcome at least one of these two major drawbacks of pure palladium. Palladium–copper alloys show, for instance, significant tolerance towards sulfur impurities, whereas palladium–silver alloys have lower phase-transition temperatures [11,13,16,17]. In addition, depending on the exact composition, palladium–copper and palladium–silver alloys have slightly higher hydrogen permeabilities than pure palladium, and alloying palladium with cheaper metals makes the membrane material, per mass, less expensive [11,13,14]. Group V metals are also less expensive than palladium, and have impressive hydrogen permeabilities way higher than that of palladium [13,18]. However, they suffer from hydrogen embrittlement and poisoning of active sites for hydrogen dissociation on the surface [12,13,18,19]. Early attempts to overcome surface poisoning of group V metal membranes were made in 1967, when Makrides, Wright, and Jewett patented membranes made from group V metals coated with thin layers of palladium on both sides [20]. Publications from the last years, on group V metal-based hydrogen membranes with palladium-based or transition metal carbide protection layers, show further development but also outline tremendous challenges, among which, attenuation of hydrogen flux over time in consequence of, for instance, interdiffusion of membrane and top-layer materials at elevated temperatures [18,19,21,22].

3.2. Membrane Design

Beside the choice of the membrane material, the membrane design is crucial when designing a reactor with integrated membrane for hydrogen feed or removal. Free standing membranes like, for instance, fabricated and examined in [18,23], have to be rather thick to be manageable and to withstand pressure differences between the retentate and permeate side. Since membrane materials are expensive, and permeance decreases with increasing thickness, thinner membranes supported on porous substrates are desirable [24,25]. Both ceramic and metallic porous substrates for metallic hydrogen separation membranes are reported in the literature (see e.g., [16,26]). Ceramic substrates have smooth surfaces with small pore sizes and narrow pore size distributions, which facilitate the fabrication of defect-free thin metal layers [24–27]. Unfortunately, ceramic supports are difficult to integrate into reactors, as joining between metal and ceramic parts is difficult [26]. By contrast, porous sinter metal membrane supports can easily be integrated into reactors, for example, via welding. However, they only show poor surface properties such as wide pore size distributions and large pore sizes [26]. Furthermore, metallic interdiffusion between the metallic support and the active metal membrane layer can deteriorate the membrane performance [24,27].

A promising approach combining both advantages from porous ceramic and metallic supports is, therefore, the fabrication of composite membranes on sinter metal supports with ceramic diffusion barrier layers, like reported, e.g., in [6,28–31]. Choosing suitable materials for the sinter metal support and the diffusion barrier layer influences the long-term stability of the fabricated composite membrane. Factors such as corrosion resistance at reaction conditions of the porous metal support, and similar thermal expansion coefficients of sinter metal, ceramic, and membrane material, are important. Kot showed that the combination of Crofer-22-APU steel and yttria stabilized zirconia (8 mol %

yttria—8YSZ) is, for example, a suitable material combination for the fabrication of substrates for palladium composite membranes [31]. The modules of the µEnhancer 2.0 design are therefore based on Crofer-22-APU as material. The composite support for palladium foils or the sprayed palladium layer is made up of Crofer-22-APU and an 8YSZ diffusion barrier layer.

3.3. *Standard Membrane Fabrication Methods*

There is a variety of techniques reported for the fabrication of metallic hydrogen separation membranes, among which include rolling [18,23,32,33], physical vapor deposition [30], electroplating [28], and electroless plating [28,30,34,35]. Extensive overviews of different fabrication methods are available in the literature (see e.g., [11,36]). Like mentioned above, almost all reported methods are iterative and time-consuming processes, demanding days to weeks to build up sufficiently thin (in the case of rolling) or thick (in the case of physical vapor deposition and electroless plating) metal layers. The most common method to prepare thin (in the range of 10 µm) palladium-based membranes is electroless plating. Using this method, the substrate must be covered with palladium seeds, for instance, via wet chemical methods or physical or chemical vapor deposition. Subsequently, the seeded substrate is immersed in a coating solution containing a stabilized palladium salt, which is then autocatalytically reduced by the palladium seeds under consumption of an added reducing agent. A detailed description of the electroless plating process is, for example, given in [29]. Depending on the properties of the substrate surface, in order to reach a dense defect-free metal layer, the coating procedure must be repeated multiple times [28]. For high surface quality asymmetric ceramic supports, single-step coating procedures for thin membranes are reported, see, for instance, [37]. For the fabrication of palladium alloy membranes, different metals are deposited subsequently, and annealed under elevated temperature (see, e.g., [16,38]). Simultaneous deposition of palladium and silver via electroless plating is also reported, see, for instance, [37,39]. Another drawback of electroless plating, besides the time-consuming iterative layer build up, is the production of metal-loaded wastewater.

A novel and promising technique for the cost-effective fabrication of thin metal-based hydrogen separation composite membranes is under development at IMVT in cooperation with the German Aerospace Centre (DLR) in Stuttgart, and based on industry-established plasma spray techniques. Time requirements to fabricate layers with thicknesses in the 10 µm range lie in the range of minutes. Substrates do not have to be activated like in the case of electroless plating, and no metal-loaded liquid waste is produced [6,30].

3.4. *Plasma Spraying Techniques for Membrane Preparation*

3.4.1. General Process Description

Plasma spraying is an industry-established process, which is typically used for the fabrication of thermal barrier layers on gas turbine blades, as well as corrosion and wear protection layers on tube walls and bearings. A schematic sketch of a plasma spray facility is shown in Figure 4. Typically, particles are injected into a plasma of a plasma torch with the help of a carrier gas. The particles injected into the plasma melt and hit a surface, where they form either dense or porous layers, depending on the process parameters. Both ceramic and metal particles can be processed. Large areas can be coated in a short time with impressive deposition rates up to 1 mm/min [40].

If dense layers with thicknesses in the range of only a few microns shall be fabricated, very small particles with mean particle sizes smaller than 1 µm must be injected into the plasma. The injection of such small particles is impossible with a carrier gas, since the particle's momentum is too small to enable the particles to penetrate the plasma. Instead, suspensions containing the particles are injected into the plasma. The solvent of the injected suspension evaporates when reaching the core of the plasma. The remaining particles melt, and form a layer on the coated surface [41].

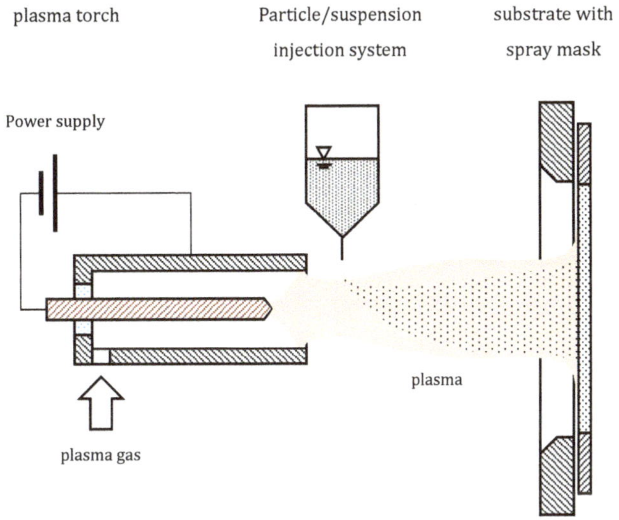

Figure 4. Schematic sketch of a plasma spray facility to produce metallic membrane layers on porous substrates. Plasma is lighted in a plasma torch, and carried out by the plasma carrier gas. Metal particles, or a suspension of metal particles, are injected into the plasma. The molten metal particles hit the substrate surface with high velocity, and form a dense metal layer. Sketch after [6].

Plasma spray processes, and especially the suspension plasma spray process, are very complex, depending on a variety of variables, such as the particle size and agglomeration state of the particles (in the suspension or powder), the solvent used when suspensions are injected, the injection speed and rate, the plasma parameters, such as plasma gas composition and flow rate, plasma current, spraying distance, and spraying atmosphere, and the substrate temperature and surface properties [6,40,41].

3.4.2. Development and State of the Art

Although there are several groups reporting the fabrication of dense membranes via plasma spraying, for example, the fabrication of oxygen ion transport membranes via suspension plasma spraying (see [42]), the authors are only aware of the Dittmeyer group fabricating palladium-based hydrogen separation membranes via plasma spraying. A first attempt was reported 2007, by Dittmeyer and Huang, using atmospheric plasma spraying with a powder feed. They were using a commercial palladium powder with particle size <45 µm and, as substrates, porous stainless steel with plasma-sprayed yttria-stabilized zirconia diffusion barrier layers. The diffusion barrier layers were maintained with their rough surface after spraying. Regarding future prospects, Dittmeyer and Huang argued that a rough surface would improve membrane adhesion. Other groups, using plasma-sprayed ceramic diffusion barrier layers, sand blast the layers to smoothen the surface and facilitate the application of defect-free thin palladium layers (see, e.g., [34]). Despite the thicknesses of 30 µm and 70 µm of the palladium layers deposited by Dittmeyer and Huang, the fabricated composite membranes had no satisfactory permselectivity, due to remaining open porosity. Dittmeyer and Huang wrote that smaller particles injected into the plasma could result in thinner and denser metal layers. [30]

A second attempt at fabricating palladium composite membranes via plasma spraying techniques was reported 2014 by Boeltken et al. Like Dittmeyer and Huang, Boeltken et al. were using porous stainless steel substrates with sprayed yttria-stabilized zirconia diffusion barrier layers. The surface of the diffusion barrier layers in the experiments of Boeltken et al. also remained in its rough initial state. Palladium particles, with a size between 250 nm and 550 nm, were injected into the plasma dispersed in dethylene glyocol monobutyl ether stabilized with ethyl cellulose. Boeltken et al.

fabricated much thinner layers of palladium than Dittmeyer and Huang (9.5 μm instead of 30 μm and 70 μm, respectively) but the permselectivity of the fabricated composite membranes was still not satisfactory; it only reached a value of 60. For the future, Boeltken et al. proposed to use substrates with smoother surfaces and smaller pore sizes to reach higher permselectivities at smaller palladium layer thicknesses [6].

In 2017, a third attempt to fabricating palladium-based composite membranes for hydrogen separation started in the cooperation network of the Institute for Energy and Climate Research (IEK-1) at Research Centre Jülich (FZJ), DLR, in Stuttgart, and IMVT in Karlsruhe. Since the substrate surface quality was of great importance in the experiments of Boeltken et al., the project focused, at first, on the fabrication of high-quality composite membrane substrates. Detailed results of the substrate fabrication and characterization, conducted in 2017, are reported in [43]. Crofer-22-APU sinter metal plates, provided by IEK-1, were welded into dense metal frames for good integrability into test reactors, and coated via dip-coating with an yttria-stabilized zirconia layer. The resulting membrane substrates had a smooth defect-free surface. Figure 5 shows scanning electron microscope (SEM) images of the surface, and a cross-section of the fabricated substrates. The surface-weighted pore size distribution was determined from SEM images, and followed an approximately log–normal distribution, with a geometric mean pore size of 98 nm, and a geometric standard deviation of 0.617. The nitrogen permeance at room temperature of the substrates was determined to be 7.9 ± 1.2 $\mu mol/m^2/s/Pa$. Hydrogen permeance at room temperature was 22 ± 3 $\mu mol/m^2/s/Pa$. The literature reports substrates with room temperature nitrogen permeances more than twice as high as the value determined in the current work at IMVT (see e.g., [28]). Further investigations showed that the main transport resistance in the fabricated substrates did not lie in the ceramic diffusion barrier layer, but in pores of the sinter metal substrates blocked with coating suspension formed during dip-coating. This shows a potential for improvement of the permeance of the substrates. The palladium layer was deposited at DLR in Stuttgart via suspension plasma spraying. One focus of the coating process development was the fabrication of stable palladium suspensions departing from commercially available palladium powder with a mean particle size of ca. 100 nm. First, results were not yet satisfactory, since the suspension stability and deagglomeration state of palladium particles in suspension were not optimal, and the deposited layers were very thin (1–4 μm) and with a remaining open porosity, but work is ongoing.

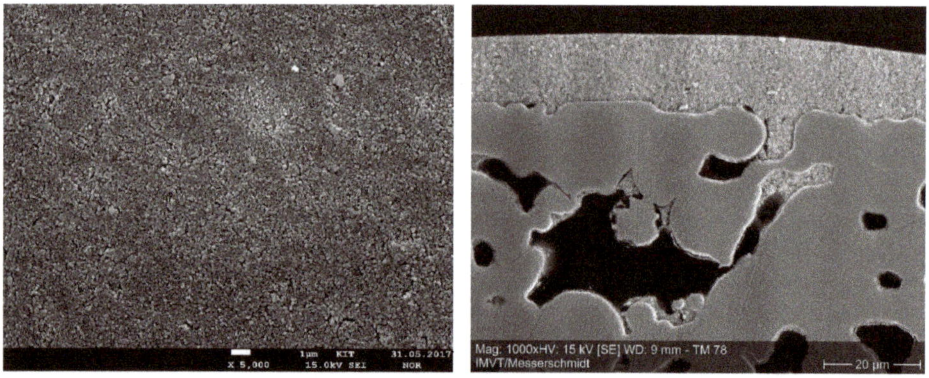

Figure 5. Surface (**left**) and cross-sectional view (**right**) of substrates currently fabricated at IEK-1 in Jülich and Institute for Micro Process Engineering (IMVT) in Karlsruhe.

4. Process Intensification in LOHC Dehydrogenation Using Pd-Based Membranes

Liquid organic hydrogen carriers are recently discussed substances that could give an answer to the question of safe and handy storage and distribution of hydrogen [44,45]. Especially with regard to future energy systems fed from renewable sources, the suitability of this technology will be investigated [46,47]. The hydrogen is produced via electrolysis supplied by regenerative electricity to provide further applications independent of location and time. The extraction of hydrogen in locations, where a lot more hydrogen is available than needed, makes this technology feasible. These locations are mostly sunny and windy regions or places where no customers can be found. Conventional solutions store hydrogen physically, by compression or liquefaction, to increase the low volumetric energy density compared to atmospheric conditions. LOHC technology, on the other hand, propagates the reversible chemical binding of hydrogen to an organic liquid. Due to a high cycle stability, the use of the carrier substance should resemble a "H_2 deposit bottle". A suitable LOHC is also characterized by high temperature stability, low toxicity, high storage density, good availability and, ultimately, thermodynamic conditions that enable profitable technical implementation. The decisive advantage of this technology lies in the secure and long-term stable storage, as well as the use of existing infrastructure for the distribution of the liquid. A comparison of the mentioned storage technologies and the potential LOHC material systems is discussed, in detail, by Preuster et al. [48].

Brückner et. al [49] have shown that the isomer mixture of dibenzyltoluene (0H-DBT, Marlotherm SH® from Sasol), which is commercially available as a heat transfer oil, partially or completely fulfils all these criteria and is, therefore, suitable as LOHC. The LOHC can store 6.2 wt % hydrogen when fully loaded, which corresponds to a storage density of 17.5 L_{LOHC}/kg_{H2}. A study on environmental and health impact attributes a high potential of acceptance in the population by the proposed material, since the handling of organic liquids, such as diesel, is already known [50]. Furthermore, 0H-DBT and the hydrogenated form, perhydro-dibenzyltoluene (18H-DBT), have a considerably lower risk potential than fuels used today—they are neither flammable nor volatile. In addition, no carcinogenic effects have been proven [51,52]. The system 18H-DBT/0H-DBT is, therefore, currently the state of the art in LOHC research. Various applications, considering 18H-DBT/0H-DBT as carrier system, were evaluated on a thermodynamic and techno-economic basis [53–55]. Rüde et al. predict high reliability and robustness to the LOHC unit with upstream electrolysis and a downstream fuel cell [5,56]. Dynamic operation of a dehydrogenation unit has already been demonstrated [57].

In order to be able to describe the substance system, which has not yet been sufficiently researched for this purpose, firstly, material properties had to be determined [58–62]. Secondly, suitable analytics had to be found to determine the degree of (de)hydrogenation. A comparison of different methods has shown that refractometry is the simplest, and nuclear magnetic resonance (NMR) spectroscopy the most accurate method to determine the hydrogen load of the carrier [63,64]. Recent developments show that, despite high boiling points (0H-DBT $\approx$ 390 °C) of the substances, gas chromatography can also be used [65].

Considering thermodynamics, favorable conditions result for hydrogenation at high pressures (15–30 bar) and low temperatures (150–250 °C), and for dehydrogenation at low pressures (1–5 bar) and high temperatures (280–330 °C). Consequently, the superimposed phase and reaction equilibrium can only be determined with great inaccuracies in these multiphase reactions [66]. Do et al. investigated the reaction path, and found that the reaction takes place via two stable intermediates (12H-DBT, 6H-DBT) resulting from consecutive C6-cycle dehydrogenation [67]. In total, the 0H-DBT/18H-DBT can absorb/release nine H_2 molecules, whereby 71 kJ/mol_{H2} heat is released/consumed per molecule [49]. The strong heat tone means that, in case of decentralized applications, a lot of waste heat is needed. Therefore, Preuster et al. investigated the power generation by linking the dehydrogenation unit with a solid oxide fuel cell [68].

In general, dehydrogenation is technically more difficult to implement than hydrogenation. In addition to the thermodynamically unfavorable conditions, the enormous gas production makes the reaction control more difficult. Assuming that the residence time of the gas and liquid phase is equal, the reaction chamber is almost exclusively filled by gas. Depending on the reactor concept, the liquid can be "pressed" through the apparatus by gas bubbles with short contact times at the catalyst. Including wetting properties between catalyst and LOHC, this flow guidance can lead to enormous losses in catalyst-related productivity. The high gas content additionally causes LOHC to evaporate due to the vapor–liquid equilibrium (VLE), which is detectable despite subsequent condensation of the product gas. In addition to traces of LOHC, other low-boiling components could also be determined, which result from side reactions, e.g., from decomposition or transfer hydrogenation [69]. In order to provide hydrogen in sufficient quality for other applications, such as fuel cells, purification of the product gas is, therefore, still indispensable.

4.1. State of the Art: LOHC Dehydrogenation

Hydrogenious Technologies GmbH, a young company from Erlangen, Germany, offers commercial container-based solutions for loading/unloading liquid hydrogen carriers in different performance classes [52]. As reactor geometry for dehydrogenation, a horizontal pipe, half-filled with catalyst particles, is used. The liquid flows in from below and covers the catalyst. This allows the resulting hydrogen to be removed from the top [70]. This approach prevents a multiphase flow, and allows the liquid to almost follow the hydrodynamic law.

Another concept is the so-called "one reactor", which makes it possible to switch between hydrogenation and dehydrogenation reactions in the same container with the same catalyst and similar temperature level, purely by changing the reaction pressure. A benefit, especially for stationary applications, is the possibility to recuperate the heat obtained from the hydrogenation reaction. In addition, investment costs are significantly reduced. The reactor consists of a vessel with vertically arranged tubes filled with catalyst, which are tempered from the outside by a thermostat. In the dehydrogenation configuration, the loaded LOHC flows from bottom to top, so that the resulting gas bubbles leave the pipes at the top outlet [71].

Both concepts presented, however, cannot deliver high-purity hydrogen, due to low-boiling by-products and traces of evaporated LOHC. A purification can be managed, e.g., by a pressure swing adsorption plant (PSA) which, however, becomes unprofitable with low hydrogen capacities. One solution to this problem could be the integration of Pd-based membranes as a purification step.

4.2. Multi-Staged Approach Using Pd Membranes

Pd-based membrane reactors have already been developed for the dehydrogenation of LOHC methylcyclohexane, and the steam reforming of methane, where promising results could be achieved [3,8,9,72–75]. For adaptation to the multiphase system 18H-DBT/0H-DBT/H_2, a multi-stage reactor concept with intermediate separation of hydrogen was developed. The use of consecutive reactors and membrane separators is not new, however, the concept has so far only been applied for reactions in the gas phase. Due to the unknown interaction between LOHC and the membrane surface, the integrated form, the membrane reactor, was avoided in this work. In order to shed light on the influence of the intermediate separation, a reactor and a membrane separator were modeled and sequentially arranged (see Figure 6). The simulations were carried out under strong simplified conditions, in order to show the positive influence of the gas separation via Pd-based membranes from the system. For the reactor, a radial flow reactor was assumed, which behaves like a plug flow reactor (PFR). This is a 1-dimensional problem which was solved integrally by the CSTR (continuous stirred tank reactor) cascade using MATLAB software. The resulting product stream from the reactor stage is fed to the membrane separator, where the hydrogen is completely separated. This also means that the potential wetting of the membrane surface by LOHC, which leads to a reduction of the effective membrane area, has not been taken into account. For the calculations, no experimentally determined kinetics were used for the reaction—rather, the kinetics were adjusted such that a relevant degree of dehydrogenation could be achieved. A detailed description of the modeling, the reactors used, and the experimental determined kinetics of the system, can be found in [Wunsch 2018, manuscript will be submitted for the special issue].

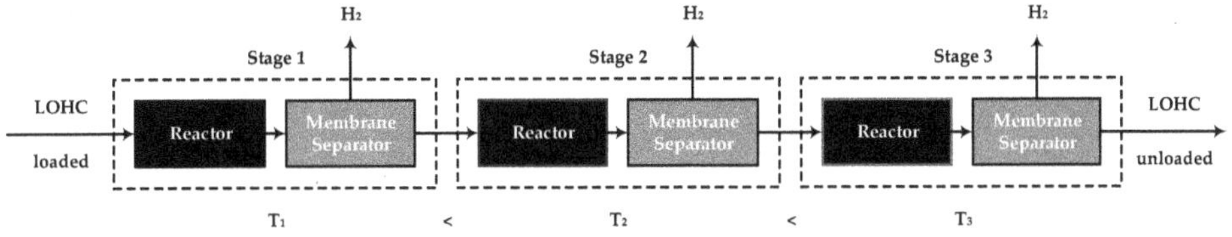

Figure 6. Schematic of the multi-staged reactor concept with intermediate hydrogen separation.

To characterize the reactor, the ratio of the residence time between the gas phase and the liquid phase must first be defined. The least efficient case was assumed, which means that the arising gas phase stays the same time as the liquid phase:

$$\tau_{liq} = \tau_{gas}. \tag{1}$$

In order to describe the influence of the high gas production on the effectively used catalyst mass, a correction factor, α, will be introduced:

$$m_{Cat}^{eff} = \alpha \cdot m_{Cat}. \tag{2}$$

The correction factor primarily describes the poorer wetting of the catalyst with LOHC under the condition that the gas and liquid phases have the same residence time. For a simplified description of this property, a relationship between correction factor α and liquid phase fraction ε_L was constructed:

$$\alpha = \ln\left(1 + \varepsilon_L^b \cdot (e - 1)\right). \tag{3}$$

This relationship is not determined experimentally; it is simply a construction that could describe the unknown efficiency loss by reduced wetting of the catalyst surface area with decreasing liquid phase fraction in the reactor. As the conversion increases, the liquid phase fraction decreases and, finally, the effectively used catalyst mass decreases (see Figure 7). This results in the limiting case $\alpha = 1$, when the evolved gas has no impact on the efficiency, and the limiting case $\alpha \approx \varepsilon_L$, when the efficiency decreases nearly linear with gas production. This relationship can be controlled using parameter b.

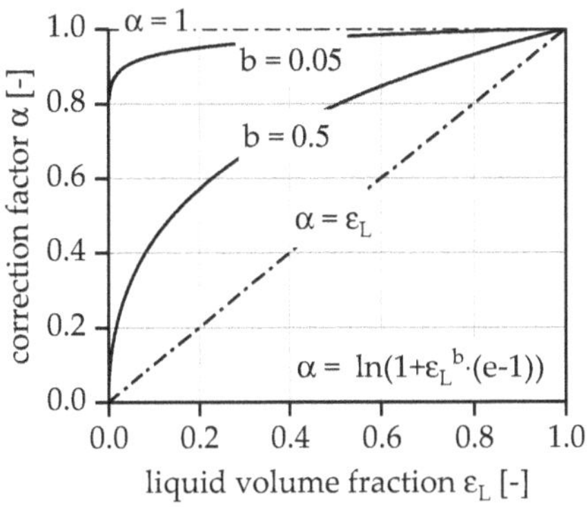

Figure 7. Constructed relationship between correction factor α and liquid volume factor ε_L.

As mentioned, we assumed that an increased degree of dehydrogenation is present in each reactor stage, and that the membrane completely separates the hydrogen from the product gas:

$$\dot{V}_{H2,\ Retentate} = 0. \tag{4}$$

Figure 8 shows the degree of dehydrogenation depending on the number of reactors used. A constant catalyst mass is distributed over a different number of reactors (1–5). A reactor is divided into N equal pieces, and the hydrogen is removed at these points via a Pd-based membrane. It is immediately apparent that the degree of dehydrogenation can be almost doubled by the more frequent intermediate separation alone, irrespective of the residence time distribution between gas and liquid phase. As expected, the total degree of dehydrogenation increases with decreasing impact of the gas phase on the effectively used catalyst mass ($b = 0.05$).

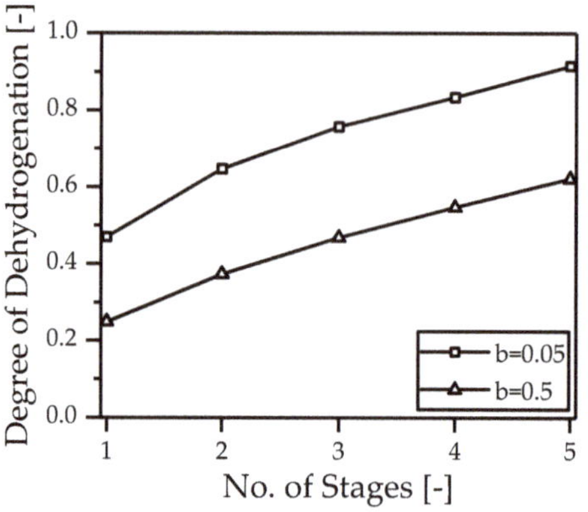

Figure 8. Degree of dehydrogenation as a function of the number of reactors used with intermediate separation at constant catalyst mass and varying influence of the gas phase on the effectively used catalyst mass.

A comparison of the degree of dehydrogenation between five reactor stages, with intermediate separation and one reactor stage, is shown in Figure 9. The reactor volume/catalyst mass was also kept constant. It can be seen that already after 40% of the reactor length, and one intermediate separation stage, the same degree of dehydrogenation is achieved as after the entire reactor length without separation.

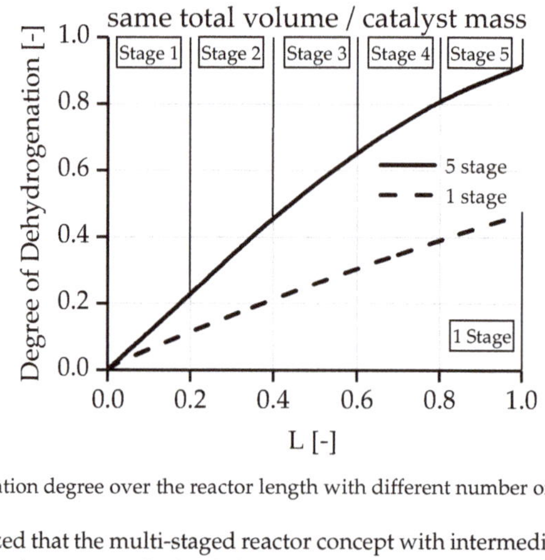

Figure 9. Dehydration degree over the reactor length with different number of separation stages.

It can be summarized that the multi-staged reactor concept with intermediate hydrogen separation can be used to intensify the process of LOHC dehydrogenation of 18H-DBT. In addition to the actual goal, namely, hydrogen purification, the intermediate separation increases the efficiency of the process by reducing the residence time of the liquid.

5. Conclusions

In this article, we described the general concept for ultra-compact microstructured membrane reactors for hydrogen production. The benefits of the reactor concept are the very large membrane surface area per catalyst volume (ca. 103–106 m^{-1}) and negligible mass transport resistance towards membrane, even for high-flux membranes. By the integration of the heating module in the compact device, efficient heating by hot gas or catalytic combustion of retentate with air is ensured. Altogether, the design convinces by its high compactness, low weight, and modular plant design.

Furthermore, an alternative method, suspension plasma spraying, for the fabrication of Pd membranes in relation to the commonly used membranes, either thin metal foils or layers fabricated by electroless plating, has been shown. As supports, porous sinter metallic supports made of Crofer-22 APU with a YSZ diffusion barrier layer have been used. An important advantage of the Crofer-22 APU supports is the possibility for gas-tight integration into the compact module by laser welding.

As an example for the application of these compact modules, besides on-site methane reforming (see [8]), the dehydrogenation of liquid organic hydrogen carriers is described. Here, the advantages, like the equilibrium shift of the reaction and the integrated purification of the released hydrogen, are presented in a multi-staged reactor concept. This concept can be realized in a single compact device by stacking of multiple combined reaction and separation modules.

One important aspect of the application of compact membrane reactors for the catalytic reaction are suitable preparation methods for the catalyst in the microstructured reactors. These developments and results of the preparation of catalyst layers by inkjet printing in the thickness of 10–15 μm are not in the focus of this article, being reported elsewhere [8,76]

Looking ahead, the first commercial applications of membrane reactors may appear in small-capacity hydrogen production for industrial uses via on-site reforming, where hydrogen is at a relatively low pressure (<3 bar) and moderate purity (max 99.5%), and for hydrogen generation from LOHC in the context of hydrogen logistics, rather than in large-scale reforming or water gas shift (WGS). This modular concept is a flexible concept for a variety of applications producing hydrogen, like in reforming reactions and LOHC dehydrogenation. Small-capacity hydrogen supply is an excellent opportunity for the extension of new technologies, such as compact membrane reactors, as they are more energy-efficient compared to large-scale processes, and require a lower OPEX (operating expenditure), and by the simplification of the process scheme, a reduced CAPEX (capital expenditure) is needed. However, for the commercialization of this technology, there is some indispensable research work that is still ongoing at our institute.

Author Contributions: Conceptualization, P.K. and A.W.; methodology, P.K., A.W. and P.P.; software, A.W. and P.P.; validation, A.W., P.P.; formal analysis, P.K. and A.W.; investigation, M.M., A.W. and P.K..; resources, IMVT; data curation, P.K. and A.W.; writing—original draft preparation, K.H.S., A.W. and P.K.; writing—review and editing, K.H.S.; supervision, P.P. and R.D.; funding acquisition, P.P. and R.D.

Funding: The authors gratefully acknowledge funding by the German Federal Ministry of Education and Research (BMBF) within the Kopernikus Project P2X: Flexible use of renewable resources—exploration, validation and implementation of 'Power-to-X' concepts.

Acknowledgments: All colleagues from the micro fabrication and materials engineering groups at the institute IMVT for their dedication and professional work for major contributions to the design and manufacturing of the various reactors described in this paper by the colleagues are gratefully acknowledged. The cooperation with Sayed-Asif Ansar and Dirk Ullmer, at the German Aerospace Center (DLR) in Stuttgart and Martin Bram and his group at the Institute of Energy and Climate Research (IEK-1), Forschungszentrum Jülich GmbH, Germany is also gratefully acknowledged.

Conflicts of Interest: The authors declare no conflict of interest.

References

1. Basile, A.; Gallucci, F.; Tosti, S. Synthesis, Characterization, and Applications of Palladium Membranes. In *Inorganic Membranes: Synthesis, Characterization and Applications*; Mallada, R., Menéndez, M., Eds.; Elsevier: Amsterdam, The Netherlands; Oxford, UK, 2013; pp. 255–323.
2. Peters, T.A.; Rørvik, P.M.; Sunde, T.O.; Stange, M.; Roness, F.; Reinertsen, T.R.; Ræder, J.H.; Larring, Y.; Bredesen, R. Palladium (Pd) Membranes as Key Enabling Technology for Pre-combustion CO_2 Capture and Hydrogen Production. *Energy Procedia* **2017**, *114*, 37–45. [CrossRef]
3. Dittmeyer, R.; Boeltken, T.; Piermartini, P.; Selinsek, M.; Loewert, M.; Dallmann, F.; Kreuder, H.; Cholewa, M.; Wunsch, A.; Belimov, M.; et al. Micro and micro membrane reactors for advanced applications in chemical energy conversion. *Curr. Opin. Chem. Eng.* **2017**, *17*, 108–125. [CrossRef]
4. Gallucci, F.; van Sint Annaland, M. *Process Intensification for Sustainable Energy Conversion*; Wiley: Chichester, UK, 2015.
5. Rahimpour, M.R.; Samimi, F.; Babapoor, A.; Tohidian, T.; Mohebi, S. Palladium membranes applications in reaction systems for hydrogen separation and purification: A review. *Chem. Eng. Process. Process Intensif.* **2017**, *121*, 24–49. [CrossRef]
6. Boeltken, T.; Soysal, D.; Lee, S.; Straczewski, G.; Gerhards, U.; Peifer, P.; Arnold, J.; Dittmeyer, R. Perspectives of suspension plasma spraying of palladium nanoparticles for preparation of thin palladium composite membranes. *J. Membr. Sci.* **2014**, *468*, 233–241. [CrossRef]
7. Fernandez, E.; Helmi, A.; Medrano, J.A.; Coenen, K.; Arratibel, A.; Melendez, J.; de Nooijer, N.C.A.; Spallina, V.; Viviente, J.L.; Zuñiga, J.; et al. Palladium based membranes and membrane reactors for hydrogen production and purification: An overview of research activities at Tecnalia and TU/e. *Int. J. Hydrog. Energy* **2017**, *42*, 13763–13776. [CrossRef]
8. Boeltken, T.; Wunsch, A.; Gietzelt, T.; Pfeifer, P.; Dittmeyer, R. Ultra-compact microstructured methane steam reformer with integrated Palladium membrane for on-site production of pure hydrogen: Experimental demonstration. *Int. J. Hydrog. Energy* **2014**, *39*, 18058–18068. [CrossRef]
9. Boeltken, T.; Belimov, M.; Pfeifer, P.; Peters, T.A.; Bredesen, R.; Dittmeyer, R. Fabrication and testing of a planar microstructured concept module with integrated palladium membranes. *Chem. Eng. Process. Process Intensif.* **2013**, *67*, 136–147. [CrossRef]
10. Snelling, W.O. Apparatus for Separating Gases. U.S. Patent 1174631, 7 March 1916.
11. Paglieri, S.N.; Way, J.D. Innovations in Palladium Membrane Research. *Sep. Purif. Methods* **2006**, *31*, 1–169. [CrossRef]
12. Mundschau, M. Hydrogen Separation Using Dense Composite Membranes Part I: Fundamentals. In *Inorganic Membranes for Energy and Environmental Applications*; Bose, A.C., Ed.; Springer: New York, NY, USA, 2009; pp. 125–153.
13. Conde, J.J.; Maroño, M.; Sánchez-Hervás, J.M. Pd-Based Membranes for Hydrogen Separation: Review of Alloying Elements and Their Influence on Membrane Properties. *Sep. Purif. Rev.* **2016**, *46*, 152–177. [CrossRef]
14. O'Brien, C.P.; Howard, B.H.; Miller, J.B.; Morreale, B.D.; Gellman, A.J. Inhibition of hydrogen transport through Pd and Pd47Cu53 membranes by H2S at 350 °C. *J. Membr. Sci.* **2010**, *349*, 380–384. [CrossRef]
15. Calles, J.A.; Sanz, R.; Alique, D.; Furones, L. Thermal stability and effect of typical water gas shift reactant composition on H2 permeability through a Pd-YSZ-PSS composite membrane. *Int. J. Hydrog. Energy* **2014**, *39*, 1398–1409. [CrossRef]
16. Pizzi, D.; Worth, R.; Giacinti Baschetti, M.; Sarti, G.C.; Noda, K.-I. Hydrogen permeability of 2.5 μm palladium–silver membranes deposited on ceramic supports. *J. Membr. Sci.* **2008**, *325*, 446–453. [CrossRef]
17. She, Y.; Emerson, S.C.; Magdefrau, N.J.; Opalka, S.M.; Thibaud-Erkey, C.; Vanderspurt, T.H. Hydrogen permeability of sulfur tolerant Pd–Cu alloy membranes. *J. Membr. Sci.* **2014**, *452*, 203–211. [CrossRef]
18. Alimov, V.N.; Hatano, Y.; Busnyuk, A.O.; Livshits, D.A.; Notkin, M.E.; Livshits, A.I. Hydrogen permeation through the Pd–Nb–Pd composite membrane. *Int. J. Hydrog. Energy* **2011**, *36*, 7737–7746. [CrossRef]
19. Fuerst, T.F.; Zhang, Z.; Hentges, A.M.; Lundin, S.-T.B.; Wolden, C.A.; Way, J.D. Fabrication and operational considerations of hydrogen permeable Mo2C/V metal membranes and improvement with application of Pd. *J. Membr. Sci.* **2018**, *549*, 559–566. [CrossRef]

20. Makrides, A.C.; Wright, M.A.; Jewett, D.N. Separation of Hydrogen by Permeation. U.S. Patent 3350846, 7 November 1967.
21. Fuerst, T.F.; Petsalis, E.P.; Wolden, C.A.; Way, J.D. Application of TiC in Vanadium-Based Hydrogen Membranes. *Ind. Eng. Chem. Res.* **2018**. [CrossRef]
22. Viano, D.M.; Dolan, M.D.; Weiss, F.; Adibhatla, A. Asymmetric layered vanadium membranes for hydrogen separation. *J. Membr. Sci.* **2015**, *487*, 83–89. [CrossRef]
23. Zhang, Y.; Ozaki, T.; Komaki, M.; Nishimura, C. Hydrogen permeation of Pd–Ag alloy coated V–15Ni composite membrane. *J. Membr. Sci.* **2003**, *224*, 81–91. [CrossRef]
24. Ma, Y.H. Composite Pd and Pd/Alloy membranes. In *Inorganic Membranes for Energy and Environmental Applications*; Bose, A.C., Ed.; Springer: New York, NY, USA, 2009; pp. 251–254.
25. Zheng, L.; Li, H.; Xu, H. "Defect-free" interlayer with a smooth surface and controlled pore-mouth size for thin and thermally stable Pd composite membranes. *Int. J. Hydrog. Energy* **2016**, *41*, 1002–1009. [CrossRef]
26. Mardilovich, I.P.; Engwall, E.; Ma, Y.H. Dependence of hydrogen flux on the pore size and plating surface topology of asymmetric Pd-porous stainless steel membranes. *Desalination* **2002**, *144*, 85–89. [CrossRef]
27. Guo, Y.; Wu, H.; Fan, X.; Zhou, L.; Chen, Q. Palladium composite membrane fabricated on rough porous alumina tube without intermediate layer for hydrogen separation. *Int. J. Hydrog. Energy* **2017**, *42*, 9958–9965. [CrossRef]
28. Dittmar, B.; Behrens, A.; Schödel, N.; Rüttinger, M.; Franco, T.; Straczewski, G.; Dittmeyer, R. Methane steam reforming operation and thermal stability of new porous metal supported tubular palladium composite membranes. *Int. J. Hydrog. Energy* **2013**, *38*, 8759–8771. [CrossRef]
29. Huang, Y.; Dittmeyer, R. Preparation and characterization of composite palladium membranes on sinter-metal supports with a ceramic barrier against intermetallic diffusion. *J. Membr. Sci.* **2006**, *282*, 296–310. [CrossRef]
30. Huang, Y.; Dittmeyer, R. Preparation of thin palladium membranes on a porous support with rough surface. *J. Membr. Sci.* **2007**, *302*, 160–170. [CrossRef]
31. Kot, A.J. Entwicklung eines Metallbasierten Substratkonzeptes für Energieeffiziente Gasmembranen. Ph.D. Thesis, Ruhr-Universität, Bochum, Germany, 2015.
32. Böltken, T. Microstructured Reactors with Integrated Metallic Membranes for on-Site Production of Pure Hydrogen. Ph.D. Thesis, Karlsruhe Institute of Technology, Karlsruhe, Germany, 25 May 2015.
33. Tosti, S. Rolled thin Pd and Pd–Ag membranes for hydrogen separation and production. *Int. J. Hydrog. Energy* **2000**, *25*, 319–325. [CrossRef]
34. Sanz, R.; Calles, J.A.; Alique, D.; Furones, L.; Ordóñez, S.; Marín, P.; Corengia, P.; Fernandez, E. Preparation, testing and modelling of a hydrogen selective Pd/YSZ/SS composite membrane. *Int. J. Hydrog. Energy* **2011**, *36*, 15783–15793. [CrossRef]
35. Straczewski, G.; Völler-Blumenroth, J.; Beyer, H.; Pfeifer, P.; Steffen, M.; Felden, I.; Heinzel, A.; Wessling, M.; Dittmeyer, R. Development of thin palladium membranes supported on large porous 310L tubes for a steam reformer operated with gas-to-liquid fuel. *Chem. Eng. Process. Process Intensif.* **2014**, *81*, 13–23. [CrossRef]
36. Yun, S.; Ted Oyama, S. Correlations in palladium membranes for hydrogen separation: A review. *J. Membr. Sci.* **2011**, *375*, 28–45. [CrossRef]
37. Melendez, J.; Fernandez, E.; Gallucci, F.; van Sint Annaland, M.; Arias, P.L.; Pacheco Tanaka, D.A. Preparation and characterization of ceramic supported ultra-thin (~1 μm) Pd-Ag membranes. *J. Membr. Sci.* **2017**, *528*, 12–23. [CrossRef]
38. Qiao, A.; Zhang, K.; Tian, Y.; Xie, L.; Luo, H.; Lin, Y.S.; Li, Y. Hydrogen separation through palladium–copper membranes on porous stainless steel with sol–gel derived ceria as diffusion barrier. *Fuel* **2010**, *89*, 1274–1279. [CrossRef]
39. Zeng, G.; Shi, L.; Liu, Y.; Zhang, Y.; Sun, Y. A simple approach to uniform PdAg alloy membranes: Comparative study of conventional and silver concentration-controlled co-plating. *Int. J. Hydrog. Energy* **2014**, *39*, 4427–4436. [CrossRef]
40. Heimann, R.B. *Plasma Spray Coating. Principles and Applications*, 2., Completely rev. and Enlarged ed.; Wiley-VCH: Weinheim, Germany, 2008.
41. Fauchais, P.; Rat, V.; Coudert, J.-F.; Etchart-Salas, R.; Montavon, G. Operating parameters for suspension and solution plasma-spray coatings. *Surf. Coat. Technol.* **2008**, *202*, 4309–4317. [CrossRef]

42. Fan, E.S.C.; Kesler, O. Deposition of Lanthanum Strontium Cobalt Ferrite (LSCF) Using Suspension Plasma Spraying for Oxygen Transport Membrane Applications. *J. Therm. Spray Technol.* **2015**, *24*, 1081–1092. [CrossRef]
43. Kant, P. Palladiumkompositmembranen Durch Suspensionsplasmaspritzen auf Planaren Substraten und deren Integration in Kompakte Membranreaktoren. Master's Thesis, Karlsruhe Institute of Technology, Karlsruhe, Germany, 2017.
44. Teichmann, D.; Arlt, W.; Wasserscheid, P.; Freymann, R. A future energy supply based on Liquid Organic Hydrogen Carriers (LOHC). *Energy Environ. Sci.* **2011**, *4*, 2767–2773. [CrossRef]
45. Aakko-Saksa, P.T.; Cook, C.; Kiviaho, J.; Repo, T. Liquid organic hydrogen carriers for transportation and storing of renewable energy—Review and discussion. *J. Power Sources* **2018**, *396*, 803–823. [CrossRef]
46. Teichmann, D.; Arlt, W.; Wasserscheid, P. Liquid Organic Hydrogen Carriers as an efficient vector for the transport and storage of renewable energy. *Int. J. Hydrog. Energy* **2012**, *37*, 18118–18132. [CrossRef]
47. Preuster, P.; Papp, C.; Wasserscheid, P. Liquid Organic Hydrogen Carriers (LOHCs): Toward a Hydrogen -free Hydrogen Economy. *Acc. Chem. Res.* **2017**, *50*, 74–85. [CrossRef] [PubMed]
48. Preuster, P.; Alekseev, A.; Wasserscheid, P. Hydrogen Storage Technologies for Future Energy Systems. *Annu. Rev. Chem. Biomol. Eng.* **2017**, *8*, 445–471. [CrossRef] [PubMed]
49. Bruckner, N.; Obesser, K.; Bosmann, A.; Teichmann, D.; Arlt, W.; Dungs, J.; Wasserscheid, P. Evaluation of Industrially Applied Heat-Transfer Fluids as Liquid Organic Hydrogen Carrier Systems. *ChemSusChem* **2014**, *7*, 229–235. [CrossRef] [PubMed]
50. Markiewicz, M.; Zhang, Y.Q.; Bosmann, A.; Bruckner, N.; Thoming, J.; Wasserscheid, P.; Stolte, S. Environmental and health impact assessment of Liquid Organic Hydrogen Carrier (LOHC) systems—Challenges and preliminary results. *Energy Environ. Sci.* **2015**, *8*, 1035–1045. [CrossRef]
51. IFA-Institut-für-Arbeitsschutz-der-Deutschen-Gesetzlichen-Unfallversicherung. *Material Safety Data Sheet (MSDS)—Dibenzyltoluene, Isomers*; GESTIS-Stoffdatenbank: Sankt Augustin, Germany, 2018.
52. Hydrogenious-Technologies-GmbH. *Material Safety Data Sheet (MSDS)—Verordnung (EG) Nr. 1907/2006 Version 1.1*; Hydrogenious-Technologies-GmbH: Erlangen, Germany, 2017.
53. Eypasch, M.; Schimpe, M.; Kanwar, A.; Hartmann, T.; Herzog, S.; Frank, T.; Hamacher, T. Model-based techno-economic evaluation of an electricity storage system based on Liquid Organic Hydrogen Carriers. *Appl. Energy* **2017**, *185*, 320–330. [CrossRef]
54. Reuss, M.; Grube, T.; Robinius, M.; Preuster, P.; Wasserscheid, P.; Stolten, D. Seasonal storage and alternative carriers: A flexible hydrogen supply chain model. *Appl. Energy* **2017**, *200*, 290–302. [CrossRef]
55. Adametz, P.; Potzinger, C.; Muller, S.; Muller, K.; Preissinger, M.; Lechner, R.; Bruggemann, D.; Brautsch, M.; Arlt, W. Thermodynamic Evaluation and Carbon Footprint Analysis of the Application of Hydrogen-Based Energy-Storage Systems in Residential Buildings. *Energy Technol.* **2017**, *5*, 495–509. [CrossRef]
56. Rude, T.; Bosmann, A.; Preuster, P.; Wasserscheid, P.; Arlt, W.; Muller, K. Resilience of Liquid Organic Hydrogen Carrier Based Energy-Storage Systems. *Energy Technol.* **2018**, *6*, 529–539. [CrossRef]
57. Fikrt, A.; Brehmer, R.; Milella, V.O.; Muller, K.; Bosmann, A.; Preuster, P.; Alt, N.; Schlucker, E.; Wasserscheid, P.; Arlt, W. Dynamic power supply by hydrogen bound to a liquid organic hydrogen carrier. *Appl. Energy* **2017**, *194*, 1–8. [CrossRef]
58. Aslam, R.; Muller, K. Adsorption Isotherm of Dibenzyl Toluene and its Partially Hydrogenated Forms Over Phenyl Hexyl Silica. *Mod. Chem. Appl.* **2017**, *5*, 221. [CrossRef]
59. Aslam, R.; Muller, K.; Ant, W. Experimental Study of Solubility of Water in Liquid Organic Hydrogen Carriers. *J. Chem. Eng. Data* **2015**, *60*, 1997–2002. [CrossRef]
60. Aslam, R.; Muller, K.; Muller, M.; Koch, M.; Wasserscheid, P.; Arlt, W. Measurement of Hydrogen Solubility in Potential Liquid Organic Hydrogen Carriers. *J. Chem. Eng. Data* **2016**, *61*, 643–649. [CrossRef]
61. Heller, A.; Rausch, M.H.; Schulz, P.S.; Wasserscheid, P.; Froba, A.P. Binary Diffusion Coefficients of the Liquid Organic Hydrogen Carrier System Dibenzyltoluene/Perhydrodibenzyltoluene. *J. Chem. Eng. Data* **2016**, *61*, 504–511. [CrossRef]
62. Muller, K.; Stark, K.; Emel'yanenko, V.N.; Varfolomeev, M.A.; Zaitsau, D.H.; Shoifet, E.; Schick, C.; Verevkin, S.P.; Arlt, W. Liquid organic hydrogen carriers: Thermophysical and thermochemical studies of benzyl- and dibenzyl-toluene derivatives. *Ind. Eng. Chem. Res.* **2015**, *54*, 7967–7976. [CrossRef]

63. Muller, K.; Aslam, R.; Fischer, A.; Stark, K.; Wasserscheid, P.; Arlt, W. Experimental assessment of the degree of hydrogen loading for the dibenzyl toluene based LOHC system. *Int. J. Hydrog. Energy* **2016**, *41*, 22097–22103. [CrossRef]
64. Aslam, R.; Minceva, M.; Muller, K.; Arlt, W. Development of a liquid chromatographic method for the separation of a liquid organic hydrogen carrier mixture. *Sep. Purif. Technol.* **2016**, *163*, 140–144. [CrossRef]
65. Modisha, P.M.; Jordaan, J.H.L.; Bosmann, A.; Wasserscheid, P.; Bessarabov, D. Analysis of reaction mixtures of perhydro-dibenzyltoluene using two-dimensional gas chromatography and single quadrupole gas chromatography. *Int. J. Hydrog. Energy* **2018**, *43*, 5620–5636. [CrossRef]
66. Westermeyer, M.; Muller, K. Probability density distribution in the prediction of reaction equilibria. *Fluid Phase Equilib.* **2017**, *437*, 96–102. [CrossRef]
67. Do, G.; Preuster, P.; Aslam, R.; Bosmann, A.; Muller, K.; Arlt, W.; Wasserscheid, P. Hydrogenation of the liquid organic hydrogen carrier compound dibenzyltoluene—Reaction pathway determination by H-1 NMR spectroscopy. *React. Chem. Eng.* **2016**, *1*, 313–320. [CrossRef]
68. Preuster, P.; Fang, Q.P.; Peters, R.; Deja, R.; Nguyen, V.N.; Blum, L.; Stolten, D.; Wasserscheid, P. Solid oxide fuel cell operating on liquid organic hydrogen carrier-based hydrogen—Making full use of heat integration potentials. *Int. J. Hydrog. Energy* **2018**, *43*, 1758–1768. [CrossRef]
69. Geburtig, D.; Preuster, P.; Bosmann, A.; Muller, K.; Wasserscheid, P. Chemical utilization of hydrogen from fluctuating energy sources—Catalytic transfer hydrogenation from charged Liquid Organic Hydrogen Carrier systems. *Int. J. Hydrog. Energy* **2016**, *41*, 1010–1017. [CrossRef]
70. Preuster, P. Entwicklung eines Reaktors zur Dehydrierung Chemischer Wasserstoffträger als Bestandteil eines Dezentralen, Stationären Energiespeichers. Ph.D. Thesis, University of Erlange, Erlangen, Germany, 7 March 2017.
71. Jorschick, H.; Preuster, P.; Durr, S.; Seidel, A.; Muller, K.; Bosmann, A.; Wasserscheid, P. Hydrogen storage using a hot pressure swing reactor. *Energy Environ. Sci.* **2017**, *10*, 1652–1659. [CrossRef]
72. Kreuder, H.; Muller, C.; Meier, J.; Gerhards, U.; Dittmeyer, R.; Pfeifer, P. Catalyst development for the dehydrogenation of MCH in a microstructured membrane reactor-For heat storage by a Liquid Organic Reaction Cycle. *Catal. Today* **2015**, *242*, 211–220. [CrossRef]
73. Kreuder, H.; Boeltken, T.; Cholewa, M.; Meier, J.; Pfeifer, P.; Dittmeyer, R. Heat storage by the dehydrogenation of methylcyclohexane—Experimental studies for the design of a microstructured membrane reactor. *Int. J. Hydrog. Energy* **2016**, *41*, 12082–12092. [CrossRef]
74. Cholewa, M.; Durrschnabel, R.; Boukis, N.; Pfeifer, P. High pressure membrane separator for hydrogen purification of gas from hydrothermal treatment of biomass. *Int. J. Hydrog. Energy* **2018**, *43*, 13294–13304. [CrossRef]
75. Cholewa, M.; Zehner, B.; Kreuder, H.; Pfeifer, P. Optimization of membrane area to catalyst mass in a microstructured membrane reactor for dehydrogenation of methylcyclohexane. *Chem. Eng. Process.* **2018**, *125*, 325–333. [CrossRef]
76. Lee, S.; Boeltken, T.; Mogalicherla, A.K.; Gerhards, U.; Pfeifer, P.; Dittmeyer, R. Inkjet printing of porous nanoparticle-based catalyst layers in microchannel reactors. *Appl. Catal. A Gen.* **2013**, *467*, 69–75. [CrossRef]

membranes

MDPI

Article

Application of Pd-Based Membrane Reactors: An Industrial Perspective

Emma Palo [1], Annarita Salladini [2], Barbara Morico [2], Vincenzo Palma [3], Antonio Ricca [3] and Gaetano Iaquaniello [1,*]

1 KT—Kinetics Technology S.p.A., Viale Castello della Magliana 27, 00148 Rome, Italy; e.palo@kt-met.it
2 Processi Innovativi srl, Via di Vannina 88, 00156 Rome, Italy; salladini.a@processiinnovativi.it (A.S.); morico.b@processiinnovativi.it (B.M.)
3 Department of Industrial Engineering, University of Salerno, via Giovanni Paolo II 132, 84084 Fisciano, SA, Italy; vpalma@unisa.it (V.P.); aricca@unisa.it (A.R.)
* Correspondence: g.iaquaniello@kt-met.it; Tel.: +39-06-6021-6231

Received: 13 September 2018; Accepted: 23 October 2018; Published: 1 November 2018

Abstract: The development of a chemical industry characterized by resource efficiency, in particular with reference to energy use, is becoming a major issue and driver for the achievement of a sustainable chemical production. From an industrial point of view, several application areas, where energy saving and CO_2 emissions still represent a major concern, can take benefit from the application of membrane reactors. On this basis, different markets for membrane reactors are analyzed in this paper, and their technical feasibility is verified by proper experimentation at pilot level relevant to the following processes: (i) pure hydrogen production; (ii) synthetic fuels production; (iii) chemicals production. The main outcomes of operations in the selected research lines are reported and discussed, together with the key obstacles to overcome.

Keywords: Pd-based membrane; hydrogen; closed architecture; open architecture; gas to liquid; propylene

1. Introduction

Membrane reactors are currently increasingly recognized as an effective way to replace conventional separation, process, and conversion technologies for a wide range of applications. In particular, by taking benefit from advanced membrane materials development, they are able to provide enhanced efficiency, are very adaptable, and may have great economic potential.

On the basis of their flexibility, membrane reactors can be employed in a wide range of applications where energy saving and the enhancement of performance in terms of reactants conversion and products selectivity can lead to improved economics.

Among the different membrane reactors reported in the scientific literature [1–3], the palladium-based ones are the most commonly used when hydrogen is the product to be separated [4–9].

Steam reforming is the most widely application for that, since this process is strongly energy-intensive; however, the whole hydrocarbon processing industry, in principle, could take benefit from this technology, since, while enabling a substantial energy saving, it can be helpful also for the production of a highly concentrated CO_2 stream, ready for valorization [8].

Accordingly, several application areas of Pd-based membrane reactors have been considered by KT—Kinetics Technology (KT): (i) pure hydrogen production; (ii) gas-to-liquid (GTL) processes; (iii) propylene production.

Natural gas (NG) steam reforming is the most widely used hydrogen production process. Currently, around 50% of the worldwide hydrogen yearly production results from this technology.

The main reaction occurring in NG steam reforming is endothermic ($CH_4 + H_2O = CO + 3H_2$) and limited by chemical equilibrium, thereby significant hydrogen yields are achieved only with operation at high temperatures (850–900 °C). As a consequence, a portion of methane feedstock must be burned in furnaces in order to sustain the reaction heat duty. This can be responsible for a reduction of the overall process efficiency, an increase of greenhouse gas (GHG) emissions, and a dependence of the hydrogen production cost on the natural gas cost. Accordingly, the coupling of the reaction unit with hydrogen-selective membranes can represent a promising way to enhance hydrogen yield at lower temperatures, since the continuous selective removal of hydrogen from the reaction environment allows to maintain the gas mixture composition far from equilibrium, so that equilibrium conditions are not achieved, and the endothermic reactions can be carried out at a lower temperature.

In addition, the lower thermal duty could also allow the operation of the membrane reformer with a cleaner energy source or with waste heat available from another process, instead of the high-temperature flue gases used in the furnace. Moreover, the lower operation temperature makes possible to use cheaper steel alloys for the fabrication of the reforming tube instead of the expensive materials currently used to withstand the high operating temperatures of conventional steam reforming plants.

With reference to the GTL processes, the main challenge of monetizing gas resources is logistical. Natural gas reserves close to markets are usually transported via pipelines. Where this is not feasible, the gas can be transported in alternative forms, such as compressed natural gas (CNG), liquefied natural gas (LNG), and GTL products, which all address this challenge by densifying the gas and reducing the transportation costs. Natural gas-to-liquid technologies have been a matter of study for many years but are considered economically not convenient, owing to the high costs of natural gas. Nevertheless, the current prospects in shale gas production as well as an increase in oil price have determined a significant difference between oil and gas prices, thus improving the economic benefits from GTL processes application and making this technology the most promising alternative for the valorization of natural gas assets, in particular in North America [10,11]. The GTL process has three main steps: (i) feedstock preparation and syngas production; (ii) Fisher–Tropsch (FT) synthesis; (iii) product upgrading. Syngas production typically involves steam reforming or an autothermal reforming reaction with pure oxygen from an Air Separation Unit (ASU); product upgrading typically involves hydrocracking processes of syncrude. The core of the technology is represented by Fischer Tropsch (FT) synthesis which requires an H_2/CO ratio of about 2.0.

The syngas production step is the most expensive of the three processes, accounting for up to 50% of the Capital Expenditure (CAPEX). However, feed consumption is responsible for up to more than 80% of all operating costs and more than 60% of the cost of production. Therefore, there is a significant incentive for developing new technologies to decrease the capital and operating cost of syngas production units. Furthermore, since developing and constructing a large-scale GTL plant is very capital-intensive and takes years, with significant market timing and hence economic risks involved, the possibility to think of a modular design of smaller scale GTL plants is opening up opportunities to reduce risks and, at the same time, for the use of natural gas in both offshore and remote on-shore locations [12]. The development of innovative process schemes for the production of synthesis gas at lower temperatures than the traditional ones, without affecting natural gas conversion and, at the same time, with saving in terms of feed consumption and plant complexity, is crucial to assess the potentiality of distributed GTL plants. In particular, the use of membrane reactors coupled with novel routes for syngas production such as Catalytic Partial Oxidation (CPO) can be considered the basis for the development of a novel process scheme suitable for GTL applications.

With reference to the last area of the studied applications, propylene is one of the most important derivatives in the petrochemical industry, after ethylene. Worldwide, propylene is mostly produced as a co-product in steam crackers (>55%) or as by-product in Fluid Catalytic Cracking (FCC) units (around 35%). However, the high reaction temperatures of such processes, as well as the high instability of hydrocarbons, lead to formation of coke and the occurrence of side reactions that significantly impact propylene yield [13]. In order to meet the increase in the market demand, being estimated as 5% per year until 2018, several "on-purpose" technologies have been proposed as propylene sources, such as propane dehydrogenation (PDH), methanol-to-propylene (MTP) conversion, and olefin metathesis. It is widely recognized that, in the longer term, "on-purpose" propylene production technologies should be able to stabilize the supply–demand balance.

PDH (C_3H_8 = C_3H_6 + H_2) is a highly endothermic reaction, accordingly favored at high temperatures of operation. In these conditions, other side reactions usually can be observed, responsible for the production of lighter hydrocarbons or coke deposited on the catalyst, thereby forcing to carry out a periodic catalyst regeneration in order to recover its activity after deactivation. The two targets of decreased coke formation and increased propylene yield could be in principle achieved by integrating a low-temperature dehydrogenation reaction step with a H_2-selective membrane [14,15] able to remove the produced hydrogen from the system and thus promoting a chemical equilibrium towards the production of propylene [16,17]. Ideally, a membrane reactor would enable the operation at a lower temperature, avoiding: (i) the search for a trade-off between conversion and selectivity; (ii) too many frequent regeneration cycles. Pd-based membranes appear as the most promising solution for the integration with the PDH reaction, owing to their significant selectivity/flux ratio and to the conventional operating temperature (400–500 °C) that is aligned with the PDH reaction operating conditions [18].

In this paper, it is reported the experience gained by KT in the design and testing of pilot facilities where membranes for hydrogen separation play a key role for the overall process performance. The results reported in the paper, being far from presenting a detailed characterization of the adopted membrane in terms of flux and permeance, aim to give a sketch of the influence of membrane integration on the overall process performance, evidencing that the experimentation at pilot level is necessary to fully understand that the operation of an industrial catalytic membrane reactor is closely linked not only to the selection of active and selective membranes and catalysts but also to the individuation of reliable procedures for a correct operation and maintenance.

2. Experimental: Pilot Plants Description

A selective membrane can be integrated with the reaction environment in two different configurations, with different benefits and drawbacks: (i) the selective membrane is in direct contact with the reaction environment/catalyst, and the reaction product is continuously removed simultaneously to its production (Integrated Membrane Reactor (IMR) or closed architecture); (ii) the selective membrane is not in direct contact with the reaction environment/catalyst but installed outside the reactor and followed downstream by another reaction unit, where the overcoming of chemical equilibrium is observed (Staged Membrane Reactor (SMR), or Reactor and Membrane Module (RMM), or open architecture) [19,20].

The level of integration of catalyst and membrane is actually important to determine the overall process yield, with the main benefit of an open architecture lying in the possibility to keep the membrane and reaction environment separate. In this way, the operating temperatures in the reactor and in the separator can be managed and optimized separately. Accordingly, for each of the studied applications, the selection between open and closed architecture was carried out on the basis of the reaction characteristics, in particular the eventual occurrence of more stressful conditions that can be detrimental for the membrane lifetime.

2.1. Pure Hydrogen Production

For the pure hydrogen production case, both architectures were taken into account.

The open architecture was tested at pilot level and at a capacity of 20 Nm^3/h of pure hydrogen (facility available at Chieti Scalo, Italy, Italian FISR project). The process scheme of the pilot unit is reported in Figure 1 [21–24].

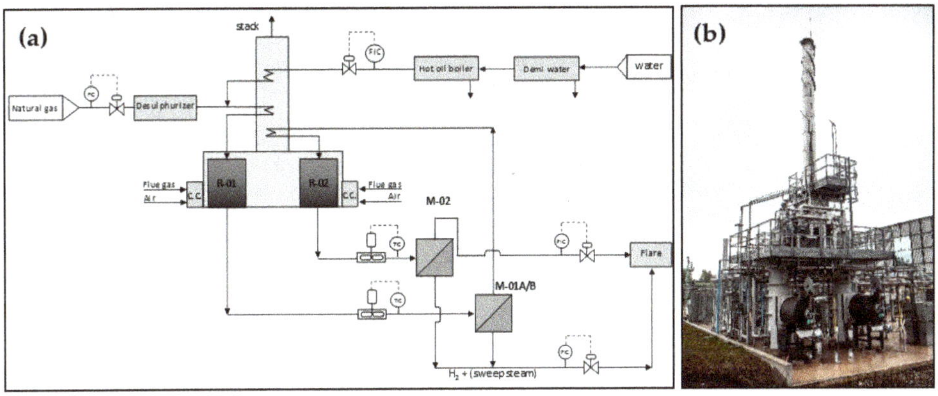

Figure 1. Pure hydrogen production membrane reactor in open architecture (**a**) Process scheme; (**b**) Bird-eye view.

The plant is characterized by two-stages reformers and two membrane modules operated in the temperature range 550–650 °C and 400–450 °C, respectively. Natural gas, available at battery limits at 12 barg, was subjected to desulphurization and then mixed with the process steam. The process steam was produced separately by a real hot oil boiler, superheated in the plant convection section and routed to the first reforming reactor. Each reforming module is characterized by two main sections: (i) a radiant tube, loaded with the catalyst, and (ii) a convection section, where the recovery of heat from the flue gases, available at a temperature of 650–700 °C, was carried out for: (i) feed preheating, (ii) steam superheating.

The syngas produced in the first reformer R-01 was cooled down to the temperature suitable for membrane operation and fed to the first separation module M01A/B (0.4 and 0.6 m^2 respectively). The retentate was recycled to the second reformer R-02, at the outlet of which a mixture with an increased amount of H_2 in consequence of the higher feed conversion was produced. The syngas from the second reformer stage was cooled down from 650 °C to the temperature suitable for membrane operation and routed to the second separation module (M-02, 0.13 m^2). H_2 recovered from both membrane modules was collected and routed to the final cooling and condensate separation. The retentate from R-02 reformer was sent to the flare.

Both R-01 and R-02 were loaded with a structured foam-shaped catalyst, based on noble metals; the main reactors and catalyst characteristics are reported in Table 1.

Table 1. Main characteristics of steam reforming reactors and catalyst.

Element	R-01/R-02
Type	Single tube
Tube Nominal Diameter, in	2 $^{1/2}$
Tube Active length, m	3
Reactor Volume, L	9
Catalyst shape	Cylinder
Catalyst ID-L, mm-mm	58–150
Catalyst Support material, -	SiC
Catalyst Support porosity, %	87
Catalyst Support pore density, ppi	15

Three different membrane modules, Pd- and Pd/Ag -based, able to operate at high temperatures (480 °C for M-01/A and M-02, 500 °C for M-01/B), were installed in the pilot unit. Their main features are reported in Table 2.

Table 2. Main characteristics of the employed membranes.

Membrane	Supplier	Support	Membr. Selective Layer	Thick. Selective Layer, μm	Memb. Area, m^2	Geometry	Production Method	Permeance, $Nm^3/(hm^2bar^{0.5})$ @ 400 °C
M-01/A	ECN	Al_2O_3	Pd	3–6	0.4	Tubular	Electroless deposition	32–35
M-01/B	MRT	SS	Pd/Ag	25	0.6	Planar	Proprietary	8–11
M-02	Japanese Company	Al_2O_3	Pd/Ag	2–3	0.13	Tubular	Electroless deposition	22–24

Table 3 reports the main operating conditions adopted during the experimental test on open architecture.

Table 3. Main operating conditions for two-stage reaction and separation configuration.

Description	I Separation Stage	II Separation Stage
Syngas Flowrate		
Overall flowrate, kg/h	25–50	25–50
Syngas Composition		
CH_4, mol%	7–15%	4–8%
CO, mol%	1–3%	1.5–3.5%
CO_2, mol%	6–7%	7–8.5%
H_2, mol%	30–36%	27–36%
H_2O, mol%	42–56%	48–56%
Pressure (P)		
P feed side, barg	10.2–10.8	10–9.8
P permeate side, barg	0.4–0.6	0.4–0.6
Temperature		
Membrane temperature, °C	350–440	380–440

The closed architecture was tested at pilot level and at a capacity of 3 Nm^3/h of pure hydrogen (facility available at ENEA Casaccia, Italy, EU Comethy project). The process scheme of the pilot unit is reported in Figure 2 [25–27].

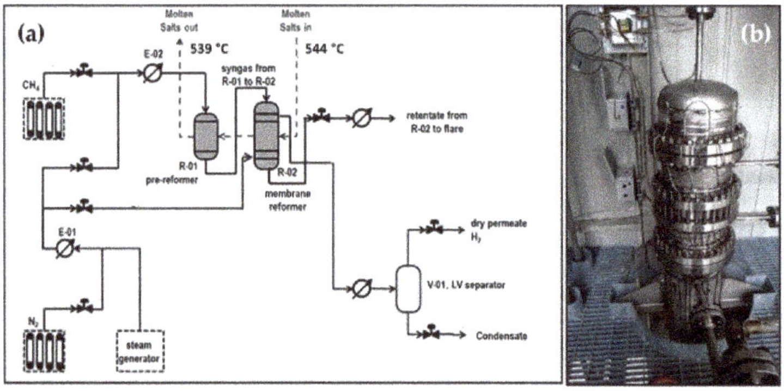

Figure 2. Pure hydrogen production membrane reactor in closed architecture (**a**) Process scheme; (**b**) Reactor assembly.

The plant architecture is based on a first pre-reformer stage (R-01) followed by an integrated membrane reactor (R-02). The main characteristics of the reactors are reported in Table 4. Methane is supplied by cylinders, while process steam and sweep steam are generated by a dedicated electrical boiler. The reaction heat was supplied through a molten salt mixture fed to R-01 at a maximum inlet temperature of 550 °C and further routed to the R-01. Thanks to the partial conversion carried out in R-01, syngas fed to R-02 contained a certain amount of hydrogen, allowing the membrane to be active just at the entrance of the reactor. R-01 is organized as a shell-and-tube configuration, where the molten salts mixture flows in the shell side supplying reaction heat, and the catalyst is installed inside the tubes. R-02 is also arranged in a shell-and-tube configuration, with the molten salts flowing on the shell side. Catalyst and membrane are arranged according to a tube-in-tube configuration, with the catalyst in the annular section around the membranes tube. The latter is equipped with an inner tube to allow the sweep gas to flow in the permeate side. The permeate stream collected from R-02 was cooled down in order to easily separate the sweep gas as a condensate.

Table 4. Main characteristics of closed-architecture membrane reactors.

Element	R-01	R-02
Type	Shell & Tube	Shell & Tube
Nominal Tube Diameter, in	1	1 $^{1/2}$
Active length, mm	400	800
N. tube	7	10
Total Reactor Volume, L	1.6	10
Catalyst shape	Cylinder	2-half hollow cylinder
Support material, -	SiC	SiC
Catalyst, -	Pt-Rh	Pt-Ni

The nickel noble metal-based catalysts deposited on silicon carbide foam were shaped in the form of a cylinder and an annular cylinder for R-01 and R-02, respectively. The catalysts were prepared at the ProCeed Lab of the University of Salerno. A total of 10 Pd-based membranes on ceramic supports were arranged in R-02, developing an overall area of about 0.35 m^2, whose main characteristics are reported in Table 5. Each membrane had an outside diameter of 14 mm and a length of 80 cm.

Table 5. Main characteristics of membranes tested under an integrated reactor.

Membrane	Supplier	Support	Membr. Selective Layer	Thick. Selective Layer, m	Membr. Area, m^2	Geometry	Production Method	Permeance, $Nm^3/(hm^2bar^{0.5})$ @ 400–450 °C
M (R-02)	ECN	Al_2O_3	Pd	3–6	0.35	Tubular	Electroless Deposition	10–15

To improve the separation efficiency, superheated steam was employed as sweep gas in a countercurrent configuration. R-01, R-02, and piping in contact with the molten salts were electrically traced in order to assure a temperature above the salts' freezing point during the start-up and shut-down procedures.

The main operating conditions of the catalytic tests carried out with the integrated membrane reactor are reported in Table 6.

Table 6. Main operating conditions for the integrated membrane reactor.

Description	Value
Flowrate	
CH_4 inlet pre-reforming reactor R-01, kg/h	0.3–1.5
H_2O inlet pre-reforming reactor R-01, kg/h	4.5–6.0
H_2O Sweep gas, kg/h	0–2.0
Molten salts, kg/h	1250–1650
Pressure	
P feed side, barg	9.8–9.5
P permeate side, barg	0.4
Temperature	
T range molten salts, °C	450–550

2.2. Synthetic Fuel Production

The overall concept was tested in open architecture at pilot level and at a capacity of 20 Nm^3/h of pure hydrogen (facility available at Chieti Scalo, Italy, EU NEXT-GTL project). The process scheme of the pilot unit is reported in Figure 3 [28–30].

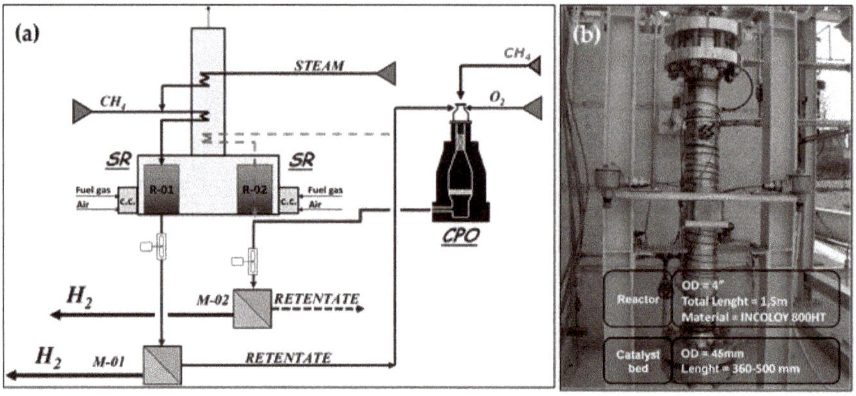

Figure 3. Syngas/pure hydrogen production membrane reactor for gas-to-liquid (GTL) process (**a**) Process scheme; (**b**) Catalytic partial oxidation (CPO) reactor assembly.

After desulphurization, the natural gas was mixed with the process steam and preheated in the convective section of the reformer. The preheated steam was further routed to the reforming reactor operated at an outlet temperature of 550–600 °C. The syngas produced was cooled at a temperature in the range of 450–480 °C before entering the first Pd-based membrane module M-01, where a permeate and a retentate stream were produced. The retentate, poor in hydrogen, was routed to the CPO reactor properly mixed inside the reactor with a stream of pure oxygen from the gas cylinders. The hot syngas available at the outlet of the CPO reactor was cooled down to 450–480 °C in a gas–gas heat exchanger and routed to the second Pd-based membrane separator M-02 for a further hydrogen recovery step. The resulted retentate was a syngas, whose composition could be adjusted on the basis of the membrane hydrogen recovery factor. As reported in Figure 3, the CPO reactor could be operated in a standalone mode (with an external CH_4 stream) or fully integrated with the membrane, accordingly fed with retentate, as described above. Structured catalysts in the forms of honeycomb monolith and pellets, both based on noble metals, were used in the CPO reactor. The main characteristics of the CPO reactor are reported in Table 7.

Table 7. Main characteristics of the CPO reactor.

Element	Description
Type	Reactor
Nominal Diameter, in	4
Active length, mm	500
Total Active Reactor Volume, l	0.9
Catalyst shape	Cylinder/pellets
Support material, -	Confidential
Catalyst, -	Confidential

The membrane-based GTL process was operated at the operating conditions reported in Table 8.

Table 8. Main operating conditions of the membrane-based GTL process.

Description	Value
Gas Flowrate	
CH_4 inlet reforming reactor R-01, kg/h	6–10
H_2O inlet reforming reactor R-01, kg/h	18–30
O_2 inlet CPO reactor, kg/h	1.5–7
Pressure	
P, barg	10.5–10.2
T inlet CPO reactor, °C	280–300
T outlet CPO reactor, °C	650–750

2.3. Propylene Production

The overall concept was tested in open architecture at pilot level and at a capacity of 0.05 kg/h of propylene (facility available at the University of Salerno, Italy, EU CARENA project). The process scheme of the pilot unit is reported in Figure 4 [31–34].

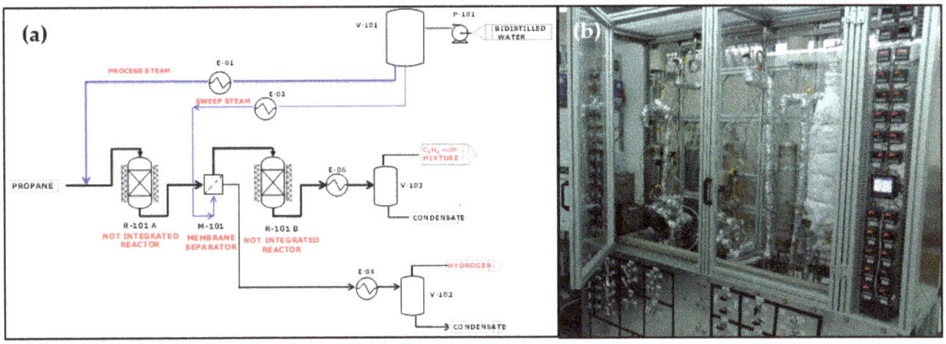

Figure 4. Propylene production membrane reactor (**a**) Process scheme; (**b**) Pilot unit assembly.

The experimental apparatus is constituted by two tubular catalytic reactors (R-101 and R-102) and, between them, a selective separation unit (membrane, M-101): such arrangement defines the "open architecture" of the membrane-based process. The catalytic units consist of a catalytic reactor, characterized by a tubular shape and made of AISI 310 stainless steel (SS); the reactor was loaded with a platinum–tin-based catalyst. The catalytic bed loading was also optimized with inters in order to minimize the undesired side reactions occurring in the homogeneous phase, both before and after the catalytic bed (on reactants and products streams). The separation unit is constituted by a Pd-based membrane prepared on an SS porous tube and having an overall permeation surface of 0.01 m^2. The temperature of the three process devices was controlled by means of electrical heaters, driven by a controller–programmer. The units were then wrapped in a thick layer of insulating mat to minimize heat losses. The catalytic activity tests were carried out in the following operating conditions: Weight Hourly Space Velocity (WHSV) ($kg_{C3H8}/h \cdot kg_{catalyst}$) of 8 h^{-1}, steam/propane ratio of 0.25 mol/mol, $T_{reaction}$ = 500 °C, P = 5 barg, $T_{membrane}$ = 370 °C.

3. Results and Discussion

3.1. Pure Hydrogen Production

The behavior of the two-stages reaction and separation based configuration was deeply investigated. The experimental results in terms of membrane stability and feed conversion confirmed, from a technical point of view, the feasibility of the proposed architecture. The development of proper start-up and shut-down procedures, especially with respect to the heating-up and cooling-down sequences, assured a stable operation for the three membrane modules.

The results clearly evidenced the effect of membrane separation on shifting the reaction towards the product, once hydrogen was removed from the syngas stream, and a second reaction stage was foreseen.

Figure 5a reports the experimental data collected on a two-stage membrane reforming when the M-01A membrane was in operation. With a hydrogen recovery factor in the order of 30–35%, the two-stage configuration allowed to overcome the equilibrium conversion by 20–25%. Hydrogen purity of at least 99.9% mol was detected for all membranes. Because of the modular concept characterizing the open architecture, the performance for a higher number of reaction and separation stages could be easily extrapolated, as shown in Figure 5b. With a number of reaction stages up to six, for example, feed conversion increased up to 70% and 90% at a reformer outlet temperature of 650 °C and 600 °C, respectively. Starting from the experimental results, the behavior of the system for a higher membrane area per stage as well for different values of sweep gas flow rate and reaction pressure could be extrapolated.

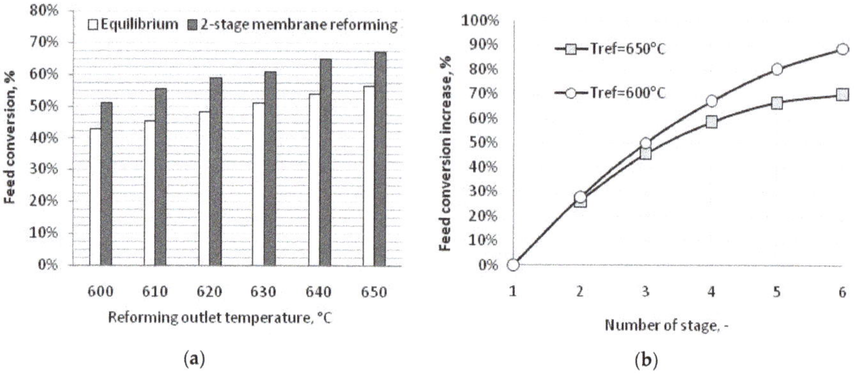

Figure 5. (**a**) Feed conversion for different reforming outlet temperatures (P = 10 barg, Steam/Carbon ratio S/C = 4.3 mol; Tmem = 410–420 °C, Amem = 0.4 m^2, Molar sweep gas/feed ratio = 0); (**b**) Multistage membrane steam reforming performance (P = 10 barg, S/C 4.3, Tmem = 410–420 °C, Amem = 0.4 m^2, Molar Sweep gas/feed ratio = 0).

An experimental campaign on closed-architecture membrane steam reforming was carried out in order to check the overall system performance. The reforming outlet temperature was regulated by adjusting the inlet molten salts temperature. The latter was limited to 550 °C because of the thermal degradation phenomena occurring in the molten salt mixture at a higher temperature.

Despite the low reaction temperature, a very high feed conversion was observed due to the effectiveness of hydrogen removal from the reaction environment. Hydrogen purity of at least 99.8%mol was detected during the experimental test.

As reported in Figure 6a, the integrated closed configuration allowed to overcome the thermodynamic equilibrium conversion up to 150% under sweep gas condition. The latter plays a major role in the integrated configuration, where H_2 partial pressure on the reaction side is quite low because of continuous withdrawals. The experimental results showed that, under a molar sweep gas/feed ratio equal to 1.5, the feed conversion increased up to 135–150% with respect to an operating condition without sweep gas. It derived that, by properly adjusting the steam to carbon (S/C) ratio and sweep gas flow rate, a very high feed conversion could be obtained. In a scenario based on the integration of a Concentrated Solar Power (CSP) plant and steam reforming, the operating temperature window of the solar salts mixture (550–290 °C) accounts for a large steam production, thus allowing to optimize the integrated membrane reforming by increasing both the steam/carbon ratio and the sweep gas/feed ratio, without reducing the overall plant efficiency as for a conventional steam reforming [35].

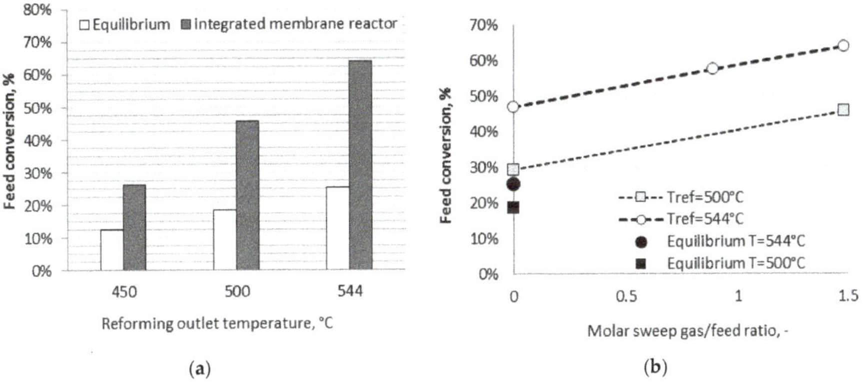

Figure 6. (**a**) Feed conversion at different reforming outlet temperatures (P = 9 barg, S/C = 3 mol and (**b**) Effect of the sweep gas on the integrated membrane reforming plant (P = 9 barg, S/C = 3 mol).

The low reaction temperatures combined with a high feed conversion achievable with the integrated configuration allows to minimize the CO content in the retentate side, thus avoiding any post-shift reaction and accounts for a final retentate stream reach in CO_2 and under pressure. With respect to a conventional steam reformer where the overall produced CO_2 is available diluted and at atmospheric pressure in the flue gas stream, the proposed architecture allows for a less energy-intensive CO_2 capture, due to the fact that it is available at s higher partial pressure.

Although more complex from a technologic point view, requiring a new reactor design with respect to conventional one, being able to house a catalyst, a membrane, and sweep gas, the integrated membrane configuration benefits from a higher effectiveness in equilibrium shifting. The contextual hydrogen removal and production allow for a higher hydrogen recovery factor characterizing the integrated membrane architecture, which definitively means a higher feed conversion.

3.2. Synthetic Fuel Production

The most relevant results in this application are reported in Figure 7a,b.

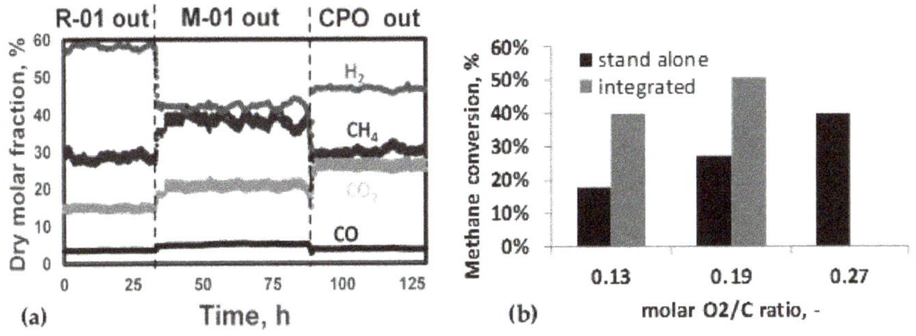

Figure 7. Performance of the membrane-based GTL process: (**a**) Stability, (**b**) Oxygen consumption.

Figure 7a shows the product composition on a dry basis at steady-state conditions measured at the outlet of reformer, the membrane separation, and the CPO reactor, respectively. The data were collected when operating the reformer at an outlet temperature of 590 °C, P = 11 atm, with a hydrogen

Recovery Factor (HRF) of 45% and with a ratio $O_2/(CO + CH_4)$ at the inlet of the CPO reactor of 0.16. The oxygen to carbon ratio was evaluated by taking into account the carbon contribution of CO and CH_4 in the retentate stream.

A significant reduction in hydrogen content was observed in the retentate stream due to hydrogen recovery carried out by the first-stage membrane, whereas the concentration of other components increased, since the mixture became more concentrated. In addition, the performance of the membrane system resulted stable for more than 100 h of continuous operation.

Looking at the economics of the novel process for syngas production, it can be observed that the solution consisting of the CPO reactor integrated with the reformer and the membrane enables for a reduction in oxygen consumption, since a portion of feed conversion is achieved in the upstream reformer stage. The removal of hydrogen in the first membrane module has the double role to favor both reactions of partial oxidation of methane and steam reforming inside the CPO reactor. The second membrane module installed downstream of the CPO reactor enables for an additional recovery of hydrogen and might be also used to adjust the final H_2/CO ratio. In Figure 7b, the comparison of oxygen consumption is reported, observed with both standalone and integrated configuration. The oxygen consumption was referred to the natural gas available at the inlet of the reaction scheme in order to have a direct comparison.

It is possible to observe that, if a total feed conversion of 40% is taken into account, the integrated architecture allows to reduce the oxygen consumption over 50%. This can be translated into a lower operating temperature for the CPO section, accordingly with a difference in the outlet gas composition. The achieved reduction of oxygen consumption enables for a reduction of about 10% in the variable operating cost.

3.3. Propylene Production

To proper investigate the membrane-based propylene production process, a dedicated computational model was developed in Matlab environment, aiming to combine the kinetics expression relevant to the main reaction as derived from the literature [36], heat and material balance, and hydrogen permeance law. In Figure 8a, the results of the numerical simulation of propane dehydrogenation reaction coupled with membrane separation are reported, whereas, in Figure 8bc, the experimental results at pilot level are reported. With reference to Figure 8a, the results are reported by indicating, for different values of the membrane permeance, the evolution of the conversion of propane along the catalytic bed. It can be observed that, with a membrane permeance of 40 $Nm^3/(m^2hbar^{0.5})$ at 550 °C, it was possible to increase the propane conversion of 35%. The improvement of such performance became 48% when the membrane permeance was increased up to 80 $Nm^3/(m^2hbar^{0.5})$.

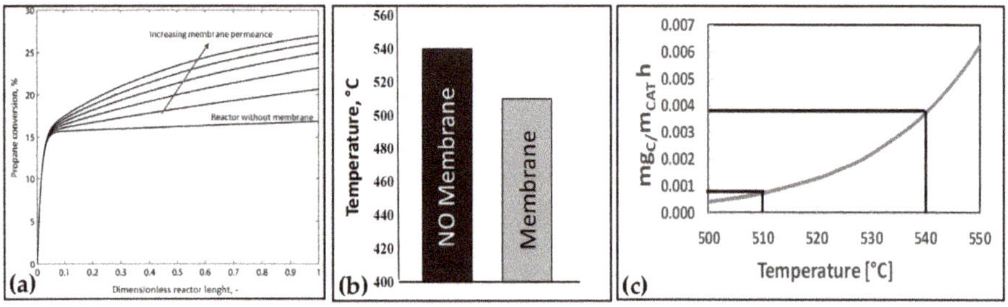

Figure 8. Performance of membrane-based propane dehydrogenation: (**a**) numerical simulation of propane conversion at T = 550 °C and different membrane permeances 40–80 $Nm^3/hm^2bar^{0.5}$, (**b**) experimental operating temperature at fixed propane conversion (X_{C3H8} = 10%, 5 barg, S/C3 = 0.25), (**c**) coke amount as a function of the operating temperature.

The results reported in Figure 8b show the temperature at which it was possible to reach a propane conversion of 10% without and with the membrane. In particular, in the absence of the membrane, it was necessary to reach 540 °C to achieve such conversion level. This temperature decreased of about 30 °C when a membrane reactor was used. This achievement could be translated into a lower deposition of carbon matter on the catalytic bed. Indeed, under the assumption of a deactivating model for the catalyst, developed on the basis of the experimental results achieved, it was possible to evaluate that the coke deposited was reduced five times when the membrane was applied (Figure 8c).

The overall results reported up to now, show the effective applicability of Pd-based membrane reactors in a very large number of chemical processes, even if the actual industrial implementation is strongly related to the possibility of having a stable membrane performance for a long period (at least about 10,000 h) and a cheap cost.

Indeed, while in the case of hydrogen production (both pure and for GTL applications) the application is in a more advance phase, even if it is still necessary to demonstrate the membrane stability on the long term, for propane dehydrogenation, potential applications might require a further developmental step. In this latter case, in fact, it is much more desirable to limit the membrane operation at temperatures not higher than 400 °C, after a decline in hydrogen flux is observed, owing to the tendency of propylene to form oligomers and ring structures on the membrane that can reduce the performance [37]. This would mean that an industrial application of this concept should take into account such requirement.

4. Conclusions

The development of a chemical industry characterized by resource efficiency, in particular with reference to energy use, is becoming a major issue and driver for the achievement of a sustainable chemical production. From an industrial point of view, several application areas, where energy saving and CO_2 emissions still represent a major concern, can take benefit of the application of membrane reactors.

Different markets for membrane reactors applications have been analyzed in this paper, and the technical feasibility verified by proper experimentation at pilot level for: (i) pure hydrogen production; (ii) synthetic fuels production; (iii) chemicals production. The achieved results showed that membrane reactors can be effectively used in all mentioned applications.

In most of the proposed solutions, the concept of membrane reactor is based on the application of a sequence of reaction–separation–reaction units rather than on the application of a reactor in which catalyst and membrane are installed in a same vessel according to the concept of process intensification. Indeed, the former solution is still an efficient solution for the overcoming of chemical equilibrium but is characterized by a lower compactness degree compared to the latter. However, in particular applications where the optimal operating conditions for catalyst and membrane operation are so different from each other and the engineering solution to cope with them may become more complex, thereby impacting also on the operation and maintenance of the reaction system, the former solution might represent a faster approach to boost industrial acceptance in the first phase of transition to the novel catalytic membrane reactors technology.

Nevertheless, it is worth to mention that, in order to progress with the industrial implementation of membrane reactors, a few issues need to be solved, such as the reduction of the fabrication costs, the improvement of membrane stability under poisoning, the current lack of industry-produced, commercial-scale units, and the identification of acceptable accelerated ageing tests, without which the membrane stability could be only assessed by carrying out tests for thousands of hours.

5. Patents

Iaquaniello, G., Salladini, A. Method for hydrogen production, EP Patent Application 11150491.6, 10 January 2011.

Iaquaniello, G., Salladini, A., Morico, B. Method and system for the production of hydrogen. US Patent US9493350B2, 15 November 2016 (priority date 16 March 2012).

Palo, E., Iaquaniello, G. Method for olefins production. US Patent 9,776,935B2, 3 October 2017 (Priority date 29 March 2011)

Author Contributions: All authors contributed to writing and correcting the paper.

Funding: Part of this work was carried out within the framework of the project "Pure hydrogen from natural gas reforming up to total conversion obtained by integrating chemical reaction and membrane separation", financially supported by MIUR (FISR DM 17/12/2002)-Italy. Part of this work received a financial support by the European Community throughout the Next-GTL project "Innovative Catalytic Technologies & Materials for the next Gas to Liquid processes" (Contract NMP3-LA-2009-229183). CoMETHy project has received funding from the European Union's Seventh Framework Programme (FP7/2007-2013) for the Fuel Cells and Hydrogen Joint Technology Initiative under grant agreement n. 279075. The research leading to a portion of these results has received funding from the European Union Seventh Framework Programme FP7-NMP-2010-Large-4, under Grant Agreement no. 263007 (acronym CARENA).

Conflicts of Interest: The authors declare no conflict of interest.

References

1. Gallucci, F.; Basile, A.; Hai, F.I. Introduction—A review of Membrane reactors. In *Membranes for Membrane Reactors: Preparation, Optimization and Selection*; Basile, A., Gallucci, F., Eds.; Wiley: Hoboken, NJ, USA, 2011; pp. 1–62. ISBN 9780470746523.
2. Gallucci, F.; Medrano, J.A.; Roses, L.; Brunetti, A.; Barbieri, G.; Viviente, J.L. Process Intensification via Membrane Reactors, the DEMCAMER Project. *Processes* **2016**, *4*, 16. [CrossRef]
3. Tan, X.; Li, K. *Inorganic Membrane Reactors: Fundamentals and Applications*; Wiley: Hoboken, NJ, USA, 2014; ISBN 978-1-118-67284-6.
4. van Delft, Y.C.; Overbeek, J.P.; Saric, M.; de Groot, A.; Dijkstra, J.W.; Jansen, D. Towards application of palladium membrane reactors in large scale production of hydrogen. In Proceedings of the 8th World Congress on Chemical Engineering, Montreal, QC, Canada, 23–27 August 2009.
5. Basile, A. Hydrogen Production Using Pd-based Membrane Reactors for Fuel Cells. *Top. Catal.* **2008**, *51*, 107. [CrossRef]
6. Fernandez, E.; Helmi, A.; Medrano, J.A.; Coenen, K.; Arratibel, A.; Melendez, J.; de Nooijer, N.C.A.; Spallina, V.; Viviente, J.L.; Zuñiga, J.; et al. Palladium based membranes and membrane reactors for hydrogen production and purification: An overview of research activities at Tecnalia and TU/e. *Int. J. Hydrogen Energy* **2017**, *42*, 13763–13776. [CrossRef]
7. Salladini, A.; Iaquaniello, G.; Palo, E. Membrane Reforming Pilot Testing: KT Experiences. In *Membrane Reforming Pilot Testing: Applications for a Greener Process Industry*; Wiley: Hoboken, NJ, USA, 2016; ISBN 978-1-118-90680-4.
8. De Falco, M.; Salladini, A.; Palo, E.; Iaquaniello, G. Pd-Alloy Membrane Reactor for Natural Gas Steam Reforming: An Innovative Process Design for the Capture of CO2. *Ind. Eng. Chem. Res.* **2015**, *54*, 6950–6958. [CrossRef]
9. Shu, J.; Grandjean, B.P.A.; Van Neste, A.; Kaliaguine, S. Catalytic palladium-based membrane reactors: A review. *Can. J. Chem. Eng.* **1991**, *69*, 1036–1060. [CrossRef]
10. Salehi, E.; Nel, W.; Save, S. Viability of GTL for the North America Gas Market. *Hydrocarbon Process.* **2013**, *92*, 41.
11. Baliban, R.C.; Elia, J.A.; Floudas, C.A. A Novel Natural Gas to Liquids Processes: Process Synthesis and Global Optimization Strategies. *AIChE J.* **2013**, *59*, 505–531. [CrossRef]
12. Roberts, K. Modular design of smaller-scale GTL plants. *Petrol. Technol. Quaterly* **2013**, *18*, 101–103.
13. Ibarra, A.; Veloso, A.; Bilbao, J.; Arandes, J.M.; Castaño, P. Dual coke deactivation pathways during the catalytic cracking of raw bio-oil and vacuum gasoil in FCC conditions. *Appl. Catal. B* **2016**, *182*, 336–346. [CrossRef]
14. Shelepova, E.V.; Vedyagin, A.A.; Mishakov, I.V.; Noskov, A.S. Simulation of hydrogen and propylene coproduction in catalytic membrane reactor. *Int. J. Hydrogen Energy* **2015**, *40*, 3592–3598. [CrossRef]
15. Quicker, P.; Hollein, V.; Dittmeyer, R. Catalytic dehydrogenation of hydrocarbons in palladium composite membrane reactors. *Catal. Today* **2000**, *56*, 21–34. [CrossRef]

16. Moparthi, A.; Uppaluri, R.; Gill, B.S. Economic feasibility of silica and palladium composite membranes for industrial dehydrogenation reactions. *Chem. Eng. Res. Des.* **2010**, *88*, 1088–1101. [CrossRef]
17. Gbenedio, E.; Wu, Z.T.; Hatim, I.; Kingsbury, B.F.K.; Li, K. A multifunctional Pd/alumina hollow fibre membrane reactor for propane dehydrogenation. *Catal. Today* **2010**, *156*, 93–99. [CrossRef]
18. Ockwig, N.W.; Nenoff, T.M. Membranes for Hydrogen Separation. *Chem. Rev.* **2007**, *107*, 4078–4110. [CrossRef] [PubMed]
19. Iaquaniello, G.; De Falco, M.; Salladini, A. Palladium-based Selective Membranes for Hydrogen Production. In *Membrane Engineering for the Treatment of Gases*; Drioli, E., Barbieri, G., Eds.; RSC Publishing: Cambridge, UK, 2011; Volume 2, pp. 110–136. ISBN 978-1-84973-239-0.
20. De Falco, M.; Iaquaniello, G.; Salladini, A. Steam reforming of natural gas in a reformer and membrane modules test plant: Plant design criteria and operating experience. In *Membrane Reactors for Hydrogen Production Processes*; De Falco, M., Marrelli, L., Iaquaniello, G., Eds.; Springer: New York, NY, USA, 2011; pp. 201–224. ISBN 978-0-85729-150-9.
21. Iaquaniello, G.; Salladini, A. Method for Hydrogen Production. EP Patent Application 11150491.6, 10 January 2011.
22. De Falco, M.; Iaquaniello, G.; Salladini, A. Experimental tests on steam reforming of natural gas in a reformer and membrane modules (RMM) plant. *J. Membr. Sci.* **2011**, *368*, 264–274. [CrossRef]
23. De Falco, M.; Salladini, A.; Iaquaniello, G. Reformer and membrane modules (RMM) for methane conversion: Experimental assessment and perspectives of said innovative architecture. *ChemSusChem* **2011**, *4*, 1157–1165. [CrossRef] [PubMed]
24. Iaquaniello, G.; Palo, E.; Salladini, A.; Cucchiella, B. Using palladium membrane reformers for hydrogen production. In *Palladium Membrane Technology for Hydrogen Production, Carbon Capture and Other Applications*; Doukelis, A., Panopoulos, K., Koumanakos, A., Kakara, E., Eds.; Woodhead Publishing: Sawston, UK, 2014; pp. 287–301. ISBN 9781782422341.
25. Gentile, A.; Morico, B.; Iaquaniello, G.; Giaconia, A.; Caputo, G. Development, testing and experimental results evaluation of a membrane reactor for solar steam reforming using molten salts as heat transfer fluid. In Proceedings of the ISCRE24, Minneapolis, MN, USA, 12–15 June 2016.
26. Iaquaniello, G.; Salladini, A.; Morico, B. Method and System for the Production of Hydrogen. US Patent US9493350B2, 15 November 2016.
27. Iaquaniello, G.; Palo, E.; Salladini, A. Application of membrane reactors: An industrial perspective. In Proceedings of the MR4PI2017, Villafrance de Verona, Italy, 9–10 March 2017.
28. Capoferri, D.; Cucchiella, B.; Mangiapane, A.; Abate, S.; Centi, G. Catalytic partial oxidation and membrane separation to optimise the conversion of natural gas to syngas and hydrogen. *ChemSusChem* **2011**, *4*, 1787–1795. [CrossRef] [PubMed]
29. Iaquaniello, G.; Salladini, A.; Palo, E.; Centi, G. Catalytic Partial Oxidation Coupled with Membrane Purification to Improve Resource and Energy Efficiency in Syngas Production. *ChemSusChem* **2015**, *8*, 717–725. [CrossRef] [PubMed]
30. Iaquaniello, G.; Palo, E.; Salladini, A. Membrane assisted syngas production for gas-to-liquid processes. In *Membrane Engineering for the Treatment of Gases: Volume 2: Gas-Separation Issues Combined with Membrane Reactors*; Drioli, E., Barbieri, G., Brunetti, A., Eds.; Royal Society of Chemistry: London, UK, 2017; pp. 247–272. ISBN 978-1-78262-875-0.
31. Palo, E.; Iaquaniello, G. Method for Olefins Production. US Patent 9,776,935B2, 3 October 2017.
32. Ricca, A.; Montella, F.; Iaquaniello, G.; Palo, E.; Salladini, A.; Palma, V. Membrane assisted propane dehydrogenation: Experimental investigation and mathematical modelling of catalytic reactions. *Catal. Today* **2018**. [CrossRef]
33. Ricca, A.; Palma, V.; Iaquaniello, G.; Palo, E.; Salladini, A. Highly selective propylene production in a membrane assisted catalytic propane dehydrogenation. *Chem. Eng. J.* **2017**, *330*, 1119–1127. [CrossRef]
34. Ricca, A.; Truda, L.; Iaquaniello, G.; Palo, E.; Palma, V. Pd-membrane Integration in a Propane Dehydrogenation Process for Highly Selective Propylene Production. *Int. J. Membr. Sci. Technol.* **2018**, *5*, 1–15. [CrossRef]
35. Morico, B.; Gentile, A.; Iaquaniello, G. Molten Salt Solar Steam Reforming: Process Schemes Analysis. In *Membrane Reactor Engineering Applications for a Greener Process Industry*; Basile, A., De Falco, M., Centi, G., Iaquaniello, G., Eds.; Wiley: Hoboken, NJ, USA, 2016; ISBN 9781118906804.

36. Lobera, M.P.; Tellez, C.; Herguido, J.; Menendez, M. Transient kinetic modelling of propane dehydrogenation over a Pt–Sn–K/Al_2O_3 catalyst. *Appl. Catal. A Gen.* **2008**, *349*, 156–164. [CrossRef]
37. Peters, T.A.; Polfus, J.M.; van Berkel, F.P.F.; Bredesen, R. Interplay between propylene and H_2S co-adsorption on the H_2 flux characteristics of Pd-alloy membranes employed in propane dehydrogenation (PDH) processes. *Chem. Eng. J.* **2016**, *304*, 134–140. [CrossRef]

membranes MDPI

Article

New Insight to the Effects of Heat Treatment in Air on the Permeation Properties of Thin Pd77%Ag23% Membranes

Nicla Vicinanza [1], Ingeborg-Helene Svenum [1], Thijs Peters [2], Rune Bredesen [2] and Hilde Venvik [1,*]

[1] Department of Chemical Engineering, Norwegian University of Science and Technology, 7491 Trondheim, Norway; niclavicinanza@gmail.com (N.V.); Ingeborg-Helene.Svenum@sintef.no (I.-H.S.)
[2] SINTEF Industry, P.O. Box 124 Blindern, N-0314 Oslo, Norway; Thijs.Peters@sintef.no (T.P.); Rune.Bredesen@sintef.no (R.B.)
* Correspondence: Hilde.j.Venvik@ntnu.no; Tel.: +47-7359-2831

Received: 22 August 2018; Accepted: 16 September 2018; Published: 10 October 2018

Abstract: Sputtered Pd77%Ag23% membranes of thickness 2.2–8.5 μm were subjected to a three-step heat treatment in air (HTA) to investigate the relation between thickness and the reported beneficial effects of HTA on hydrogen transport. The permeability experiments were complimented by volumetric hydrogen sorption measurements and atomic force microscopy (AFM) imaging in order to relate the observed effects to changes in hydrogen solubility and/or structure. The results show that the HTA—essentially an oxidation-reduction cycle—mainly affects the thinner membranes, with the hydrogen flux increasing stepwise upon HTA of each membrane side. The hydrogen solubility is found to remain constant upon HTA, and the change must therefore be attributed to improved transport kinetics. The HTA procedure appears to shift the transition from the surface to bulk-limited transport to lower thickness, roughly from ~5 to ≤2.2 μm under the conditions applied here. Although the surface topography results indicate that HTA influences the surface roughness and increases the effective membrane surface area, this cannot be the sole explanation for the observed hydrogen flux increase. This is because considerable surface roughening occurs during hydrogen permeation (no HTA) as well, but not accompanied by the same hydrogen flux enhancement. The latter effect is particularly pronounced for thinner membranes, implying that the structural changes may be dependent on the magnitude of the hydrogen flux.

Keywords: Pd-Ag membranes; hydrogen permeation; surface characterization; solubility; heat treatment

1. Introduction

Palladium-based membranes have been the focus of many studies due to their high hydrogen permeability and selectivity, which may find application in efficient separation technologies [1,2]. At temperature below ~300 °C and pressure below ~2 MPa, however, pure palladium undergoes the α-to-β phase transition that results in irreversible lattice strain. Over time, cycling of the temperature causes the Pd to become brittle; leading to fractures. In order to prevent hydrogen embrittlement, Pd is conveniently alloyed with other metals [3,4]. Silver is a widely used alloying element, reducing the α-to-β phase transition to below room temperature. Moreover, Pd-Ag alloys exhibit higher hydrogen permeability than pure palladium [4–6], with a maximum at ~23 wt.% of Ag [6].

Hydrogen flux can be also enhanced by heat treatment in air (HTA) procedures [7–19], which are essentially oxidation-reduction cycles. For thin membranes, typically in the range of 1–3 μm, the total hydrogen flux through the membrane can be doubled upon air thermal treatment [8–11,16], and HTA has also been used to regenerate deactivated membranes [14,20,21]. Moreover, Mejdell et al. reported a significant reduction of the CO inhibitive effects on hydrogen permeation through a ~3 μm Pd-Ag (23%) membrane after HTA [9]. The phenomena behind these advantageous effects are, however, still not clear. Different hypotheses have been suggested to explain the influence of heat treatment in air; including cleaning of the Pd-Ag surface, microstructural rearrangement, and induced segregation of Pd towards the surface. Oxidation is known to remove certain impurities and poisoning species from Pd-based surfaces [5,15,16,20]. This can facilitate and improve hydrogen flux, but is not a sufficient explanation that can fully elucidate the positive effects of HTA [14].

Microstructural changes are commonly observed in Pd-based membranes after air exposure, while the permeation properties are maintained or enhanced and even the stability/durability seems upheld. These include increase in surface roughness [10,13,14,17,22,23], as well as defect/void formation [8,11,14,17,19]. The increased surface roughness leads to an enlarged active surface area which can promote the hydrogen flux through the membrane given that surface phenomena are transport limiting. Furthermore, surface roughening is associated with grain growth in the membrane bulk [10,13,14,17,19]. As there is no clear agreement on whether larger grains inhibit [24–26] or enhance [27,28] hydrogen flux, it is not possible to irrefutably attribute hydrogen flux enhancement to formation of larger grains. Recently we established, however, a possible correlation between the solubility of hydrogen and the average grain boundary density in sputtered membranes, rendering the diffusivity practically unaffected as long as the hydrogen transport was controlled by bulk diffusion [29]. On the other hand, Zhang and co-workers [17] reported an increase in the hydrogen sorption kinetics that was attributed to higher hydrogen diffusivity for a 25 μm cold-rolled Pd-Ag 25 wt.% membrane after heat treatment in air at 300 °C for 1 day.

Another hypothesis is related to surface segregation of Pd as a result of the heat treatment in air, since H_2 dissociates over and binds to Pd but not to Ag. It is well established that Ag segregates to the membrane surface of ideal Pd-Ag alloys in absence of adsorbates; i.e., under vacuum or inert gas conditions, while a reverse segregation of Pd is induced after exposure of Pd-Ag surfaces to hydrogen and several other chemisorbing species [30–32]. The thermal treatment in air has been found to yield formation of a ~2 nm thick PdO layer [13] on the Pd-Ag surface and to an enrichment of Pd that can be involved in the enhancement of hydrogen flux through Pd-Ag membranes [13,15,16,32]. Segregation and rearrangement of the outmost atomic layers could also be linked, in the sense that this could affect in particular the transfer of hydrogen from the surface to the bulk or vice versa; processes that are particularly difficult to study experimentally.

In this work, a new approach to the heat treatment in air procedure has been applied in order to further investigate its effect on hydrogen transport properties in Pd-Ag membranes. The procedure was performed so that each side of the membranes was consecutively exposed to ambient air in order to probe the individual surface responses, which to our knowledge have not been previously addressed in the literature. A final HTA applied to both sides together was also performed in order to establish if additional effects exist. Different membrane thicknesses have been included to the investigation, as well as measurements of hydrogen solubility before and after heat treatment, to better understand the transport mechanisms involved. The membranes were characterized using atomic force microscopy (AFM) for as-grown membranes, hydrogen-stabilized, and heat-treated in air (with subsequent hydrogen stabilization), to monitor changes in surface topography.

2. Materials and Methods

2.1. Membrane Preparation

Pd77%Ag23% thin, self-supported membranes were produced by SINTEF using a unique two-step magneton sputtering technique onto silicon single crystal substrates [33,34]. Pd-Ag films with thicknesses of 2.2, 4.7, 6.9, 8.5, and 11.2 μm were studied. The membrane thickness was determined using white light interferometry. In this work, the membranes analyzed have been categorized as follows: 'as-grown' refers to Pd77%Ag23% thin film samples just pulled-off from the silicon substrate; 'hydrogen-stabilized' refers to membranes exposed to hydrogen permeation measurements only; 'air-treated' corresponds to membranes that have been heat-treated in air (HTA) and subsequently stabilized under hydrogen according to the procedures described below. The 'growth/feed side' of the membranes is the side growing during the magneton sputtering deposition and was always exposed to the feed gas during permeation experiments. The 'substrate/permeate side' refers to the membrane side facing the silicon wafer under fabrication, and was always kept to the permeate (low pressure) side of the membranes during permeation experiments.

2.2. Hydrogen Stabilization/Permeation

The permeation behavior under pure hydrogen (purity 99.999%) through the Pd77%Ag23% membranes was studied as a function of temperature and pressure with no use of sweep gas. All membranes were mounted in a microchannel configuration as depicted in Figure 1. The membranes were placed in between a polished stainless steel feed housing and a polished stainless steel plate. The feed housing had seven channels for gas flow corresponding to a total active surface area of 0.91 cm^2, in accordance with the permeate side steel plate geometry. On the permeate side, an open stainless steel housing was sealed to the perforated steel plate by a polished copper gasket. In conjunction with absence of sweep gas or dilutants, this configuration enables investigation of very thin membranes in absence of transport limitations from the gas phase (concentration polarization) or the support [35]. Membrane leakage was checked by using an Agilent 490 Micro-GC (Agilent Technologies, Santa Clara, CA, USA). No leakage could be detected during any of the permeation experiments both before and after the three-step heat treatment procedure. Experiments were performed at selected temperatures of 300, 350, and 400 °C. 300 °C was reached by ramping at 2 °C per minute with nitrogen (purity 99.999%) flushing on the feed side and argon (purity 99.999%) on the permeate side. After reaching 300 °C, nitrogen and argon were slowly removed from the system and hydrogen introduced. The permeate side was left at atmospheric pressure while a differential pressure of maximum 2 bar was applied on the retentate side. The permeate flow was measured by a bubble flow meter.

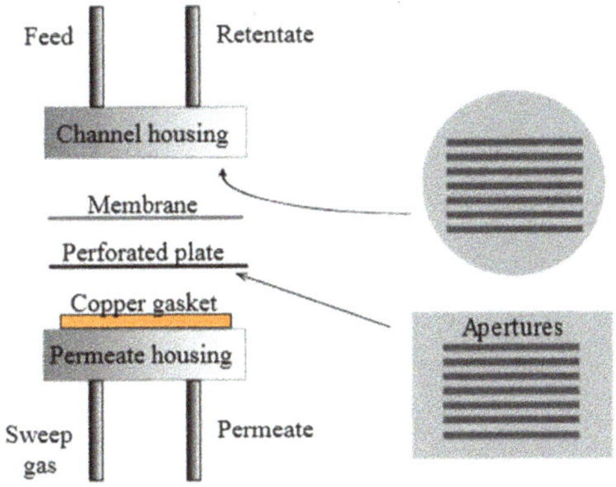

Figure 1. Sketch of the microchannel reactor configuration.

2.3. Heat Treatment in Air (HTA)

The three-step heat treatment in air (HTA) procedure was performed between hydrogen permeation experiments for three selected membrane thicknesses; 2.2, 4.7, and 8.5 µm respectively. The HTA was mainly performed using the following sequence: (i) permeate side, (ii) feed side, (iii) both sides one more time, with hydrogen permeation experiments between each step. Each HTA step was carried out at 300 °C for one hour. Before exposing the membranes to ambient air, hydrogen was removed from the system by introducing nitrogen on the feed side and argon on the permeate side for about 15 min. Nitrogen/argon were reintroduced for 15 min at the feed/permeate sides in order to flush out air. Hydrogen was then introduced again and nitrogen/argon slowly removed. As will be shown, HTA had the largest effect for the thinnest membranes. Another 2.2 μm membrane was therefore subjected to an alternative HTA sequence, denoted HTA2: (i) feed side, (ii) permeate side, (iii) both sides one more time, again with hydrogen permeation experiments between each step and procedures otherwise as described above.

2.4. Hydrogen Solubility

Equilibrium sorption measurements were carried out as described in [29] using an ASAP 2020 chemisorption analyzer (Micromeritics Instrument Corporation, Norcross, GA, USA) for 2.2 μm and 8.5 μm as-grown membranes, as well as after heat treatment in air. The heat treatment in air for the sorption samples was performed in a furnace at 300 °C under ambient air for one hour. Volumetric sorption was performed with a hydrogen pressure between 0.02 and 90.7 kPa. In every measurement, a sample mass close to 0.1 grams was used, taking into account also mass loss from degassing of the sample. The sorption measurements were carried out twice at three different temperatures: 300, 350, and 400 °C.

2.5. AFM Imaging

The surface topography was investigated by atomic force microscopy (AFM, Bruker, Boston, MA, USA) using a Multimode AFM instrument with a Veeco Multimode controller. All force spectroscopy analysis was performed in tapping mode under atmospheric conditions. Surface topography was investigated for both growth/feed side and substrate/permeate side for all the as-grown membranes, for hydrogen stabilized membranes and selected membranes after heat treatment in air with subsequent hydrogen stabilization. At least five surface scans were obtained at different locations for each sample. The first flattening order, provided by the AFM-instrument software (Nanoscope Software Version 7.2, by Veeco, Plainview, NY, USA), was performed in order to remove tilt and noise from all images. Surface roughness was estimated as the root mean square roughness (Rq) from the measured AFM images.

3. Results and Discussion

3.1. Permeability

Hydrogen permeation through Pd-Ag membranes generally follows the solution-diffusion mechanism, where Fick's law of diffusion describes the mass transport

$$J_{H_2} = \frac{P}{t}(p_1^n - p_2^n) = \frac{SD}{t}(p_1^n - p_2^n) \quad (1)$$

P is the permeability of the membrane, t the thickness, p_1 and p_2 the partial pressure of hydrogen on the high pressure side and low pressure side of the membrane, respectively, S the solubility, and D the diffusion coefficient. The n-value is determined by the rate-limiting step of the transport mechanism as explain further below. The diffusivity can be expressed as

$$D = D_0 \exp\left(-\frac{E_a}{RT}\right) \quad (2)$$

where D_0 is a pre-exponential factor and E_a the activation energy for diffusion.

Hydrogen permeance (P/t, Equation (1)) values, measured at 300 °C for untreated membranes (not subjected to HTA), are plotted in Figure 2 as a function of inverse thickness. In the calculation of the permeance, n = 0.5 was applied, which is essentially valid only with bulk diffusion as the rate-limiting step [36–41]. When the kinetics is bulk-limited, the permeance should be proportional to the inverse thickness, leading to a constant value for the permeability (a material property). In thick Pd-membranes, hydrogen generally forms a dilute solution and the transport is bulk-limited. The full line (Figure 2) refers to a bulk value of permeability equal to 1.5×10^{-8} mol·m·m^{-2}·s^{-1}·Pa$^{-0.5}$ that has been reported for a 100 μm thick Pd77%Ag23% membrane at 300 °C [5]. A previous investigation of pure Pd membranes in the thickness range 10 to 150 μm indicated that bulk diffusion was rate limiting for thicknesses above 20 μm [37]. As the thickness decreases, the transport mechanism may be controlled by a combination of surface effects and bulk diffusion ($0.5 < n < 1$) [42,43]. Eventually, surface phenomena such as adsorption/dissociation and/or association/desorption become completely rate-limiting, and the pressure exponent approaches unity [36–41]. Figure 2 indicates a bulk limited transport with permeance values somewhat higher than the bulk literature value for the thicker membranes (≤4.7 μm), while the thinnest membranes exhibit permeance values indicative of surface limitations affecting hydrogen permeation. The thickness at which this transition appears may depend both on the conditions (T, P, H_2 feed content) and the material properties as affected by the fabrication and eventual pre-treatment of the membrane. Several studies report that surface phenomena start to have an impact from 4–5 μm [10,29,40,44].

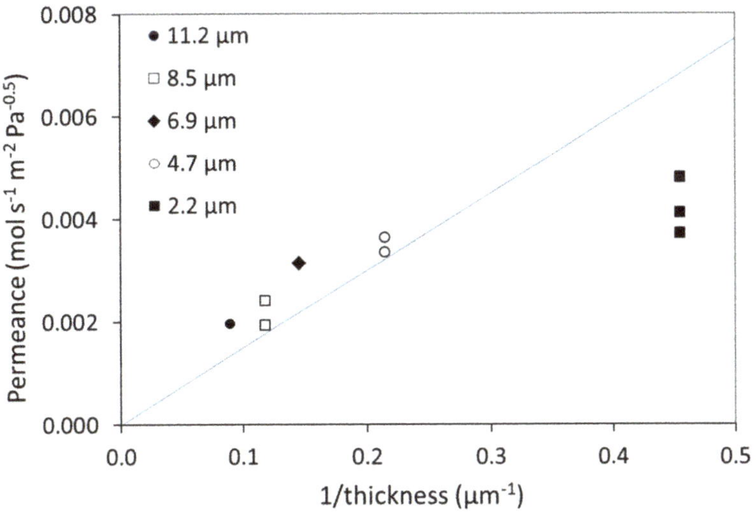

Figure 2. Measured hydrogen permeance as function of inverse thickness at 300 °C for membranes not subjected to HTA. The full line refers to values of permeance reported if bulk is the rate limiting step (1.5×10^{-8}·mol·s^{-1}·m^{-2}·Pa$^{-0.5}$) [5].

There are some variations in the measured permeance between membranes of the same thickness; larger for the 2.2 μm membrane relative to the 4.7 μm and 8.5 μm membrane. There could be several reasons for this variation, including experimental uncertainty. There may be a gradient in membrane thickness as large as ~10% across the silicon wafer, but we take care to mainly utilize the mid-sections for permeation experiments to reduce this deviation. Time and contamination during ambient storage may have an effect, as well as slight stretching of the material under total pressure difference [10]. Such sputtered membranes of similar thickness are, however, not found to exhibit major differences in microstructure, composition, or surface topography [10,45,46], but a general observation that will be further discussed below is that this variation between principally equivalent samples is also reduced by HTA [10].

After stabilization under hydrogen and measurement of the permeance, membranes with thicknesses of 2.2, 4.7, and 8.5 μm were subjected to the three-step HTA procedures as described above. The results are displayed in Figure 3 and Table 1. Figure 3a shows the hydrogen permeance plotted as a function of inverse thickness before HTA as well as after each step in the HTA procedure; n assumed equal to 0.5 also here. HTA has larger effect as the thickness decreases, with practically no effect for the 8.5 μm thick membrane. Moreover, the permeance increases after each of the first two HTA steps, i.e., upon oxidation-reduction of each side of the membrane. The first increase amounts to ~20% (HTA permeate/substrate) while the second (HTA feed/growth) is 60–40% (Table 1), depending on temperature, for the 2.2 μm membrane. There is no significant change when both sides are simultaneously exposed one more time to air. Figure 3a also implies that the permeance is inversely proportional to the membrane thickness after HTA. Hence, the oxidation-reduction cycle imposed by the HTA procedure apparently shifts the surface transport limitation to lower thickness and bulk diffusion becomes rate-limiting for the hydrogen transport over the whole thickness range and experimental conditions investigated. Moreover, the surface limitations appear to impose transport limitations on both sides initially, which can be lifted stepwise by oxidizing one side at the time.

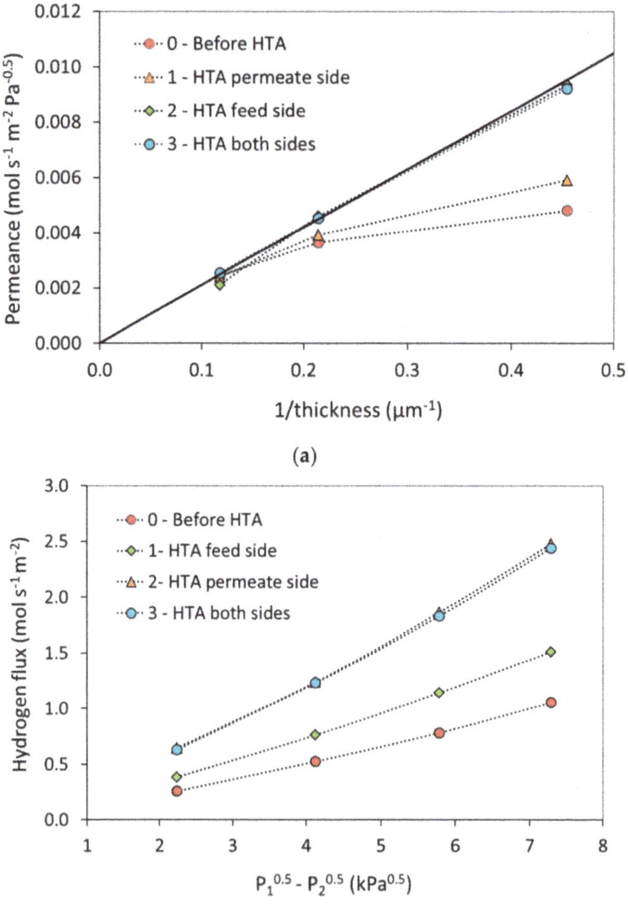

(**a**)

(**b**)

Figure 3. Permeance measured at 300 °C for each single step of heat treatment in air; (**a**) as function of inverse thickness for the main HTA membrane side sequence with the full line indicating a permeability of 2.1×10^{-8} mol·s^{-1}·m^{-2}·Pa$^{-0.5}$, and (**b**) as function of the difference in the square root of the hydrogen partial pressure for the HTA2 membrane side sequence applied to a 2.2 μm thick membrane.

In order to investigate the importance of the order of the heat treatment with respect to growth/feed or substrate/permeate side of the membrane an experiment in the opposite order (HTA2) was performed for the 2.2 μm thickness as mentioned above. The results are displayed in Figure 3b and in Table 1. The stepwise increase in permeability is definitely maintained, possibly with some differences due to the order, i.e., feed or permeate side first. The relative increases are now ~40% (HTA feed/growth) followed by 70–30% (HTA permeate/substrate) with increasing temperature, to eventually reach similar permeances values as for the main HTA sequence. Conclusions with respect to order are, however, complicated by the variation in values obtained for the hydrogen stabilization as discussed above.

Table 1. Permeability measured at 300 °C after before, between and after each of the three steps in the main HTA membrane side sequences for the 2.2 μm thick membranes.

T (°C)	Permeability 10^8 ($mol \cdot m \cdot m^{-2} \cdot s^{-1} \cdot Pa^{-0.5}$)							
	Main HTA Sequence				HTA2 Sequence			
	Before	Feed	Perm	Both	Before	Feed	Perm	Both
300	1.1	1.3	2.1	2.0	0.9	1.3	2.2	2.1
350	1.2	1.5	2.3	2.3	1.1	1.6	2.4	2.3
400	1.5	1.8	2.5	2.5	1.5	2.1	2.7	2.7

The dissociative adsorption of hydrogen over palladium surface atoms is practically non-activated [47]. It is thus difficult to ascribe the observed permeation increase on the feed side for either of the sequences in terms of adsorption properties only. Moreover, the hydrogen-stabilization as well as the HTA should have promoted Pd termination of the surface over Ag, but (temperature dependent) coverage effects may complicate the segregation behavior [48]. Nevertheless, the subsurface structure and elemental distribution may also be affected by the treatment and play a role in the transfer from the surface to the bulk and may affect both sides.

After exposure to air of both sides of the membranes, the permeability approaches a value of $2.1 \pm 0.1 \times 10^{-8} \cdot mol \cdot s^{-1} \cdot m^{-2} \cdot Pa^{-0.5}$ at 300 °C, as shown by Table 1 as well as the values for the 4.7 μm thickness (not shown). This is comparable to previous values reported in literature for similar membranes [10], for which the HTA also seems to partially diminish the variations between samples from different wafers or fabrication batches that is observed during the initial stabilization under hydrogen. The oxidation-reduction cycle may hence induce a higher degree of structural and compositional uniformity. The flux increase after HTA of both sides of the thinnest membrane corresponds to a doubling of the permeability, also in accordance with what has been previously reported in literature [10,13,14,16], although a comparison of permeabilities based on n equal to 0.5 is not valid in the strictest sense. The final permeability value at 300 °C is, however, higher than the value of $1.5 \times 10^{-8} \cdot mol \cdot m \cdot m^{-2} \cdot s^{-1} \cdot Pa^{-0.5}$ at 300 °C expected for bulk-limited transport reported in literature [5]. This could be related to differences in grain structure/density [29], grain orientation (the sputtered PdAg films contain predominantly grains oriented along <111> parallel/normal to the surface [49]), as well as purity, but requires further investigation.

3.2. *Hydrogen Solubility and Diffusivity*

In addition to hydrogen permeation experiments before and after the three-step heat treatment procedure, hydrogen sorption measurements were carried out to investigate possible variation caused by the HTA. The hydrogen solubility values were obtained before and after heat treatment in air of both sides simultaneously for 2.2 μm and 8.5 μm membranes, and the Sieverts' constants are reported in Table 2.

Table 2. Sieverts' constant measured at different temperature for 2.2 μm and 8.5 μm thick membranes before and after heat treatment in air.

Sample Thickness (μm)	Temperature (°C)	Sieverts' Constant (μmol/g·Pa$^{0.5}$)	
		Before HTA	After HTA
8.5	300	0.77	0.79
	350	0.53	0.63
	400	0.44	0.45
2.2	300	0.82	0.81
	350	0.57	0.60
	400	0.47	0.47

Solubility decreases with increasing temperature, as expected [3,50–54]. Moreover, the solubility increases as the thickness decreases as we have recently reported [29]. These thickness dependent changes in solubility were related to differences in the grain structure of the sputtered Pd-Ag films. After HTA of both sides, the values of Sieverts' constant remain basically unaffected for 300 and 400 °C. There is, however, some increase for the values at 350 °C, especially for the 8.5 μm. We are unsure whether this discrepancy can be considered an experimental error, but the data are not sufficient to conclude on significant HTA-induced changes in solubility. Previous results for a 25 μm cold-worked Pd-Ag25 wt.% indicated no change in hydrogen solubility after heat treatment in air [17,18].

Hydrogen diffusivities calculated based on reaction (1) for the 2.2 μm and 8.5 μm thick membranes before and after HTA are shown in Figure 4 as a function of inverse temperature. Estimated values for the pre-exponential factor (D_0) and activation energy (E_a) using Equation (2) are listed in Table 3. There are no significant changes found in diffusivity for the 8.5 μm membrane as both the permeability (Table 2) and solubility (Table 3) are constant before and after HTA. The solubility value obtained after HTA at 350 °C has been omitted from the fit in Figure 4, but the values remain if it is taken in. The apparent activation energy of 20 kJ/mol (before and after HTA) is comparable to results obtained for Pd-Ag with bulk-limited kinetics. Völkl and Alefeld collected data for the diffusion coefficient of hydrogen in Pd from 25 authors and estimated a mean value of E_a ~22.4 kJ/mol and D_0 ~2.9×10^{-7} m^2/s [55]. Moreover, Holleck calculated for a Pd80%Ag20% (0.08–0.20 cm thick) membrane values of E_a of ~22.3 ± 0.4 kJ/mol and D_0 ~$(2.33 \pm 0.2) \times 10^{-7}$ m^2/s [56]. This suggests that HTA does not change the hydrogen transport mechanism for thick sputtered Pd77%Ag23% membranes.

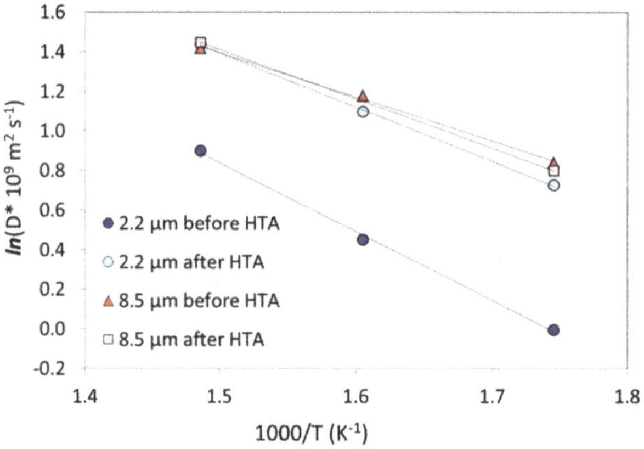

Figure 4. Arrhenius plot of the diffusivity for 2.2 μm and 8.5 μm thick membranes before and after HTA of both sides for the temperature range 300–400 °C. The diffusivity scale is presented in logarithmic form and the dotted lines are linear fits.

Table 3. Estimated values of the pre-exponential factor (D_0) and of the activation energy (E_a) for 2.2 μm and 8.5 μm thick membranes before and after the main HTA sequence.

Sample Thickness (μm)	Before HTA		After HTA	
	D_0 (m²/s)	E_a (kJ/mol)	D_0 (m²/s)	E_a (kJ/mol)
2.2	4.9×10^{-7}	29	3.6×10^{-7}	24
8.5	1.5×10^{-7}	20	1.7×10^{-7}	20

In the case of the 2.2 μm thick membrane, the permeability enhancement must be attributed to an apparent increase in the diffusivity since the solubility seems to remain constant. Both the pre-exponential factor and the activation energy changes after HTA: E_a decreases from 29 kJ/mol before HTA to 24 kJ/mol after HTA on both sides, and similar values were obtained for the opposite order of stepwise air-treatment (HTA2). Since E_a exceeds the values reported for thicker membranes in literature [56], this supports the idea that surface phenomena are rate-limiting before HTA. However, if the enhanced kinetics of hydrogen transport upon heat treatment in air is associated with a transition from surface to bulk limited transport, there should be a change in the n-value that essentially complicates the comparison by fitting. However, comprehensive analysis to extract n-values also requires a larger pressure range that what could been applied here [11,35]. Nevertheless, the activation energies associated with (associative) desorption should be generally higher than those for diffusion, but may depend on the surface composition (Pd/Ag) as well as the coverage as discussed in [29].

3.3. *Surface Topography*

AFM was used in order analyze how the hydrogen stabilization and HTA procedure affects the surface topography of the different membranes. Figures 5 and 6 show representative AFM images of 2.2 μm and 8.5 μm thick as-grown, hydrogen-stabilized, and air-treated membranes. The corresponding roughness values of all the measured membranes are reported in Table 4. The feed/growth side of as-grown samples shows an increase in the surface roughness as the thickness increases. The corresponding permeate/substrate side is very smooth (0.2–0.4 nm) with some variation in surface roughness between the different samples and thicknesses analyzed. This is in accordance with previous results for similar, sputtered membranes [10,29], and reflects the nature of the nucleation and growth phenomena involved during sputtering. A relatively high density of nuclei seem to form on the substrate, and then growth proceeds by continuing some grains while others are terminated [19].

Table 4. Surface roughness for as-grown membranes, after hydrogen stabilization and HTA with subsequent hydrogen stabilization obtained from AFM imaging analysis for both the growth/feed and substrate/permeate side. The analyzed areas are based on (5 × 5) μm² images except for the substrate/permeate side as-grown membranes where images with an area of (1 × 1) μm² are used.

Membrane Thickness (μm)	Roughness (nm)					
	As-Grown		Hydrogen Stabilization		HTA	
	Growth/ Feed	Substrate/ Permeate	Growth/ Feed	Substrate/ Permeate	Growth/ Feed	Substrate/ Permeate
2.2	8.4 ± 0.3	0.29 ± 0.02	13.8 ± 0.8	12.3 ± 0.6	20.3 ± 1.3	11.6 ± 0.6
4.7	10.7 ± 0.6	0.19 ± 0.01	18.2 ± 0.7	5.0 ± 0.4	24.6 ± 1.6	9.0 ± 0.2
6.9	11.8 ± 1.6	0.38 ± 0.04	12.0 ± 0.7	3.8 ± 0.6	-	-
8.5	10.2 ± 0.6	0.40 ± 0.03	24.0 ± 1.2	1.0 ± 0.07	26.9 ± 2.8	14.3 ± 0.6
11.2	13.2 ± 2.3	0.21 ± 0.01	20.2 ± 1.5	1.5 ± 0.3	-	-

Upon hydrogen permeation, the feed side surface roughness is found to increase moderately, while the effect of H_2 stabilization on the permeate side is found to be strongly thickness dependent. Hydrogen exposure/permeation has already been reported to increase the surface roughness of 1.3 μm sputtered Pd-Ag films [10]. For thick membranes (≥8.5 μm), hydrogen exposure causes only small changes to the surface structure. The permeate side becomes more roughened as the thickness decreases and the surface roughness of the 2.2 μm membrane becomes comparable for the two opposite sides. The reason behind this trend is not known, and it has—to our knowledge—not been reported before. The thickness dependent surface roughening on the permeate side under hydrogen permeation may possibly be connected to the fact that thinner membranes experience higher hydrogen flux than thicker membranes. In general, substantial roughening is associated with grain growth [19,29] and the higher hydrogen flux should hence facilitate restructuring and grain growth.

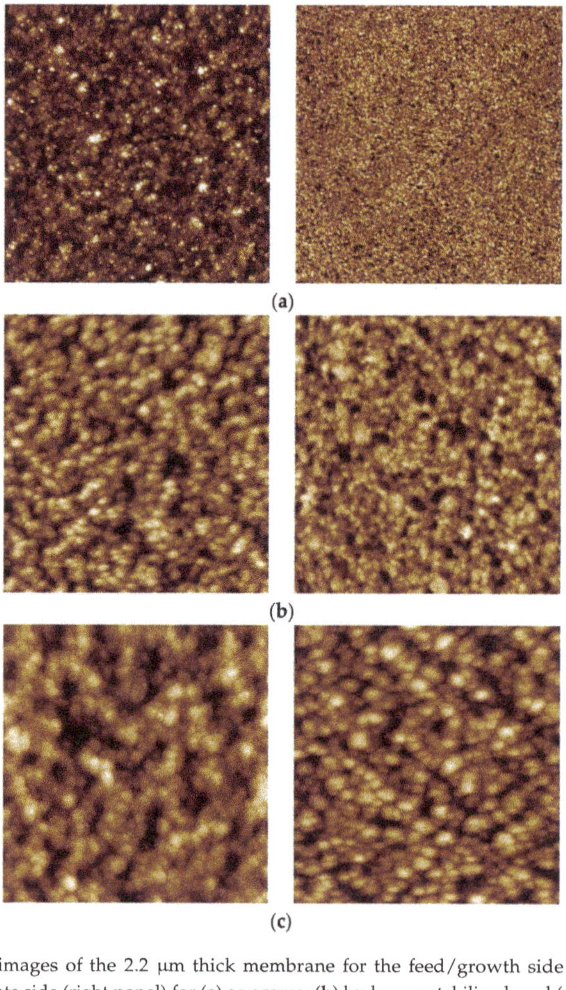

Figure 5. AFM images of the 2.2 μm thick membrane for the feed/growth side (left panel) and permeate/substrate side (right panel) for (**a**) as-grown; (**b**) hydrogen stabilized; and (**c**) HTA-subjected samples. Image areas: (**a**) right: 1 × 1 μm²; rest: 5 × 5 μm².

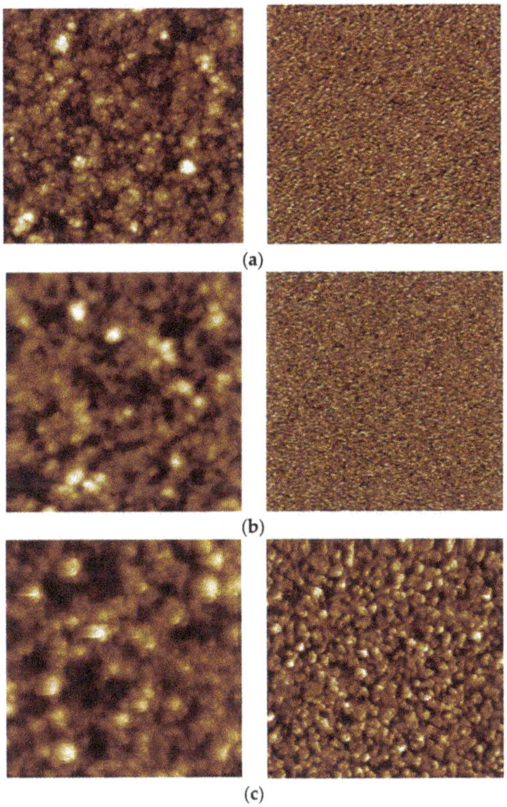

Figure 6. AFM images of the 8.5 μm thick membrane for the feed/growth side (left panel) and permeated/substrate side (right panel) for (**a**) as-grown; (**b**) hydrogen stabilized; and (**c**) HTA-subjected samples. Image areas: (**a**) right and (**b**) right: 1×1 μm^2; rest: 5×5 μm^2.

After the HTA procedure and subsequent hydrogen stabilization the surface topology undergoes further changes, depending on thickness. For the thickest (8.5 μm) membrane analyzed, the roughness of the feed side is mainly enhanced upon exposure to hydrogen while the HTA procedure has only minor effect (Figure 5). The permeate side, on the other hand, is strongly roughened after HTA. The feed side of the 4.7 μm thick membrane, already roughened during the hydrogen stabilization step, has yet another increase in roughness once oxidized. A similar behavior is also observed for the permeate side. Heat treatment in air contributes to a strong additional increase in surface roughness of feed side for the 2.2 μm membrane (Figure 4, Table 2). The permeate side roughness, instead, is enhanced only after hydrogen exposure to around 12 nm, and no further increase after HTA is observed.

Eventually, upon HTA the roughness values end up in the range of 20–30 nm on the growth/feed side and 9–15 nm on the substrate/permeate side, irrespective of the thickness dependent differences existing due to the growth process or developing during H_2 stabilization only. This is in agreement with several other investigations [10,13,14,17,22,23], reporting that the heat treatment in air helps the overall surface roughening process, thus creating new active sites and an increase of surface area. However, the roughening occurs also during hydrogen permeation—in particular for the thinner membranes—without observing an associated strong increase in flux. The permeation enhancement effect of the heat treatment in air can therefore not be fully attributed to an increase in surface area [14,19,57].

4. Conclusions

A three-step heat treatment in air has been performed on sputtered Pd77%Ag23% membranes with thickness ranging from 2.2 μm to 8.5 μm. The HTA is found to increase hydrogen permeability after each of the steps and to have a larger effect as the membrane thickness decreases. Air oxidation has no apparent effect on the hydrogen solubility, implying that enhancement in permeability may be related to an increase in diffusivity for thinner membranes ($\leq$5 μm). The data also suggest that bulk diffusion is the rate-limiting step for transport after HTA for all membranes, while surface phenomena are rate-limiting for thinner membranes prior thermal air treatment. Moreover, AFM studies reveal that the HTA increases the surface roughness of the membranes. However, significant surface roughening is already experienced upon stabilization under hydrogen, in particular for the thinner membranes, indicating that increase in permeability is not only attributable to increased surface area.

Author Contributions: Conceptualization, T.P. and H.V.; Methodology, T.P., H.V., and R.B.; Validation, N.V., H.V. and I.-H.S.; Investigation, N.V.; Resources, T.P. and N.V.; Writing—Original Draft Preparation, N.V. and I.-H.S.; Writing—Review & Editing, T.P. and H.V. and R.B.; Visualization, N.V., I.H.S.; Supervision, H.V.; Project Administration, T.P. and H.V.; Funding Acquisition, T.P. and R.B.

Funding: This research was funded by the Research Council of Norway through the RENERGI Program (190779/S60) and CLIMIT Program (215666/E20), and NTNU, SINTEF and Statoil ASA through the Gas Technology Centre NTNU-SINTEF.

Acknowledgments: Marit Stange is gratefully acknowledged for the manufacturing of the Pd-alloy films and MSc Live Nova Næss is acknowledged for contributing with some of the AFM data.

Conflicts of Interest: The authors declare no conflict of interest.

References

1. Paglieri, S.N.; Way, J.D. Innovations in palladium membrane research. *Sep. Purif. Methods* **2002**, *31*, 1–169. [CrossRef]
2. Bredesen, R.; Peters, T.A.; Stange, M.; Vicinanza, N.; Venvik, H.J. Chapter 11 Palladium-based Membranes in Hydrogen Production. In *Membrane Engineering for the Treatment of Gases: Volume 2: Gas-Separation Problems Combined with Membrane Reactors*; The Royal Society of Chemistry: Cambridge, UK, 2011; pp. 40–86.
3. Grashoff, G.J.; Pilkington, C.E.; Corti, C.W. The purification of hydrogen. A review of the technology emphasizing the current status of palladium membrane diffusion. *Platinum. Met. Rev.* **1983**, *27*, 157–169.
4. Kibria, A.K.M.F.; Sakamoto, Y. The effect of alloying of palladium with silver and rhodium on the hydrogen solubility, miscibility gap and hysteresis. *Int. J. Hydrogen Energy* **2000**, *25*, 853–859. [CrossRef]
5. Itoh, N.; Xu, W.C. Selective Hydrogenation of Phenol to Cyclohexanone Using Palladium-Based Membranes as Catalysts. *Appl. Catal. A Gen.* **1993**, *107*, 83–100. [CrossRef]
6. Uemiya, S.; Matsuda, T.; Kikuchi, E. Hydrogen Permeable Palladium Silver Alloy Membrane Supported on Porous Ceramics. *J. Membr. Sci.* **1991**, *56*, 315–325. [CrossRef]
7. Fort, D.; Farr, J.P.G.; Harris, I.R. A comparison of palladium-silver and palladium-yttrium alloys as hydrogen separation membranes. *J. Less Common Met.* **1975**, *39*, 293–308. [CrossRef]
8. Keuler, J.N.; Lorenzen, L. Developing a heating procedure to optimise hydrogen permeance through Pd–Ag membranes of thickness less than 2.2 μm. *J. Membr. Sci.* **2002**, *195*, 203–213. [CrossRef]
9. Mejdell, A.L.; Chen, D.; Peters, T.A.; Bredesen, R.; Venvik, H.J. The effect of heat treatment in air on CO inhibition of a ~3 μm Pd–Ag (23 wt.%) membrane. *J. Membr. Sci.* **2010**, *350*, 371–377. [CrossRef]
10. Mejdell, A.L.; Klette, H.; Ramachandran, A.; Borg, A.; Bredesen, R. Hydrogen permeation of thin, free-standing Pd/Ag23% membranes before and after heat treatment in air. *J. Membr. Sci.* **2008**, *307*, 96–104. [CrossRef]
11. Peters, T.A.; Stange, M.; Bredesen, R. On the high pressure performance of thin supported Pd–23%Ag membranes—Evidence of ultrahigh hydrogen flux after air treatment. *J. Membr. Sci.* **2011**, *378*, 28–34. [CrossRef]
12. Pizzi, D.; Worth, R.; Baschetti, M.G.; Sarti, G.C.; Noda, K.-I. Hydrogen permeability of 2.5 μm palladium–silver membranes deposited on ceramic supports. *J. Membr. Sci.* **2008**, *325*, 446–453. [CrossRef]

13. Ramachandran, A.; Tucho, W.M.; Mejdell, A.L.; Stange, M.; Venvik, H.J.; Walmsley, J.C.; Holmestad, R.; Bredesen, R.; Borg, A. Surface characterization of Pd/Ag23 wt.% membranes after different thermal treatments. *Appl. Surf. Sci.* **2010**, *256*, 6121–6132. [CrossRef]
14. Roa, F.; Way, J.D. The effect of air exposure on palladium–copper composite membranes. *Appl. Surf. Sci.* **2005**, *240*, 85–104. [CrossRef]
15. Yang, L.; Zhang, Z.; Gao, X.; Guo, Y.; Wang, B.; Sakai, O.; Sakai, H.; Takahashi, T. Changes in hydrogen permeability and surface state of Pd–Ag/ceramic composite membranes after thermal treatment. *J. Membr. Sci.* **2005**, *252*, 145–154. [CrossRef]
16. Yang, L.; Zhang, Z.; Yao, B.; Gao, X.; Sakai, H.; Takahashi, T. Hydrogen permeance and surface states of Pd-Ag/ceramic composite membranes. *AIChE J.* **2006**, *52*, 2783–2791. [CrossRef]
17. Zhang, K.; Gade, S.K.; Hatlevik, Ø.; Way, J.D. A sorption rate hypothesis for the increase in H_2 permeability of palladium-silver (Pd–Ag) membranes caused by air oxidation. *Int. J. Hydrogen Energy* **2012**, *37*, 583–593. [CrossRef]
18. Zhang, K.; Gade, S.K.; Way, J.D. Effects of heat treatment in air on hydrogen sorption over Pd–Ag and Pd–Au membrane surfaces. *J. Membr. Sci.* **2012**, *403–404*, 78–83. [CrossRef]
19. Tucho, W.M.; Venvik, H.J.; Walmsley, J.C.; Stange, M.; Ramachandran, A.; Mathiesen, R.H.; Borg, A.; Bredesen, R.; Holmestad, R. Microstructural studies of self-supported (1.5–10 μm) Pd/23 wt.%Ag hydrogen separation membranes subjected to different heat treatments. *J. Mater. Sci.* **2009**, *44*, 4429–4442. [CrossRef]
20. Ali, J.K.; Newson, E.J.; Rippin, D.W.T. Deactivation and regeneration of Pd-Ag membranes for dehydrogenation reactions. *J. Membr. Sci.* **1994**, *89*, 171–184. [CrossRef]
21. Peters, T.A.; Kaleta, T.; Stange, M.; Bredesen, R. Hydrogen transport through a selection of thin Pd-alloy membranes: Membrane stability, H_2S inhibition, and flux recovery in hydrogen and simulated WGS mixtures. *Catal. Today* **2012**, *193*, 8–19. [CrossRef]
22. Hideo, U.; Tsugio, O.; Kazutaka, S. New Technique of Activating Palladium Surface for Absorption of Hydrogen or Deuterium. *Jpn. J. Appl. Phys.* **1993**, *32*, 5095.
23. Wang, D.; Clewley, J.D.; Flanagan, T.B.; Balasubramaniam, R.; Shanahan, K.L. Enhanced rates of hydrogen absorption resulting from oxidation of Pd or internal oxidation of Pd-Al alloys. *J. Alloys Compd.* **2000**, *298*, 261–273. [CrossRef]
24. Bryden, K.J.; Ying, J.Y. Nanostructured palladium–iron membranes for hydrogen separation and membrane hydrogenation reactions. *J. Membr. Sci.* **2002**, *203*, 29–42. [CrossRef]
25. Natter, H.; Wettmann, B.; Heisel, B.; Hempelmann, R. Hydrogen in nanocrystalline palladium. *J. Alloys Compd.* **1997**, *253*, 84–86. [CrossRef]
26. Stuhr, U.; Striffler, T.; Wipf, H.; Natter, H.; Wettmann, B.; Janssen, S.; Hempelmann, R.; Hahn, H. An investigation of hydrogen diffusion in nanocrystalline Pd by neutron spectroscopy. *J. Alloys Compd.* **1997**, *253–254*, 393–396. [CrossRef]
27. McCool, B.A.; Lin, Y.S. Nanostructured thin palladium-silver membranes: Effects of grain size on gas permeation properties. *J. Mater. Sci.* **2001**, *36*, 3221–3227. [CrossRef]
28. Okazaki, J.; Ikeda, T.; Tanaka, D.A.P.; Suzuki, T.M.; Mizukami, F. In situ high-temperature X-ray diffraction study of thin palladium/α-alumina composite membranes and their hydrogen permeation properties. *J. Membr. Sci.* **2009**, *335*, 126–132. [CrossRef]
29. Vicinanza, N.; Svenum, I.-H.; Næss, L.N.; Peters, T.A.; Bredesen, R.; Borg, A.; Venvik, H.J. Thickness dependent effects of solubility and surface phenomena on the hydrogen transport properties of sputtered Pd77%Ag23% thin film membranes. *J. Membr. Sci.* **2015**, *476*, 602–608. [CrossRef]
30. Løvvik, O.M.; Opalka, S.M. Reversed surface segregation in palladium-silver alloys due to hydrogen adsorption. *Surf. Sci.* **2008**, *602*, 2840–2844. [CrossRef]
31. Shu, J.; Bongondo, B.E.W.; Grandjean, B.P.A.; Adnot, A.; Kaliaguine, S. Surface segregation of PdAg membranes upon hydrogen permeation. *Surf. Sci.* **1993**, *291*, 129–138. [CrossRef]
32. Svenum, I.H.; Herron, J.A.; Mavrikakis, M.; Venvik, H.J. Adsorbate-induced segregation in a PdAg membrane model system: $Pd_3Ag(1\ 1\ 1)$. *Catal. Today* **2012**, *193*, 111–119. [CrossRef]
33. Bredesen, R.; Klette, H. Method of Manufacturing Thin Metal Membranes. U.S. Patent 6,086,729, 11 July 2000.
34. Klette, H.; Bredesen, R. Sputtering of very thin palladium-alloy hydrogen separation membranes. *Membr. Technol.* **2005**, *2005*, 7–9. [CrossRef]

35. Mejdell, A.L.; Jondahl, M.; Peters, T.A.; Bredesen, R.; Venvik, H.J. Experimental investigation of a microchannel membrane configuration with a 1.4 μm Pd/Ag 23 wt.% membrane—Effects of flow and pressure. *J. Membr. Sci.* **2009**, *327*, 6–10. [CrossRef]
36. Collins, J.P.; Way, J.D. Preparation and characterization of a composite palladium-ceramic membrane. *Ind. Eng. Chem. Res.* **1993**, *32*, 3006–3013. [CrossRef]
37. Hurlbert, R.C.; Konecny, J.O. Diffusion of hydrogen through palladium. *J. Chem. Phys.* **1961**, *34*, 655–658. [CrossRef]
38. Keurentjes, J.T.F.; Gielens, F.C.; Tong, H.D.; van Rijn, C.J.M.; Vorstman, M.A.G. High-Flux Palladium Membranes Based on Microsystem Technology. *Ind. Eng. Chem. Res.* **2004**, *43*, 4768–4772. [CrossRef]
39. Moss, T.S.; Peachey, N.M.; Snow, R.C.; Dye, R.C. Multilayer metal membranes for hydrogen separation. *Int. J. Hydrogen Energy* **1998**, *23*, 99–106. [CrossRef]
40. Nam, S.-E.; Lee, S.-H.; Lee, K.-H. Preparation of a palladium alloy composite membrane supported in a porous stainless steel by vacuum electrodeposition. *J. Membr. Sci.* **1999**, *153*, 163–173. [CrossRef]
41. Ward, T.L.; Dao, T. Model of hydrogen permeation behavior in palladium membranes. *J. Membr. Sci.* **1999**, *153*, 211–231. [CrossRef]
42. Dittmeyer, R.; Höllein, V.; Daub, K. Membrane reactors for hydrogenation and dehydrogenation processes based on supported palladium. *J. Mol. Catal. A Chem.* **2001**, *173*, 135–184. [CrossRef]
43. Uemiya, S. State-of-the-art of supported metal membranes for gas separation. *Sep. Purif. Methods* **1999**, *28*, 51–85. [CrossRef]
44. Wu, L.-Q.; Xu, N.; Shi, J. Novel method for preparing palladium membranes by photocatalytic deposition. *AIChE J.* **2000**, *46*, 1075–1083. [CrossRef]
45. Mekonnen, W.; Arstad, B.; Klette, H.; Walmsley, J.C.; Bredesen, R.; Venvik, H.; Holmestad, R. Microstructural characterization of self-supported 1.6 μm Pd/Ag membranes. *J. Membr. Sci.* **2008**, *310*, 337–348. [CrossRef]
46. Peters, T.A.; Tucho, W.M.; Ramachandran, A.; Stange, M.; Walmsley, J.C.; Holmestad, R.; Borg, A.; Bredesen, R. Thin Pd–23%Ag/stainless steel composite membranes: Long-term stability, life-time estimation and post-process characterization. *J. Membr. Sci.* **2009**, *326*, 572–581. [CrossRef]
47. Padama, A.A.B.; Kasai, H.; Budhi, Y.W.; Arboleda, N.B. Ab initio Investigation of Hydrogen Atom Adsorption and Absorption on Pd(110) Surface. *J. Phys. Soc. Jpn.* **2012**, *81*, 114705. [CrossRef]
48. Fernandes, V.R.; Bossche, M.V.d.; Knudsen, J.; Farstad, M.H.; Gustafson, J.; Venvik, H.J.; Gronbeck, H.; Borg, A. Reversed Hysteresis during CO Oxidation over $Pd_{75}Ag_{25}$(100). *ACS Catal.* **2016**, *6*, 4154–4161. [CrossRef]
49. Arstad, B.; Venvik, H.; Klette, H.; Walmsley, J.C.; Tucho, W.M.; Holmestad, R.; Holmen, A.; Bredesen, R. Studies of self-supported 1.6 mu m Pd/23 wt.% Ag membranes during and after hydrogen production in a catalytic membrane reactor. *Catal. Today* **2006**, *118*, 63–72. [CrossRef]
50. Sieverts, A. Absorption of gases by metals. *Z. Metallkd.* **1929**, *21*, 37–46.
51. Hughes, D.T.; Harris, I.R. A comparative study of hydrogen permeabilities and solubilities in some palladium solid solution alloys. *J. Less Common Met.* **1978**, *61*, P9–P21. [CrossRef]
52. Picard, C.; Kleppa, O.J.; Boureau, G. High-temperature thermodynamics of the solutions of hydrogen in palladium-silver alloys. *J. Chem. Phys.* **1979**, *70*, 2710–2719. [CrossRef]
53. Bhargav, A.; Jackson, G.S. Thermokinetic modeling and parameter estimation for hydrogen permeation through $Pd_{0.77}Ag_{0.23}$ membranes. *Int. J. Hydrogen Energy* **2009**, *34*, 5164–5173. [CrossRef]
54. Fort, D.; Harris, I.R. Physical properties of some palladium alloy hydrogen diffusion membrane materials. *J. Less Common Met.* **1975**, *41*, 313–327. [CrossRef]
55. Voelkl, J.; Alefeld, G. Diffusion of hydrogen in metals. *Top. Appl. Phys.* **1978**, *28*, 321–348.
56. Holleck, G.L. Diffusion and solubility of hydrogen in palladium and palladium—Silver alloys. *J. Phys. Chem.* **1970**, *74*, 503–511. [CrossRef]
57. Tucho, W.M.; Venvik, H.J.; Stange, M.; Walmsley, J.C.; Holmestad, R.; Bredesen, R. Effects of thermal activation on hydrogen permeation properties of thin, self-supported Pd/Ag membranes. *Sep. Purif. Technol.* **2009**, *68*, 403–410. [CrossRef]

membranes

MDPI

Article

Grain Boundary Segregation in Pd-Cu-Ag Alloys for High Permeability Hydrogen Separation Membranes

Ole Martin Løvvik [1,*], Dongdong Zhao [2], Yanjun Li [2], Rune Bredesen [1] and Thijs Peters [1]

1 SINTEF Industry, N-0314 Oslo, Norway; Rune.Bredesen@sintef.no (R.B.); Thijs.Peters@sintef.no (T.P.)
2 Department of Materials Science and Engineering, Norwegian University of Science and Technology (NTNU), 7491 Trondheim, Norway; dongdong.zhao@ntnu.no (D.Z.); yanjun.li@ntnu.no (Y.L.)
* Correspondence: ole.martin.lovvik@sintef.no

Received: 25 June 2018; Accepted: 2 September 2018; Published: 12 September 2018

Abstract: Dense metal membranes that are based on palladium (Pd) are promising for hydrogen separation and production due to their high selectivity and permeability. Optimization of alloy composition has normally focused on bulk properties, but there is growing evidence that grain boundaries (GBs) play a crucial role in the overall performance of membranes. The present study provides parameters and analyses of GBs in the ternary Pd-Ag-Cu system, based on first-principles electronic structure calculations. The segregation tendency of Cu, Ag, and vacancies towards 12 different coherent Σ GBs in Pd was quantified using three different procedures for relaxation of supercell lattice constants, representing the outer bounds of infinitely elastic and stiff lattice around the GBs. This demonstrated a clear linear correlation between the excess volume and the GB energy when volume relaxation was allowed for. The point defects were attracted by most of the GBs that were investigated. Realistic atomic-scale models of binary Pd-Cu and ternary Pd-Cu-Ag alloys were created for the Σ5(210) boundary, in which the strong GB segregation tendency was affirmed. This is a starting point for more targeted engineering of alloys and grain structure in dense metal membranes and related systems.

Keywords: membrane; hydrogen; palladium alloy; grain boundary

1. Introduction

Cost-effective production of ultra-pure hydrogen can facilitate the widespread implementation of fuel cells and is one of the remaining bottlenecks before hydrogen can be introduced as an energy carrier on a large scale [1]. Dense metal membranes with high hydrogen permeance and selectivity have been identified as a promising enabling technology for efficiency improvement and cost reduction of hydrogen production. In particular, Pd-based hydrogen separation membranes are known to have 100% selectivity and high permeability, and thus allow for direct production of high purity hydrogen for use in fuel cells [2–5]. Combining these membranes with appropriate catalysts in membrane reactors to produce hydrogen from different sources has been described in numerous studies [1,6–8].

It appears that the potential of binary Pd-based membranes has been exhausted in the literature, and several groups have recently started working on ternary compounds as the next generation membrane material [9–20]. This has many possible benefits: the surface can be engineered to enhance the tolerance to impurity gases [13], the permeability can be optimized beyond what is possible with binary alloys [14], and the mechanical strength can be increased (e.g., if the self-diffusion is hindered or the morphology is changed) [2]. There is also potentially a cost reduction that is involved if expensive elements are replaced with cheaper ones [21]. The challenge with this approach is that ternary compounds are difficult to engineer when using plating, rolling, etc. as processing techniques [14]. One solution is to use a non-equilibrium process like magnetron sputtering to synthesize the active membrane material, which gives higher control of the material composition as well as the possibility of generating very thin membranes [14,22,23].

First-principles modelling has been demonstrated to be a powerful tool in the development of membrane materials [24]; as an example, density functional theory (DFT) has been used to systematically screen for novel binary intermetallic systems for hydrogen separation membranes [25]. A few studies have also investigated ternary alloys for membranes. The H_2 permeability of the $CuPd_{1-n}M_n$ system was studied by Sholl et al. [26,27] and Gao et al. [21], Pd-Ag-Cu by Ling et al. [17], while Løvvik et al. investigated sulfur adsorption on alloys in the $Pd_{1-n}AgM_n$ system [28].

Structural defects can play an important role in the kinetics of hydrogen in metals. It is well-known that hydrogen can segregate towards GBs in pure metals [29,30] and the local composition at GBs is thus of large interest for hydrogen permeation through dense metal membranes. Several DFT studies have investigated GB segregation in metals, e.g., in Ni [31], Ti [32], and Fe [33]. Other studies have focused on studying GB energies in alloys, e.g., in NiTi [34], Ti-Mo and Ti-V alloys [35], and in the ternary Ni_2MnGa [36], and Ni-Al-Co systems [37]. Some studies have also investigated GB segregation in alloys, e.g., in Cu-Ag [38], various binary V alloys [39], and in the ternary Mg-Zn-Y [40] and FeCrNi [41] alloys. We are, however, not aware of any previous studies on GB segregation in Pd alloys that are based on first-principles calculations.

This work has investigated the effect of GBs on the distribution of elements in binary and ternary Pd alloys. This was done by studying increasingly complicated models, reflected by the structure of this paper. Initially, the properties of 12 different coherent $\sum$ GBs of pure fcc Pd will be presented. The emphasis is on properties that are relevant for segregation of defects, like the excess volume and bond range. A systematic investigation of the segregation tendency of three different defects in pure Pd will then be presented: single Ag and Cu solutes, as well as vacancies. This gives a general knowledge of defect segregation in the dilute limit and is potentially relevant for all alloys in the Pd-Ag-Cu alloy system. However, real Pd alloys display concentrations of e.g., Ag and Cu far beyond the dilute limit, so these results are not necessarily valid in realistic alloy systems. In order to corroborate this, the segregation tendency of Cu in a Pd-Cu alloy with 20–25% Cu is presented next. This Cu level was selected for the following reasons: Pd-Cu alloys have excellent sulfur tolerance [14], but exhibit rather low hydrogen permeability, except around the composition $Pd_{0.6}Cu_{0.4}$, where the crystal structure is bcc [12]. However, the bcc area at $Pd_{0.6}Cu_{0.4}$ is very narrow and difficult to obtain during preparation; the present study has therefore focused on the composition around $Pd_{0.8}Cu_{0.2}$, which has a good trade-off between sulfur tolerance and permeability. The last part investigates whether the segregation trends in the binary alloy systems hold when moving to ternary alloys. It is also focused on the region around $Pd_{0.8}Cu_{0.2}$, with Ag replacing Pd.

2. Methodology

The calculations made use of the Vienna ab initio simulation package (VASP) [42,43], employing plane-wave basis functions with the projector augmented wave method [44] and the density functional theory (DFT) at the Perdew-Burke-Ernzerhof generalized gradient approximation (GGA) level [45]. The self-consistency requirement was changes in the total electronic energy less than 10^{-5} eV. The force relaxation criterion was 0.01 eV/Å, using the RMM-DIIS quasi-Newton method. Choosing a plane-wave cut-off energy of 500 eV and a k-point density of at least four points per $Å^{-1}$ gave a numerical precision of better than 1 meV for relative total electronic energies. The GB models have been implemented in the supercell scheme, and the various supercells used are shown in Figure 1.

Three different schemes were employed as volume relaxation techniques (VRT): no relaxation, full relaxation of all cell parameters (including both cell size and shape), and only the relaxation of the cell length perpendicular to the GB plane (x). The atomic positions were relaxed in all the cases. Even if the present models do not physically resemble grains in a real material (the "grains" are sheets with infinite extension perpendicular to x), we can still learn about real materials by assessing how the different VRTs correspond to limiting cases of large or small domains in different directions. With this approach, the situation when only x is allowed to relax corresponds to the kind of relaxation that would find place for infinitely large grains, i.e., one single GB. With only one semi-infinite GB, the x direction will be fully relaxed, while the infinitely large lattice perpendicular to the GB is equivalent to an infinitely stiff lattice in the other directions. The relaxation along very large grains should thus be well represented by that found in our models when only x is allowed to relax. This option is not normally available in VASP but has been facilitated in a local version of the code. The relaxation procedures when either all or none lattice parameters are allowed to relax do not correspond directly to any physical distribution of GB in a real material. However, the two methods represent the outer limits of the kind of relaxation that would take place in a material with very small grains. In this case, relaxation perpendicular to any GB (corresponding to x in our models) will be countered by nearby GBs with other orientations. Similarly, relaxation along GBs will be partially allowed due to the finite size of the GB and the short distance to neighbor grains in all directions. Thus, the relaxation taking place in a material with very small grains should be in between the relaxation that is found when keeping the lattice fixed and when it is fully relaxed. We will in the following present results for all VRTs, hence describing relaxation effects that are likely to be seen in materials with large and small grain size.

3. Results

3.1. Structure and Stability of Pure Pd GB Models

Initially, the structure and stability of pure Pd GB models are investigated. The 12 different GBs investigated in this work are listed in Table 1. They are based on the coincidence site lattice model [46] and constitute all coherent tilt GBs with Σ up to 13, thus exhibiting a quite wide range of deviations from the pure crystal. They have been represented by periodic atomistic models, as shown in Figure 1, where pure Pd models without point defects are displayed. These models range between 44 and 100 atoms (see Table 1), and their size in the x direction is between 13 Å (Σ5(310)) and 40 Å (Σ13(510)). Each model has two cancelling GBs: One at the unit cell boundary and one in the middle, identified by the dotted, red lines. The relatively large variation in size is partially due to symmetry (the models are often based on the smallest possible supercells with two such cancelling GBs) and the differing interaction range between various GBs—complex GBs typically display strain fields with larger range than simple ones.

Some inherent and calculated properties of the GB models are listed in Table 1. The excess volume per interface area V_X is defined as the difference between the DFT relaxed unit cell volume and the corresponding volume of Pd atoms in the bulk, divided by the cross-section area A of the GB plane (perpendicular to x). The unit of V_X is length and it can naively be interpreted as the accumulated extension of the lattice in the x direction due to the GB (when divided by 2, since there are two GBs per unit cell). The GB energy is defined as

$$\gamma = (E(\mathrm{GB}) - N_{\mathrm{GB}}E(\mathrm{bulk}))/2A \tag{1}$$

where the total energy E of the GB and of the bulk is that calculated by DFT, N_{GB} is the number of atoms in the GB model, and the factor 2 is due to the presence of two GBs in each unit cell. The bond range Δr_b is the difference between the longest and shortest relaxed bond within the first coordination sphere of the atoms in each model. It is centered around the DFT equilibrium Pd-Pd distance of 2.80 Å with the shortest recorded relaxed bond distance being 2.43 Å and the longest one 3.46 Å; this gives Δr_b up to 1.04 Å, which is seen for the Σ5(210) model. The distinction between the coordination spheres gets unclear for some of the models, and we have used a cut-off of 3.5 Å on the far side of this definition, which typically is a minimum between the first and second coordination spheres in these systems.

We first note that V_X depends strongly on the VRT. This is not surprising, since the calculations with no volume relaxation are at the mercy of the initial model. When the unit cell is allowed to relax partially (only x) or fully, the spread in V_X is smaller and the values of each model is consistent between the two procedures allowing for relaxation. V_X is consistently smaller in the case of full relaxation, since a part of the volume expansion then is obtained within the GB plane. Keeping in mind that the only x relaxation corresponds to large grains while the relaxation of samples with small grains should be in between that of full relaxation and no relaxation, it appears that V_X should be quite similar in magnitude, regardless of the typical grain size; the only x value is usually between the two others. The values vary between that of the $\sum3(111)$ model (0.01–0.06 Å) and the $\sum7(41\bar{5})$ model (0.70–0.90 Å). These numbers may not be fully converged due to the limited size of the unit cells but are clear indicators of the deviation from perfect stacking. The former model can be viewed merely as a stacking fault, while the latter has quite large deviations from the perfect structure along the GB (see Figure 1).

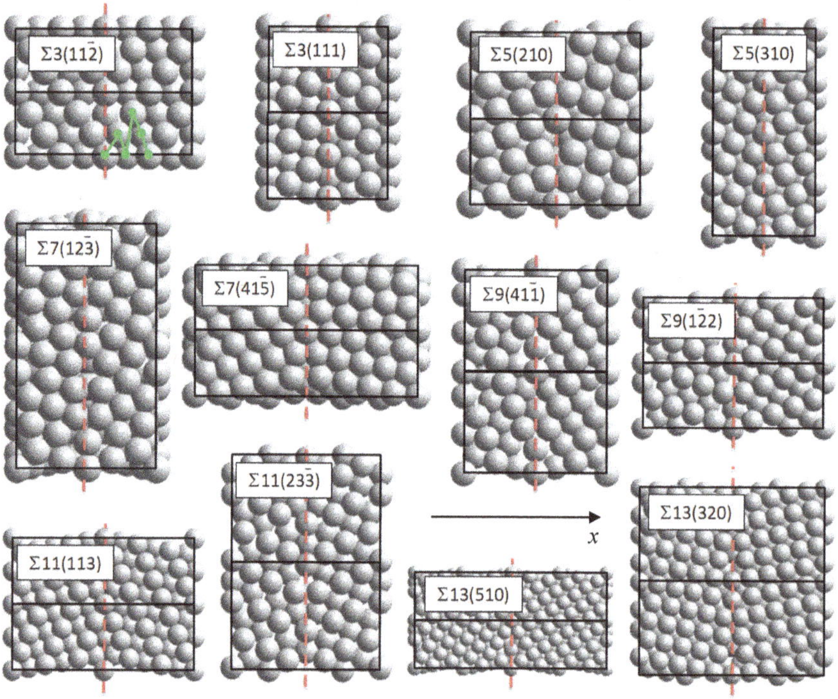

Figure 1. The different periodic grain boundary (GB) models included in this study. The dotted, red lines distinguish the GB planes. Unit cells are outlined with black lines, and the x direction is marked by the arrow. x_{gb} is defined as the horizontal distance from the GB planes. The supercell size perpendicular to the figure plane (the z direction) is listed in Table 1. The atomic positions plotted in Figure 3 have been shown as green dots connected by a solid line for the $\sum3(11\bar{2})$ model.

The same pattern is found when studying the GB energy γ; there is a clear correlation between V_X and γ for the volume relaxed models, while the same is not the case when the unit cell was not relaxed. This correspondence is plotted in Figure 2a, showing that a very clear linear relation between V_X and γ is obtained. The fit gives the empirical relation $\gamma = 1.55\ V_X$ and $\gamma = 1.19\ V_X$ for full and only x relaxation, correspondingly. The R^2 value is 0.97 and 0.93 for the two fits. This indicates that the volume misfit is an excellent predictor for the GB energy in the case of coherent defect-free GBs, which can potentially be used in experimental studies where V_X may be easier to measure than γ. A similar correlation has been found previously for other materials, e.g., Ni [47].

Table 1. The various grain boundary (GB) models investigated in this study, along with the number of atoms in their unit cells, the supercell size in the *z* direction l_z (perpendicular to the figure plane in Figure 1), their excess volume divided by area V_X, the calculated GB energy E_{GB}, and the range of bond distances Δr_b within the first coordination shell. All results are for pure Pd GBs. Three volume relaxation techniques (VRT) were used: no relaxation of the unit cell, only relaxation of the *x* lattice constant (see Figure 1 for a definition of the *x* direction), and full relaxation of all cell parameters.

GB Model	# of Atoms	l_z (Å)	Excess Volume/Area (Å)			GB Energy (J/m²)			Bond Range (Å)		
			No Relax	Only *x*	Full Relax	No Relax	Only *x*	Full Relax	No Relax	Only *x*	Full Relax
$\Sigma3(111)$	48	13.7	0.00	0.06	0.01	0.05	0.05	0.05	0.01	0.01	0.02
$\Sigma3(11\bar{2})$	44	19.4	1.61	0.48	0.38	0.99	0.67	0.67	0.69	0.75	0.75
$\Sigma5(210)$	80	17.7	0.00	0.82	0.60	1.11	0.93	0.91	1.04	0.48	0.48
$\Sigma5(310)$	72	12.5	1.25	0.71	0.57	1.01	0.89	0.89	0.72	0.92	0.71
$\Sigma7(12\bar{3})$	78	14.8	1.06	0.81	0.63	0.98	0.96	0.94	0.80	0.78	0.82
$\Sigma7(41\bar{5})$	80	25.6	1.22	0.90	0.70	1.05	1.03	1.02	0.82	0.77	0.79
$\Sigma9(1\bar{2}2)$	68	23.7	1.32	0.76	0.55	0.90	0.81	0.81	0.72	0.69	0.71
$\Sigma9(41\bar{1})$	64	16.8	1.86	0.55	0.46	1.20	0.76	0.76	0.71	0.59	0.59
$\Sigma11(113)$	88	26.2	0.00	0.35	0.19	0.32	0.30	0.27	0.26	0.24	0.22
$\Sigma11(23\bar{3})$	80	18.6	1.69	0.72	0.28	1.07	0.82	0.53	0.64	0.86	0.42
$\Sigma13(320)$	100	28.5	1.10	0.61	0.45	0.80	0.75	0.74	0.62	0.57	0.57
$\Sigma13(510)$	100	40.3	1.55	0.74	0.60	1.04	1.00	0.96	0.80	0.79	0.79

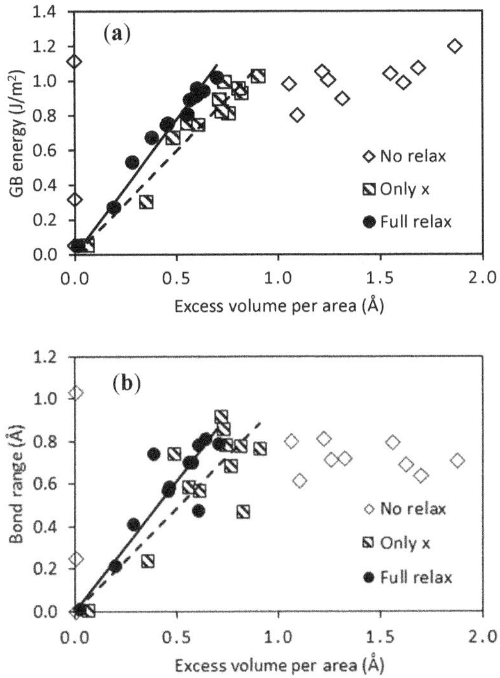

Figure 2. Relations between the excess volume per interface area V_X and the GB energy E_{GB} (**a**) or the bond range Δr_b (**b**) for the interface models listed in Table 1. The three VRTs are marked by open diamonds (no relaxation of the unit cell size), striped squares (only relaxation of the *x* axis), and filled circles (full relaxation of all degrees of freedom). Linear fits to the two latter techniques are shown as dashed and solid lines.

The bond range Δr_b is also following the excess volume and GB energy quite closely, if not as clearly as is the case between γ and V_X (Figure 2b). Nevertheless, there is a clear correlation between V_X and Δr_b—not surprising, since both parameters are a measure of the deviation from the perfect bulk crystal structure. It is perhaps more surprising that Δr_b is lower when no volume relaxation is allowed than in the relaxed case, as is seen e.g., for the $\sum 3(11\bar{2})$ model. This can be explained from the larger freedom of the volume relaxed models to accommodate strain by changing the local coordination within some of the models. Despite the spread in bond range and accompanying variation of coordination number (number of nearest neighbors) in the near vicinity of the GBs, all of the models display local atomic structures very close to fcc in between the GB regions. This ensures that close to bulk behavior can be found furthest away from the GBs in the models.

3.2. Segregation of Single Point Defects

The distribution of defects in the vicinity of GBs is determined by their relative energy at different positions and kinetics. We have only focused on energy in this work, since this is the most relevant in systems being allowed to equilibrate (which is the case in most membrane systems.) We have investigated three different defects with particular relevance for Pd membranes: substitutional Ag (Ag_{Pd}) and Cu (Cu_{Pd}), as well as vacancies (V_{Pd}). This was done with the three different VRTs described above. The energy of each vacancy was calculated at various sites with increasing distance from the GB; an example of this is shown with the green curve for the $\sum 3(11\bar{2})$ model in Figure 1. All the sites with unique x coordinates, from the GB to the midpoint between neighbor GBs were included (recall that cancelling GBs are present both at the unit size boundaries and at the red, dashed line.) The energy at the GB ($x = 0$) was used as the reference state, and the relative energy E_{gb} between that of the impurity located at $x = x_{gb}$ and $x = 0$ was calculated for all 12 models. This is plotted in Figure 3 in the case of Cu_{Pd} defects with relaxation only along the x direction. Since the number of unique sites in the x direction differs between the various models, this is also the case with the number of plot points in Figure 3. The same applies to the distance between neighbor GBs, which is why the extension of the various curves varies in Figure 3. Similar plots to the one shown in Figure 3 have been generated for the three different VRT using three different defects (Cu impurities, Ag impurities, and vacancies); in total, nine plots. For simplicity the only relax x with Cu impurities is the only series of plots shown here.

In order to quantify the tendency to segregate towards or away from the GB the segregation energy of a specific defect has been defined as follows. The lowest total energy (as calculated by DFT) of the defect among the three sites nearest the GB is taken as "the" energy at the GB, E_{GB}(def). This is compared to the average value of the three energies at the farthest distances away from the GB, defined as the "bulk" value E_{bulk}(def). The segregation energy of the defect is then defined as

$$E_{segr}(\text{def}) = E_{GB}(\text{def}) - E_{bulk}(\text{def}) \tag{2}$$

An example is shown in Figure 3: the three values with highest x_{gb} (fitted with a black, dotted line) are used to calculate $E_{bulk}(Cu_{Pd})$ for $\sum 13(510)$, and the point at $x_{gb} \approx 1$ Å (defining the lower black, dotted line) gives $E_{GB}(Cu_{Pd})$. The resulting segregation energy of this example (marked with black arrows in the figure) is $E_{segr}(Cu_{Pd}) = -0.20$ eV.

The model size should ideally be large enough for the energy to converge for large values of x_{gb}. This can be seen for some of the models in Figure 3, but not all. Computational cost restricted the use of larger unit cells to achieve better convergence with respect to unit cell size. Nevertheless, some of the models exhibit excellent convergence as the x_{gb} increases. The $\sum 13(510)$ model is one example, where the energy does not change more than 0.03 eV when x_{gb} increases from 5 to 10 Å at 16 different sites, and the difference between the three sites with largest x_{gb} is less than 1 meV. Other models fluctuate more, but most of them exhibit a clear trend when x_{gb} increases. In some cases, there is no well-defined "bulk" energy; as an example, the energy of the $\sum 3(11\bar{2})$ model varies significantly. This may be due to problems that are connected with the relatively small unit cells employed, and we have therefore in the following disregarded models where the average deviation from E_{bulk} is larger than 0.03 eV. Larger deviations than this are designating models with severe relaxation effects that are deemed as unphysical.

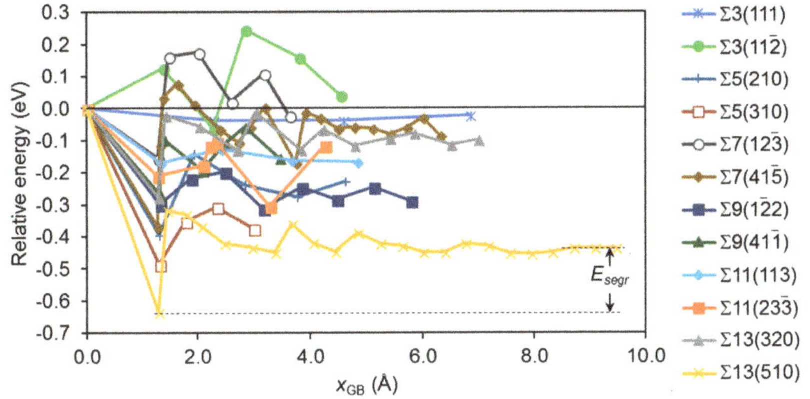

Figure 3. The relative energy E_{gb} in eV of a single Cu impurity Cu_{Pd} as a function of the distance x_{gb} to the plane of the $\sum$ boundary. The volume relaxation scheme was only relaxation of the x direction. Lines are drawn as guide to the eye. The segregation energy E_{segr} is taken as the difference between the average energy of the three sites furthest away from the GB and the lowest energy among the three nearest sites. The extension of a curve along the x_{GB} axis corresponds to half the size of the unit cell in the x direction. See the text for details.

The segregation energy was used to characterize the behavior of the three defects Cu_{Pd}, V_{Pd}, and Ag_{Pd} for all three VRTs and all 12 GB models; this has been compiled in Figure 4. A number of models displayed unphysical relaxations, and their E_{segr} values have not been included in the figure. This leaves part of the figure without data, but enough results were generated to draw some general conclusions. The most striking feature of Figure 4 is that almost all the segregation energies are negative, which indicates a tendency for all three defects to segregate towards the GB. The only small exceptions are for models where no volume relaxation was allowed, indicating that such relaxation is necessary in order to accommodate the defects at these specific GBs. The $\sum3(111)$ twin boundary is special; since the GB is merely a stacking fault, there is nothing to gain from moving a point defect towards or away from the GB. E_{segr} is thus very close to zero for all defects and VRTs for $\sum3(111)$. All other GBs exhibit significant values of E_{segr} for some or all defects and VRTs. The Cu_{Pd} defect gives the smallest absolute values of E_{segr} for most systems, indicating that the segregation tendency of Cu towards GBs in Pd is, in general, lower than that of Ag or vacancies.

Based on the computed segregation energies the equilibrium concentration at the GB of a point defect c_{def} at a given temperature T can be found from the following equation [48]:

$$c_{def} = \frac{c^0_{def}}{c^0_{def} + (1 - c^0_{def}) \exp\left(E_{segr}/k_B T\right)}, \tag{3}$$

where c^0_{def} is the overall ("bulk") equilibrium defect concentration, and k_B is Boltzmann's number. This formula assumes negligible interaction between solutes and thermodynamic equilibrium; we shall see later in this paper that the higher concentration of solutes may actually increase the anticipated equilibrium concentration in many cases. The resulting concentration has been shown in Figure 5 for the V_{Pd}, Cu_{Pd}, and Ag_{Pd} defects in Pd at T = 600 K, which is a relevant temperature for hydrogen separation membranes. The behavior is quite similar at higher temperatures (not shown), only with defect concentrations slightly closer to the bulk one (selected here to be 0.02). The negative segregation energies are reflected in defect concentrations significantly higher than c^0_{def} in most of the cases. In the example in Figure 3 ($\sum3(510)$), this means that the site at layer 2 (directly next to the GB) has approximately 50% occupancy by the Cu solutes.

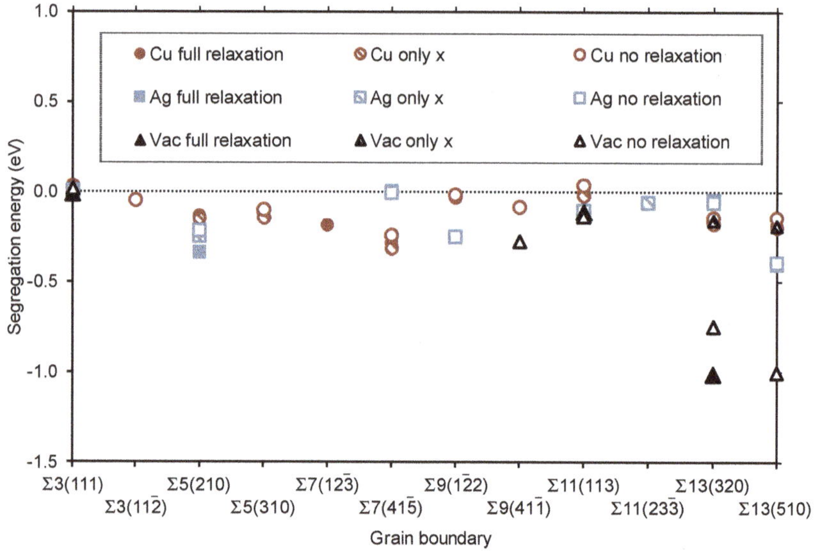

Figure 4. The segregation energy E_{segr} in eV of single Cu_{Pd}, Ag_{Pd}, and V_{Pd} defects for the various GBs. E_{segr} is shown for Cu (red circles), Ag (blue squares), and vacancy (black triangles) segregation, using full volume relaxation (filled symbols), relaxation along the x direction only (half-filled symbols), and no volume relaxation (open symbols). Only well-defined energies are included (see text).

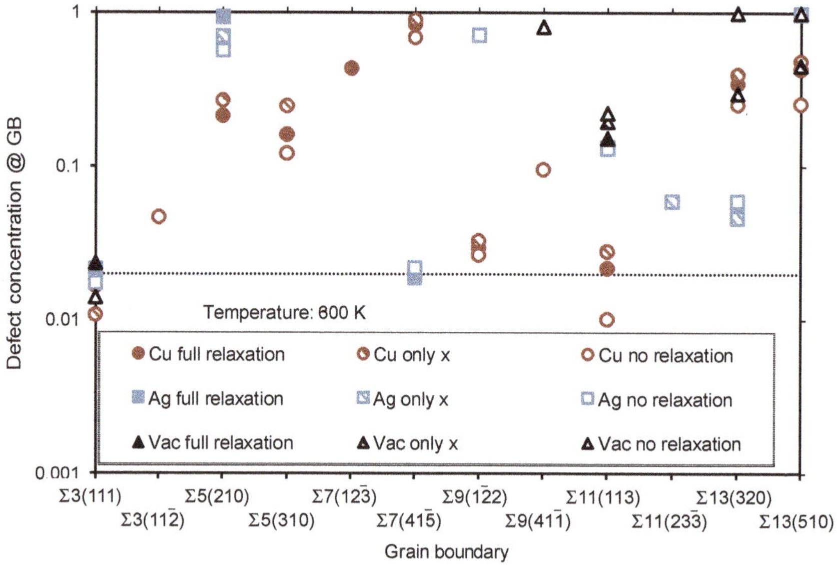

Figure 5. The defect concentration of point defects at the $\sum$ GBs listed in Table 1 and depicted in Figure 1 at T = 600 K. It is shown for the defects Cu_{Pd} (red circles), Ag_{Pd} (blue squares), and V_{Pd} (black triangles), using three VRTs: full volume relaxation (filled symbols), relaxation along the x direction only (half-filled symbols), and no volume relaxation (open symbols). The overall (bulk) defect concentration (chosen to be 0.02) is shown by the black, dotted line.

The variation of the defect concentration with temperature has been depicted for the ∑11(113) GB in Figure 6. The concentrations approach the equilibrium bulk concentration of 0.02 (black, long-dashed line) as the temperature increases. The Cu impurity does not show a strong segregation tendency. In the case of only x relaxation (which corresponds to large grains) there is a weak segregation of Cu towards the GB. When no relaxation is allowed, however, there is a similarly weak segregation of Cu away from the GB. Since the situation with small grains corresponds to an interpolation between no relaxation and full relaxation, we conclude that Cu may be weakly segregated away from the GB in this case. Ag is, on the other hand, quite strongly segregated towards the GB and there is virtually no difference between large (only x) and small (between no relaxation and full relaxation) grains. Vacancies are also strongly attracted to the GB, and the difference between materials with large and small grain size should be quite small—interpolated values between the curves based on no relaxation and full relaxation are likely to be very similar to the values of the curve based on only x relaxed. We can deduce the temperature dependence of the other GBs in Figure 5 from the behavior of the curves in Figure 6, since the segregation energy is the only parameter that determines the defect concentration in Equation (3).

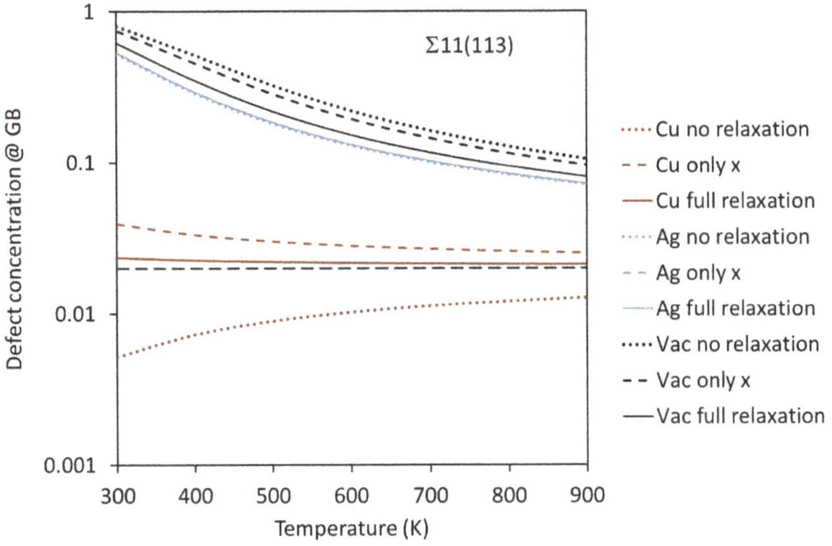

Figure 6. The defect concentration of point defects at the ∑11(113) GB as a function of temperature. The bulk equilibrium concentration (0.02) is shown as a black, long-dashed line.

3.3. Binary Systems with More Than One Impurity Atom

Even if there is a clear tendency of segregation of impurities towards the GB in many GB models, it is unclear from the above whether more than one atom can be attracted to the GBs simultaneously. Repulsive interactions between point defects may lead to lower concentrations than that predicted in Figure 5. This was therefore investigated in more detail for Cu in Pd within a ∑5(210) GB model, as shown in Figure 7. The ∑5(210) GB was selected as a typical representative of the coherent GBs investigated in this study; it has a characteristic GB energy (~1 eV), range of bond lengths Δr_b ~1 Å, as well as segregation energies of Ag (~−0.3 eV) and Cu (~−0.15 eV). This leads to a clear segregation behavior of solitary solutes with at least 21% (Cu) or 57% (Ag) equilibrium occupancy near the GB at 600 K when the overall concentration is 2%.

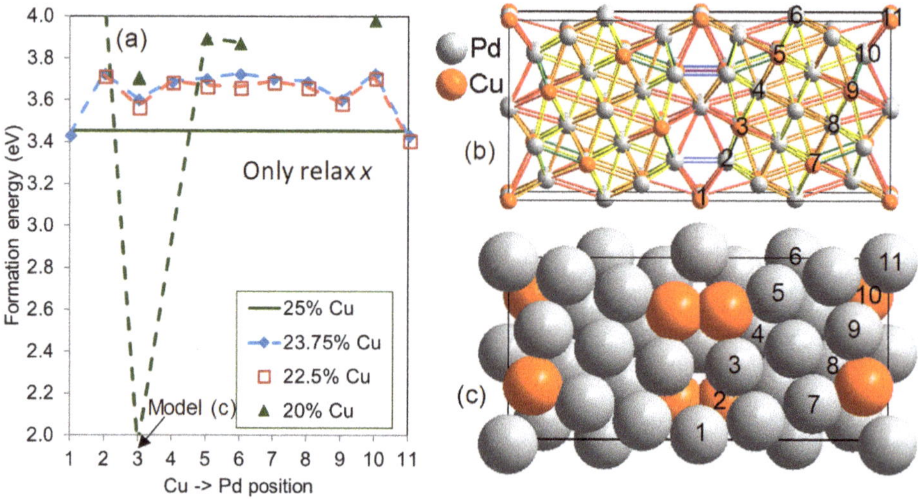

Figure 7. (**a**) The formation energy E_{form} of binary $Pd_{1-n}Cu_n$ alloys with four levels of the Cu concentration n in a $\sum5(210)$ GB model. The volume of all models is fully relaxed. The solid green line designating 25% Cu corresponds to the formation energy of perfectly ordered Pd_3Cu as shown in (**b**). Lines between atoms are rainbow colored according to the relaxed interatomic distance; violet is shortest (<2.4 Å), dark red is longest (>2.9 Å), and orange is similar to that of density functional theory (DFT) relaxed bulk Pd_3Cu (2.7 Å). Models with lower Cu content than 25% are generated by substituting Cu with Pd at various positions in the model, defined by the numbers in (**b**,**c**). The energy is drawn as a function of these positions. See the text for details about the formation energies plotted in (**a**).

Since Cu and Ag segregate towards different sites, one can expect the situation of both solutes segregating simultaneously towards the same GB. To investigate this possibility a periodic GB model with stoichiometry Pd_3Cu was generated as a starting point; a fully relaxed model is shown in Figure 7b. The bond lengths of the relaxed model are color coded in this figure. The relatively large variation of bond lengths with both elongated and contracted bonds explain why both large and small atoms may be attracted to the GB from a geometric point of view; this creates sites where atoms of various radii could be fitted geometrically. The bond lengths of this particular model vary from 2.31 to 3.29 Å, corresponding to Δr_b = 0.98 Å. This is significantly larger than Δr_b of the pure Pd $\sum5(210)$ model, which only displayed bonds between 2.51 and 2.99 Å, thus Δr_b = 0.48 Å (Table 1). This means that the addition of Cu not only decreases the smallest bond length (which can be expected when a smaller atom is added), but it also increases the largest one. This reflects a higher flexibility of the lattice when atoms with more than one size are present.

Models with the Cu content reduced below the starting point of 25% were created by replacing Cu by Pd in the model in Figure 7b. The stability of these models was assessed by their formation energy defined as

$$E_{\text{form}}(\text{Pd}_{N-n}\text{Cu}_n) = E_{\text{tot}}(\text{Pd}_{N-n}\text{Cu}_n) - (N-n)E_{\text{tot}}(\text{Pd}) - nE_{\text{tot}}(\text{Cu}), \quad (4)$$

where E_{tot} is the total electronic energy as calculated by DFT, N is the number of atoms in the GB model (listed in Table 1), and n is the number of Pd atoms being substituted by Cu. The reference energy of Pd and Cu is that of their standard state, fcc bulk. Due to the difference in standard state energy between Pd and Cu, increasing the Pd content typically increases the formation energy (it appears less stable). The Cu content is reduced to 23.75% when one Cu atom is replaced, and the most stable configuration of this model is with extra Pd placed at position 1 or 11 (Figure 7a, dashed blue curve with diamonds), i.e., at the very center of the GB. The models with less Cu also exhibit the most stable configuration when excess Pd is placed at position 1 or 11. As an example, the most stable configuration of the 22.5% Cu model is with extra Pd placed at position 1 and 11 (dashed red curve with empty squares). The latter configuration is used as the starting point for the 20% model, which displays the most stable configuration with an extra Pd placed at position 3 or 9. A number of models with 20% Cu and Pd placed according to this insight were then constructed. The most stable of those is shown in Figure 7c, where all of the Cu atoms were moved to positions 2 and 10, corresponding to E_{form} = 1.9 eV. This is marked by the arrow at position 3 in (a). In conclusion, there is a very strong thermodynamic driving force for segregation of Cu towards the coherent $\Sigma5(210)$ GB. From these results, it can be expected that all "small" sites (with short interatomic distances to the neighbor sites) close to this boundary are occupied by Cu. This can be quantified by Δr_b, which indirectly indicates the size of the smallest sites. From Table 1, it is evident that small sites exist in all of the models of this study (except the $\Sigma3(111)$ twin boundary), and in many cases to a larger degree than in the $\Sigma5(210)$ model. We can thus expect that the segregation trend of Cu is global, and that these results can be transferred to almost all GBs.

3.4. Segregation in Ternary Pd-Cu-Ag Alloys

The results of the above studies clearly show that there is a strong segregation of various point defects towards most GBs, and that this is valid even for a large density of solutes. However, it is not clear how different solutes interact with each other. From the size of defects as compared to the available sites near the GB (the "size effect"), one could expect that many GBs display a combined segregation of "small" and "large" defects (smaller and larger than the host atoms, respectively). We used a selection of Pd-Cu-Ag alloys with the composition $Pd_{0.8-\delta}Cu_{0.2}Ag_{\delta}$ to investigate this hypothesis, again using the $\Sigma5(210)$ GB as a representative model GB. The most stable models from Figure 7 were used as starting point, substituting Pd with Ag at various sites and concentrations.

The formation energy E_{form} of a Pd-Cu-Ag alloy is defined as similar to that of Pd-Cu in Equation (4):

$$E_{\mathrm{form}}(\mathrm{Pd}_{N-n-m}\mathrm{Cu}_n\mathrm{Ag}_m) = E_{\mathrm{tot}}(\mathrm{Pd}_{N-n-m}\mathrm{Cu}_n\mathrm{Ag}_m) - (N-n-m)E_{\mathrm{tot}}(\mathrm{Pd}) - (n)E_{\mathrm{tot}}(\mathrm{Cu}) - (m)E_{\mathrm{tot}}(\mathrm{Ag}). \quad (5)$$

Here, m is the number of Pd atoms that have been substituted by Ag, and the reference energy of Ag is that of the standard state, fcc bulk. This definition means that a negative E_{form} indicates a stable compound compared to the pure metals.

The formation energy is plotted for different models with Ag substituted for Pd or Cu as a function of the site (which corresponds to the distance from the GB) in Figure 8. We recognize the trend from the binary alloy; Ag is most stable at the GB (represented by the positions 1 and 11) in the model where Pd has replaced Cu at positions 1 and 3. It is also the most stable in the model where two Cu atoms in addition have been segregated to site 2 and 10, but to a smaller extent. However, in the most stable situation when all the Cu atoms are moved towards the GB and only populate sites 2 and 10 (corresponding to Figure 7c), the Ag site with lowest energy is not anymore at the GB (position 1 or 11), but rather at sites 3 and 9, which are just outside the Cu sites. This is due to relatively large local relaxations around the Cu atoms, which reduce the size of the sites at the GB (position 1 and 11). These sites are thus not significantly smaller than other sites anymore. Adding to this is the reduced affinity between Ag and Cu when compared to that between Ag and Pd. Since the latter model is the most stable one, we can expect to find the enrichment of Cu in position 2 and of Ag in position 3 close to $\Sigma5(210)$ GBs in Pd-Ag-Cu compounds. This is supported by the plot in Figure 8b, demonstrating that the formation energy decreases as the local Ag concentration increases.

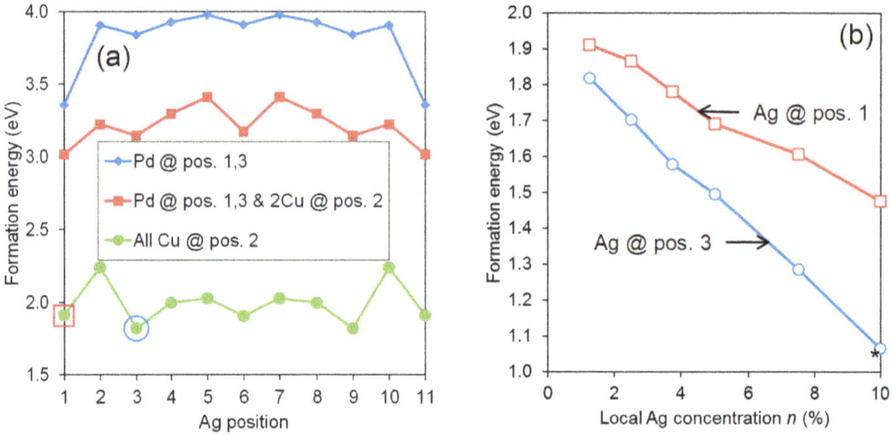

Figure 8. (**a**) The formation energy (defined in Equation (5)) of three fully relaxed ∑5(210) $Pd_{64}Cu_{16}$ models with Ag substituted at the sites defined in Figure 7c. One model was created from the periodic $Pd_{60}Cu_{20}$ model by substituting Pd for Cu at the positions 1, 3, 9, and 11 (blue line with diamonds), one had in addition moved two Cu atoms to positions 2 and 10 (red line with squares), and one had all Cu atoms located at positions 2 and 10 (green line with circles). The latter corresponds to the model shown in Figure 7c. The two most stable models from (**a**) were used to plot the formation energy as a function of local Ag concentration n in (**b**); Ag was then placed at position 1 (red circles) and position 3 (blue circles). The most stable model with n = 10% is shown in Figure 9b.

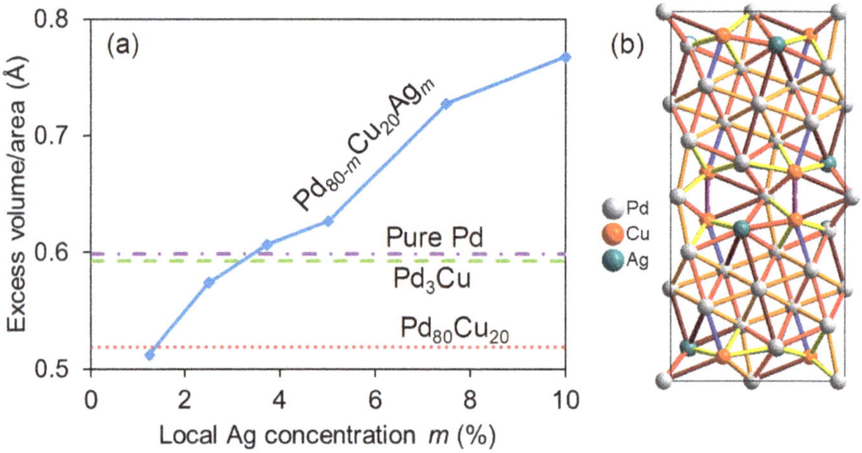

Figure 9. (**a**) The excess volume/area of an 80-atom ∑5(210) model (Figure 1) for pure Pd (dashed-dotted line), Pd_3Cu (dashed), $Pd_{80}Cu_{20}$ (dotted), and $Pd_{80-m}Cu_{20}Ag_m$ (solid). The local Ag concentration x corresponds to the concentration very close to the GB and does not necessarily reflect the overall concentration (it may have segregated towards the GB). The most stable model with m = 10% is shown in (**b**), and the interatomic distances d are indicated with rainbow colors, starting from dark red (d > 2.9 Å) to violet (d < 2.5 Å). The average bond distance in bulk Pd is approximately 2.7 Å, represented by yellow (2.6 < d < 2.7 Å) and orange (2.7 < d < 2.8 Å) bonds.

How does the addition and segregation of Ag influence the bond distances at and near the ∑5(210) GB? This may be relevant for hydrogen solubility and diffusivity since both depend strongly on the interatomic distances. Figure 9 presents how this is quantified by the excess volume divided by GB area in Figure 9a and by actual bond distances in the most stable model with the composition $Pd_{70}Cu_{20}Ag_{10}$

in Figure 9b. It is evident that an increasing amount of Ag at the GB leads to significantly increased excess volumes. Since the volumes are divided by the GB area (that of the relaxed unit cell), this primarily signifies the elongation of the GB unit cell in the x direction. The Pd-Pd-bonds in the near-GB region are thus clearly larger than those in the bulk Pd, and significantly more so than in the bulk Pd-Cu alloys, where the overall lattice constants are reduced due to the smaller size of Cu atoms.

4. Discussion

The 12 GB models depicted in Figure 1 are by no means representing all the possible GBs in fcc Pd, even when being restricted to coherent ones without compensating dislocations or other defects. The relevance of the present results is thus restricted to a selection of such interfaces, which may not govern all important properties of these compounds. Nevertheless, we have seen that the selection of models gives a broad distribution of important parameters, like the interface energy, mismatch volume, range of bond lengths, etc. This indicates that the present results should be relevant at least for all coherent GBs in Pd alloys, even those with lower angles than the present ones.

The GBs in this study are all perfectly coherent, which of course is a simplification of the real situation. Many of the GBs found in real materials are quite complex, featuring all sorts of defects that to some extent compensate geometric mismatches that are inherent to the perfect GB. Nevertheless, many of these additional defects can be relatively far apart (e.g., in the case of dislocations), leaving near-perfect GBs over large areas, as demonstrated by several microscopy studies [49,50]. We therefore assume that our coherent models may be also relevant for a number of GBs where the lack of coherence is not too severe. We expect the correspondence to fail when going to truly amorphous GBs.

The results above suggest significant segregation of a variety of point defects towards virtually all GBs. However, the absolute numbers in Figure 5, summarizing the segregation tendency, should be applied with some care due to a number of reasons. The limited size of the models means that the strain originating from the GBs is not converged to zero at any place in the super cell. This challenge has been accommodated to some extent by using the three different volume relaxation methods—they represent the outer limits of how the unit cells should realistically be relaxed in the vicinity of GBs, and the real segregation tendencies should be somewhere between those limits. So even if there is no true bulk behavior between the GBs in the various models, the local relaxation effects and the resulting segregation should be correct within the boundaries that are described by the different VRTs.

Another potential source of error in these calculations is the lack of structurally compensating defects, most notably dislocations. They can accommodate significant parts of local strain and could as such counter some of the strongest segregation tendencies that were seen in this study. However, dislocations can attract point defects themselves, so this does not necessarily hinder segregation of defects, even if the nature of the segregation might be changed.

Another reason to take the absolute numbers of Figure 4 with some care is the rather large relaxation effects that were seen for some of the models. In some cases, the entire supercell was restructured, which led to a total energy being reduced by several eV in some of the cases. This can be understood as the starting point of a full relaxation to the lowest energy structure, which is the bulk without any GB. We have disregarded the points with largest restructuring effects, but it was difficult to distinguish between reasonable relaxations and unphysical effects due to the small size of the supercells; there was a continuous range from virtually no relaxation to almost complete reorganization of the GB model. We disregarded models where the average deviation from E_{bulk} is larger than 0.03 eV, but some unphysical results due to limited unit cell size may still remain.

The calculations of the present study have all been performed without any explicit temperature being included. Temperature was included implicitly through the Arrhenius equation when obtaining equilibrium defect concentrations in Figure 5, but no other effect of temperature (thermal expansion, entropy, zero-point energy, phonon-based thermodynamics, etc.) was included. This has the potential to change the quantitative results significantly, but we expect that the qualitative trends remain unchanged.

Many of the same concerns apply when turning to the ternary compositions. We established that both larger (Ag) and smaller (Cu) atoms are attracted to a GB simultaneously, but the absolute value of the numbers and the actual sites of attraction may be different in reality than what is reported in this study. Furthermore, we did not consider simultaneous segregation of vacancies and solutes. With the knowledge from above, we expect that this is present in most boundaries, since vacancies and solutes have the ability to occupy different sites around a given GB.

The various sources of potential errors thus add up to a large and unknown uncertainty of the numbers presented in this study. The remaining main conclusion is unchanged, however: there is a clear and consistent tendency of many kinds of point defects (small and large substitutes, vacancies) to segregate towards GBs, due to the variety of local environments that are found there.

Despite the clear trends, it may be challenging to observe the segregation experimentally. It happens on the scale of single atomic layers, which makes high-resolution transmission electron microscopy the only viable way of directly probing such segregation. This relies on the ability to focus the electron beam along the GB plane, since this is where the change in concentration should be observed. It may also rely on the synthesis and heat treatment of the film; e.g., since sputtering is a non-equilibrium synthesis process, annealing (or operation under realistic conditions) may be required for the segregation to appear. Furthermore, Pd-membranes are typically aimed for hydrogen separation purposes, and the presence of hydrogen may influence the segregation behavior significantly, as has been seen in the case of segregation towards outer surfaces of similar systems [51].

Three different VRTs were compared for many of the calculations above, representing the boundaries of the likely volume relaxation regimes in real materials (from relaxation along x signifying low density of GBs to no relaxation representing very high density of GBs). However, some of the results indicate that not performing any relaxation of the volume is unreasonable. This is particularly clearly illustrated in Figure 2, where the linear trend between V_X and γ can only be seen if the partial or full relaxation of the model is allowed. The conclusion from this observation is that some relaxation of the volume is required to move away from unreasonable situations arising from the somewhat arbitrary construction of the GB models. Most of the results in this study that are based on no volume relaxation should thus be neglected, apart from serving as a far-end borderline of the values.

Which of the VRTs allowing for volume relaxation to choose is less obvious. Both partial (only x) and full relaxation of the unit cell give results in reasonable correspondence with each other, indicating that relaxation of the coordinate perpendicular to the GB plane is most important. There are some differences between the two, depending on the particulars of the GB; most notably, the $\Sigma 11(23\bar{3})$ GB exhibits large differences between only x and full relaxation, both for V_X, γ, and Δr_b (Table 1). This reflects that this particular GB displays significant relaxation of the unit cell parameters parallel to the GB plane when allowed, in contrast to the other GB models. Such relaxation is most reasonable when the real GB resembles our simulation cell: infinitely long in the directions parallel to the GB plane but with a short distance between the GBs in the x direction. Partial relaxation (only x) describes situations where the average bulk lattice constant is maintained in these parallel directions by an infinitely stiff lattice, i.e., extending far away from the GB.

5. Conclusions

Atomic-scale calculations based on density functional theory were used to investigate various properties of low-number coherent Σ grain boundaries (GBs) in fcc Pd, Pd-Cu, and Pd-Cu-Ag alloys. Their excess volumes, grain boundary energies, and ranges of bond lengths were computed while using three different volume relaxation techniques: no relaxation of volume, full relaxation of all lattice parameters, and partial relaxation of volume only allowing for one lattice constant to change. A linear correlation between the excess volume and grain boundary energy was found in pure Pd when partial or full relaxation of the volume was allowed. The tendency to segregate towards the GBs was assessed for three different point defects: Cu_{Pd}, Ag_{Pd}, and V_{Pd}. Virtually all GBs exhibited strong segregation tendencies for all defects, quantified by the segregation energy and the corresponding equilibrium concentration of the defects at or near the GBs. An amplified segregation tendency was observed when increasing the local solute concentration, indicating that many of these sites might be nearly fully occupied by substitutional defects. This also demonstrated that the initial study at the dilute limit is clearly relevant for alloys with higher concentrations of the alloying element. Ternary Pd-Cu-Ag compounds were finally investigated, and simultaneous segregation of both Cu and Ag towards the $\sum$5(210) GB was observed; i.e., the results also hold for more complex alloy systems. The most stable model furthermore displayed a pronounced increase in the excess volume, indicating significantly increased local lattice parameters. In summary, this study demonstrates how insight from first principles calculations can be used to understand the complex behavior of point defects at and around grain boundaries of metals and alloys.

Author Contributions: Conceptualization, O.M.L. and T.P.; Methodology, O.M.L., D.Z. and Y.L.; Software, O.M.L.; Validation, O.M.L. and T.P.; Formal Analysis, O.M.L.; Investigation, O.M.L.; Resources, X.X.; Data Curation, O.M.L.; Writing-Original Draft Preparation, O.M.L.; Writing-Review & Editing, O.M.L., D.Z., Y.L., R.B. and T.P.; Visualization, O.M.L.; Project Administration, T.P.; Funding Acquisition, O.M.L., R.B. and T.P.

Funding: Financial support from the Research council of Norway through the CLIMIT program (Contract No. 215666/E20) is gratefully acknowledged. The computational part was executed with a grant from the Notur metacenter for supercomputing.

Acknowledgments: The authors appreciate useful discussions with Bernhard Dam, Sarmila Dutta, Marit Stange, and Patricia Carvalho.

Conflicts of Interest: The authors declare no conflict of interest. The funders had no role in the design of the study; in the collection, analyses, or interpretation of data; in the writing of the manuscript, and in the decision to publish the results.

References

1. Gallucci, F.; Fernandez, E.; Corengia, P.; Annaland, M.v.S. Recent advances on membranes and membrane reactors for hydrogen production. *Chem. Eng. Sci.* **2013**, *92*, 40–66. [CrossRef]
2. Al-Mufachi, N.A.; Rees, N.V.; Steinberger-Wilkens, R. Hydrogen selective membranes: A review of palladium-based dense metal membranes. *Renew. Sustain. Energy Rev.* **2015**, *47*, 540–551. [CrossRef]
3. Li, P.; Wang, Z.; Liu, Z.Y.; Cao, X.; Li, W.; Wang, J.; Wang, S. Recent developments in membranes for efficient hydrogen purification. *J. Membr. Sci.* **2015**, *495*, 130–168. [CrossRef]
4. Lu, G.Q.; Diniz da Costa, J.C.; Duke, M.; Giessler, S.; Socolow, R.; Williams, R.H.; Kreutz, T. Inorganic membranes for hydrogen production and purification: A critical review and perspective. *J. Colloid Interface Sci.* **2007**, *314*, 589–603. [CrossRef] [PubMed]
5. Bredesen, R.; Peters, T.A.; Boeltken, T.; Dittmeyer, R. Pd-Based Membranes in Hydrogen Production for Fuel cells. In *Process Intensification for Sustainable Energy Conversion*; John Wiley & Sons, Ltd.: West Sussex, UK, 2015; pp. 209–242.
6. Peters, T.A.; Stange, M.; Bredesen, R. Development of thin Pd-23%Ag/Stainless Steel composite membranes for application in Water Gas Shift membrane reactors. In *Carbon Dioxide Capture for Storage in Deep Geological Formations*; Eide, L.I., Ed.; CPL Press and BP: Berkshire, UK, 2010; pp. 135–155.
7. Uemiya, S. State-of-the-Art of Supported Metal Membranes for Gas Separation. *Sep. Purif. Rev.* **1999**, *28*, 51–85. [CrossRef]
8. Oertel, M.; Schmitz, J.; Weirich, W.; Jendryssek-Neumann, D.; Schulten, R. Steam reforming of natural gas with intergrated hydrogen separation for hydrogen production. *Chem. Eng. Technol.* **1987**, *10*, 248–255. [CrossRef]
9. Coulter, K.E.; Way, J.D.; Gade, S.K.; Chaudhari, S.; Sholl, D.S.; Semidey-Flecha, L. Predicting, Fabricating, and Permeability Testing of Free-Standing Ternary Palladium-Copper-Gold Membranes for Hydrogen Separation. *J. Phys. Chem. C* **2010**, *114*, 17173–17180. [CrossRef]
10. Galipaud, J.; Martin, M.H.; Roue, L.; Guay, D. Pulsed Laser Deposition of PdCuAu Alloy Membranes for Hydrogen Absorption Study. *J. Phys. Chem. C* **2015**, *119*, 26451–26458. [CrossRef]
11. Kim, D.W.; Park, Y.J.; Woo, B.I.; Kang, S.M.; Park, J.S. Study on the perm-selectivity of thin Pd-Cu-Ni ternary alloy membrane for hydrogen purification and separation. *Jpn. J. Appl. Phys.* **2010**, *49*, 018003. [CrossRef]

12. Nayebossadri, S.; Speight, J.; Book, D. Effects of low Ag additions on the hydrogen permeability of Pd-Cu-Ag hydrogen separation membranes. *J. Membr. Sci.* **2014**, *451*, 216–225. [CrossRef]
13. Peters, T.A.; Kaleta, T.; Stange, M.; Bredesen, R. Development of ternary Pd-Ag-TM alloy membranes with improved sulphur tolerance. *J. Membr. Sci.* **2013**, *429*, 448–458. [CrossRef]
14. Peters, T.A.; Kaleta, T.; Stange, M.; Bredesen, R. Development of thin binary and ternary Pd-based alloy membranes for use in hydrogen production. *J. Membr. Sci.* **2011**, *383*, 124–134. [CrossRef]
15. Ryi, S.K.; Park, J.S.; Kim, S.H.; Cho, S.H.; Kim, D.W.; Um, K.Y. Characterization of Pd-Cu-Ni ternary alloy membrane prepared by magnetron sputtering and Cu-reflow on porous nickel support for hydrogen separation. *Separ. Purif. Technol.* **2006**, *50*, 82–91. [CrossRef]
16. Chen, F.L.; Kinari, Y.; Sakamoto, Y. The hydrogen miscibility gaps in hydrogenated Pd-Y-In(Sn, Pb) ternary alloys. *J. Alloys Compd.* **1994**, *205*, 119–124. [CrossRef]
17. Ling, C.; Semidey-Flecha, L.; Sholl, D.S. First-principles screening of PdCuAg ternary alloys as H-2 purification membranes. *J. Membr. Sci.* **2011**, *371*, 189–196. [CrossRef]
18. Pati, S.; Jat, R.A.; Anand, N.S.; Derose, D.J.; Karn, K.N.; SMukerjee, K.; Parida, S.C. Pd-Ag-Cu dense metallic membrane for hydrogen isotope purification and recovery at low pressures. *J. Membr. Sci.* **2017**, *522*, 151–158. [CrossRef]
19. Tarditi, A.M.; Cornaglia, L.M. Novel PdAgCu ternary alloy as promising materials for hydrogen separation membranes: Synthesis and characterization. *Surf. Sci.* **2011**, *605*, 62–71. [CrossRef]
20. Zhao, L.F.; Goldbach, A.; Bao, C.; Xu, H.Y. Structural and Permeation Kinetic Correlations in PdCuAg Membranes. *ACS Appl. Mater. Interfaces* **2014**, *6*, 22408–22416. [CrossRef] [PubMed]
21. Gao, M.C.; Ouyang, L.; Dogan, O.N. First principles screening of B2 stabilizers in CuPd-based hydrogen separation membranes: (1) Substitution for Pd. *J. Alloys Compd.* **2013**, *574*, 368–376. [CrossRef]
22. Bredesen, R.; Klette, H. Method of Manufacturing Thin Metal Membranes. US Patent 6,086,729, 11 July 2000.
23. Peters, T.A.; Stange, M.; Bredesen, R. 2-Fabrication of palladium-based membranes by magnetron sputtering. In *Palladium Membrane Technology for Hydrogen Production, Carbon Capture and Other Applications*; Woodhead Publishing: Cambridge, UK, 2015; pp. 25–41.
24. Nicholson, K.M.; Chandrasekhar, N.; Sholl, D.S. Powered by DFT: Screening Methods That Accelerate Materials Development for Hydrogen in Metals Applications. *Acc. Chem. Res.* **2014**, *47*, 3275–3283. [CrossRef] [PubMed]
25. Chandrasekhar, N.; Sholl, D.S. Large-Scale Computational Screening of Binary Intermetallics for Membrane-Based Hydrogen Separation. *J. Phys. Chem. C* **2015**, *119*, 26319–26326. [CrossRef]
26. Kamakoti, P.; Sholl, D.S. Towards first principles-based identification of ternary alloys for hydrogen purification membranes. *J. Membr. Sci.* **2006**, *279*, 94–99. [CrossRef]
27. Semidey-Flecha, L.; Ling, C.; Sholl, D.S. Detailed first-principles models of hydrogen permeation through PdCu-based ternary alloys. *J. Membr. Sci.* **2010**, *362*, 384–392. [CrossRef]
28. Lovvik, O.M.; Peters, T.A.; Bredesen, R. First-principles calculations on sulfur interacting with ternary Pd-Ag-transition metal alloy membrane alloys. *J. Membr. Sci.* **2014**, *453*, 525–531. [CrossRef]
29. Kirchheim, R.; Kownacka, I.; Filipek, S.M. Hydrogen segregation at grain-boundaries in nanocrystalline nickel. *Scr. Metall. Mater.* **1993**, *28*, 1229–1234. [CrossRef]
30. Lemier, C.; Weissmueller, J. Grain boundary segregation, stress and stretch: Effects on hydrogen absorption in nanocrystalline palladium. *Acta Mater.* **2007**, *55*, 1241–1254. [CrossRef]
31. Siegel, D.J.; Hamilton, J.C. Computational study of carbon segregation and diffusion within a nickel grain boundary. *Acta Mater.* **2005**, *53*, 87–96. [CrossRef]
32. Aksyonov, D.A.; Lipnitskii, A.G.; Kolobov, Y.R. Grain boundary segregation of C, N and O in hexagonal close-packed titanium from first principles. *Model. Simul. Mater. Sci. Eng.* **2013**, *21*, 12. [CrossRef]
33. Lejcek, P.; Sob, M. An analysis of segregation-induced changes in grain boundary cohesion in bcc iron. *J. Mater. Sci.* **2014**, *49*, 2477–2482. [CrossRef]
34. Yazdandoost, F.; Mirzaeifar, R. Tilt grain boundaries energy and structure in NiTi alloys. *Comput. Mater. Sci.* **2017**, *131*, 108–119. [CrossRef]
35. Yan, J.Y.; Ehteshami, H.; Korzhavyi, P.A.; Borgenstam, A. Sigma 3(111) grain boundary of body-centered cubic Ti-Mo and Ti-V alloys: First-principles and model calculations. *Phys. Rev. Mater.* **2017**, *1*, 023602. [CrossRef]

36. Zhao, Z.M.; Wan, J.F.; Wang, J.N. Ab-Initio Study of Electronic Structure of Martensitic Twin Boundary in Ni_2MnGa Alloy. *Mater. Trans.* **2016**, *57*, 477–480. [CrossRef]
37. Kim, Y.K.; Jung, W.S.; Lee, B.J. Modified embedded-atom method interatomic potentials for the Ni-Co binary and the Ni-Al-Co ternary systems. *Model. Simul. Mater. Sci. Eng.* **2015**, *23*, 055004. [CrossRef]
38. Kiyohara, S.; Mizoguchi, T. Investigation of Segregation of Silver at Copper Grain Boundaries by First Principles and Empirical Potential Calculations. *AIP Conf. Proc.* **2016**, *1763*, 040001.
39. Zhang, P.B.; Li, X.J.; Zhao, J.J.; Zheng, P.F.; Chen, J.M. Atomic investigation of alloying Cr, Ti, Y additions in a grain boundary of vanadium. *J. Nucl. Mater.* **2016**, *468*, 147–152. [CrossRef]
40. Basha, D.A.; Sahara, R.; Somekawa, H.; Rosalie, J.M.; Singh, A.; Tsuchiya, K. Interfacial segregation induced by severe plastic deformation in a Mg-Zn-Y alloy. *Scr. Mater.* **2016**, *124*, 169–173. [CrossRef]
41. Piochaud, J.B.; Becquart, C.S.; Domain, C. Ab initio and Atomic kinetic Monte Carlo modelling of segregation in concentrated FeCrNi alloys. In Proceedings of the Sna + Mc 2013—Joint International Conference on Supercomputing in Nuclear Applications + Monte Carlo, Paris, France, 27–31 October 2013.
42. Kresse, G.; Hafner, J. Ab initio molecular-dynamics for liquid-metals. *Phys. Rev. B* **1993**, *47*, 558–561. [CrossRef]
43. Kresse, G.; Furthmuller, J. Efficient iterative schemes for ab initio total-energy calculations using a plane-wave basis set. *Phys. Rev. B* **1996**, *54*, 11169. [CrossRef]
44. Kresse, G.; Joubert, D. From ultrasoft pseudopotentials to the projector augmented-wave method. *Phys. Rev. B* **1999**, *59*, 1758. [CrossRef]
45. Perdew, J.P.; Burke, K.; Ernzerhof, M. Generalized gradient approximation made simple. *Phys. Rev. Lett.* **1996**, *77*, 3865–3868. [CrossRef] [PubMed]
46. Ranganathan, S. On geometry of coincidence-site lattices. *Acta Crystallogr.* **1966**, *21*, 197. [CrossRef]
47. Olmsted, D.L.; Foiles, S.M.; Holm, E.A. Survey of computed grain boundary properties in face-centered cubic metals: I. Grain boundary energy. *Acta Mater.* **2009**, *57*, 3694–3703. [CrossRef]
48. Dontsova, E.; Rottler, J.; Sinclair, C.W. Solute-defect interactions in Al-Mg alloys from diffusive variational Gaussian calculations. *Phys. Rev. B* **2014**, *90*, 174102. [CrossRef]
49. Mekonnen, W.; Arstad, B.; Klette, H.; Walmsley, J.C.; Bredesen, R.; Venvik, H.; Holmestad, R. Microstructural characterization of self-supported 1.6 μm Pd/Ag membranes. *J. Membr. Sci.* **2008**, *310*, 337–348. [CrossRef]
50. Peters, T.A.; Tucho, W.M.; Ramachandran, A.; Stange, M.; Walmsley, J.C.; Holmestad, R.; Borg, A.; Bredesen, R. Thin Pd-23%Ag/stainless steel composite membranes: Long-term stability, life-time estimation and post-process characterisation. *J. Membr. Sci.* **2009**, *326*, 572–581. [CrossRef]
51. Lovvik, O.M.; Opalka, S.M. Reversed surface segregation in palladium-silver alloys due to hydrogen adsorption. *Surf. Sci.* **2008**, *602*, 2840–2844. [CrossRef]

membranes

MDPI

Article

"Modified" Liquid–Liquid Displacement Porometry and Its Applications in Pd-Based Composite Membranes

Lei Zheng [1,2], Hui Li [1,*], Haijun Yu [1], Guodong Kang [1], Tianying Xu [1], Jiafeng Yu [1], Xinzhong Li [3,*] and Hengyong Xu [1,*]

1 Dalian National Laboratory for Clean Energy, Dalian Institute of Chemical Physics, Chinese Academy of Sciences, Dalian 116023, China; zlei@dicp.ac.cn (L.Z.); yuhj@dicp.ac.cn (H.Y.); kangguod@dicp.ac.cn (G.K.); xutianying@dicp.ac.cn (T.X.); yujf@dicp.ac.cn (J.Y.)

2 University of Chinese Academy of Sciences, Beijing 100049, China

3 School of Materials Science and Engineering, Harbin Institute of Technology, Harbin 150001, China

* Correspondence: hui.li@dicp.ac.cn (H.L.); hitlxz@163.com (X.L.); xuhy@dicp.ac.cn (H.X.)

Received: 14 May 2018; Accepted: 6 June 2018; Published: 8 June 2018

Abstract: For H_2 separation by Pd-based composite membranes, the pore mouth size distribution of the porous support immediately affects the quality of the deposited layer, including continuity and defect/pinhole formation. However, there is a lack of convenient and effective methods for characterization of pore mouth size of porous supports as well as of defect distribution of dense Pd-based composite membranes. Here we introduce a novel method by modifying conventional liquid–liquid displacement porometry. When the pore tunnels are filled with Liquid B and the outer surface is occupied by Liquid A, the reopening of the pore mouth depends on the pressure of Liquid B and the interfacial tension at the position of the pore mouth, from which the pore mouth size can be determined according to the Young–Laplace equation. Our experimental tests using this method with model samples show promising results, which are well supported by those obtained using FESEM (fild emission scanning electron microscope), AFM (atomic force microscope), and conventional liquid–liquid displacement porometry. This novel method can provide useful information for not only surface coatings on porous substrates but also for modification of dense membrane defects; thus, broad utilizations of this technique can be expected in future study.

Keywords: MLLDP; porous membrane; pore mouth size distribution; dense Pd membrane; defect distribution

1. Introduction

Membrane technology has been extensively investigated in energy- and environment-related issues, such as H_2 separation, natural gas purification, water treatment, etc. The functional layers of either inorganic or organic materials are often supported on a porous substrate such as α-alumina, zeolites, or stainless steel. Within this asymmetric structure, the porous substrate provides mechanical support and, thus, the thickness of functional layers can be significantly reduced. Conventional Pd metal tubes have been applied in the semiconductor and electronics industries, but are at least 100 micrometers. By forming a composite membrane on a porous alumina or stainless steel substrate, the thickness of the palladium layer can be reduced to several micrometers, which greatly lowers the cost and improves the hydrogen permeability [1,2]. For porous materials used as a membrane substrate, the size of the pore mouth is more of a concern than that of the pore throat [3], as it immediately determines the quality of the deposited layer including continuity and defect/pinhole formation [4–9]. For Pd composite membranes, major defects of the porous substrate lead to increased thickness as well as crack/pinhole formations [1,10]. On the other hand, the exfoliation of the Pd layer from the substrate can easily occur at a pore mouth size below ca. 20 nm due to a weak adhesion effect [11].

Currently, there exist several techniques for the determination of the pore size distribution of porous materials, as elaborated in Table 1, but there is still lack of efficient methods for pore mouth size characterization. Direct observation methods including SEM [12], FESEM [13], ESEM (enviromental scanning electron microscope) [14], and AFM [15] can provide general surface information of porous materials directly, but they are expensive and time consuming. Moreover, they provide only local information of a specific area (ca. 100 μm^2) and require broken pieces of membranes.

Table 1. Comparison of existing pore size characterization methods.

	Methods	Equation	Pore Size Information	Ref.
Direct	SEM	-	Pore mouth	[12]
	AFM	-	Pore mouth	[15]
	ESEM	-	Pore mouth	[14]
	FESEM	-	Pore mouth	[13]
Indirect	GAD	Kelvin	Average	[16]
	Permporometry	Kelvin	Pore throat	[17]
	Nanopermporometry	Kelvin	Kelvin diameter	[18]
	EP	Kelvin	Average	[13]
	Thermoporometry	Gibbs–Thompson	Average	[19]
	BPM	Young–Laplace	Pore throat	[20]
	MBPM	Young–Laplace	Pore mouth	[3]
	LLDP	Young–Laplace	Pore throat	[21]
	Mercury intrusion porometry	Young–Laplace	Pore throat	[16]
	MLLDP	Young–Laplace	Pore mouth	This work

Recently Krantz et al. [13] reported a detailed description of indirect methods, which can be separated into three groups, i.e., GAD (gas adsorption/desorption) [22], permporometry [17], and EP (evapoporometry) [13] based on the Kelvin equation; thermoporometry [19] based on the Gibbs–Thompson equation; and mercury intrusion porometry [16], BPM (bubble point method) [20], MBPM ("modified" bubble point method) [3], and LLDP (liquid–liquid displacement porometry) [21] based on the Young–Laplace equation. GAD, EP, and thermoporometry appear effective for average pore size measurement. GAD detects not only continuous pores but also dead-end pores. EP is a simple approach based on gravimetric measurement which does not require any assumed model for the pore geometry. Permporometry is based on capillary condensation of vapor and the blocking effect of permeation of a noncondensable gas, which measures pore throat size distribution. Nanopermorometry was also reported and is a method based on the Kelvin equation to characterize the Kelvin diameter of porous membranes [18], where the results denote a bimodal membrane structure described by a dense matrix and highly permeable regions.

Mercury intrusion porosimetry [16] provides pore throat size information, but it detects not only the continuous pores but also the dead-end pores. BPM [20] and LLDP [21], due to their convenience, have been widely applied in pore size measurements. These two methods work via a straightforward mechanism and the pressure required to reopen the pore depends on the capillary force in the pore. With the increase of pressure, the pores reopen from big to small ones gradually. Usually, the maximum capillary force throughout the pore is at the pore throat along the pore tunnel, and, thus, these two methods measure the pore throat size distribution. Huang et al. [3] reported MBPM to determine the pore mouth size distribution, a method which is modified from conventional BPM. Liquid is added to the pore mouth while the pore tunnel is purged with gas under pressure, and the closure of the pore depends on the capillary force at the pore mouth when gradually decreasing the pressure. However, this method requires relatively high pressures to measure small pore mouth sizes due to high gas–liquid surface tensions, e.g., 2.9 MPa for a pore size of 0.1 μm when using water as the impregnating agent. This increases the sealing difficulty and, in addition, the "entrainment phenomenon" (bubbling of liquid due to high-pressure gas) during experiments leads to large errors in pore size analysis, especially in case of high gas fluxes.

In this work, we introduce convenient "modified" liquid–liquid displacement porometry (MLLDP) to measure the pore mouth size. This method can operate under reasonably low pressures for a wide spectrum of pore sizes due to the relatively lower liquid–liquid interfacial tensions than gas–liquid surface tensions. In addition, the "entraining phenomenon" can be eliminated in the MLLDP method. This novel technique is especially suitable for pore mouth size analysis of multichannel membranes due to the recyclability of the testing liquid. The defect distribution of supported palladium membranes can also be characterized by this novel method, assuming straight defect pores in the thin dense layer (Figure 1).

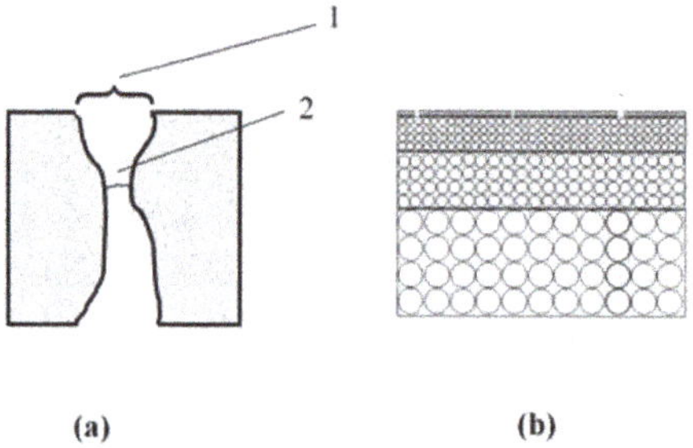

Figure 1. (**a**) Schematic of a pore tunnel. 1, pore mouth; 2, pore throat; (**b**) Schematic of dense membrane defects.

Figure 2 shows a comparison between conventional and "modified" liquid–liquid displacement porometry. In conventional liquid–liquid displacement porometry, the pore tunnels of porous samples are first filled with Liquid A, and then purged with immiscible Liquid B. With the increase of pressure, Liquid B is gradually pushed outwards until the liquid–liquid interface reaches the pore throat. Once the pressure is high enough to overcome the interfacial tension of the liquid at the pore throat, the pore tunnel would be reopened.

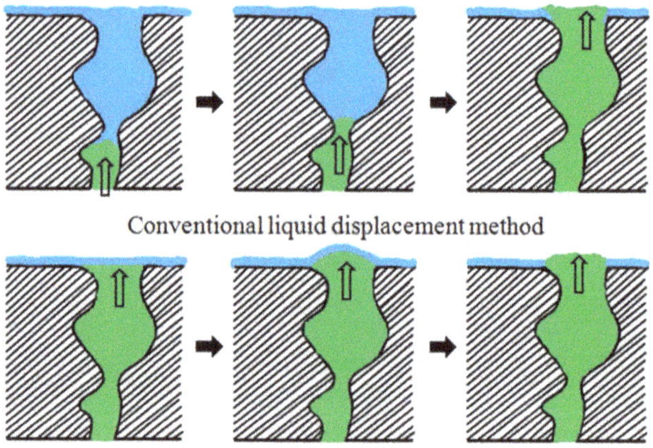

Figure 2. Schematic of the conventional and "modified" liquid–liquid displacement methods (Liquid A in blue, Liquid B in green).

In "modified" liquid–liquid displacement porometry, however, the pore tunnels are first filled with Liquid B and then the outer surface is occupied by Liquid A. Therefore, the liquid–liquid interface appears at the pore mouth instead of the pore throat. Once the pressure of Liquid B is high enough to overcome the interfacial tension at the pore mouth, the pore would be reopened. It should be noted that the measurement errors in conventional liquid–liquid displacement porometry (LLDP) due to resistance in the sublayer [21] do not apply to this "modified" liquid–liquid displacement porometry (MLLDP). According to the Young–Laplace equation, due to lower liquid–liquid interfacial tensions used in the MLLDP method, it is possible to measure relatively small pores with reasonably low pressure which does not damage the porous material. In addition, the low-pressure operation relaxes the requirements for equipment sealing.

2. Methods, Materials, and Apparatus

2.1. "Modified" Liquid–Liquid Displacement Porometry (MLLDP)

MLLDP is an indirect method based on the Young–Laplace equation, and the pore diameter can be calculated as [3]

$$d = \frac{4\gamma cos\theta}{P} \tag{1}$$

where γ is the interfacial tension coefficient between the liquid pair, θ is the contact angle of the penetrating agent on the pore wall, and P is the critical pressure to reopen the pore mouth. When the wetting effect is perfect, the contact angle can be assumed as 0. The pore mouths will be reopened from bigger to smaller ones with increasing pressure of the penetrating agent. A pressure–flux curve can be obtained by monitoring the pressure and flux of the penetrating agent through the pore tunnels. The theoretical derivation of the distribution curve from experimental pressure–flux data was introduced by Grabar et al. [23] and later applied by McGuire et al. [24] and Piątkiewicz et al. [25]. This is based on the assumption of cylindrical and separated pores with various diameters for the real pore structure of the porous membranes. In addition, a continuous distribution function $f(D)$ is assumed for the varying sizes of the pores.

Based on the Hagen–Poiseuille equation, the liquid flow through pore tunnels can be described as

$$Q = \frac{n\pi r^4 \Delta P}{8\mu l \tau} \tag{2}$$

Then, the pore mouth size distribution can be calculated using the following equation [3,16,20]:

$$f(r) = \left(\frac{dQ}{d(\Delta P)} - \frac{Q}{\Delta P}\right)\frac{1}{r^5 C_2} \tag{3}$$

where Q is the liquid flux, μ is the viscosity of the penetrating agent, r is the pore radius, l is the thickness of the porous material, τ is the pore curvature, n is the number of pores that can be opened, ΔP is the operation pressure, and C_2 is a constant number.

2.2. Materials

A wide range of pore sizes can be determined with different immiscible liquid systems depending on the interfacial tensions. Table 2 shows that the operation pressures required for MBPM are one or two orders of magnitude higher than that required for MLLDP, due to higher gas–liquid surface tensions than liquid–liquid interfacial tensions.

Well-defined membranes with a regular noninterconnected pore structure and narrow pore size distribution can be considered as standard to check the feasibility of pore size characterization methods. In this study, 100 nm Anopore^{TM} (Whatman, Maidstone, UK) membranes were adopted for validating the MLLDP method. These membranes comprise a thin layer of symmetric structure without any support layer. Therefore, the pressure difference was kept below 1 bar during MLLDP analysis to maintain integrity. MLLDP was also extended to characterization of commercially available symmetric PVDF membranes (polyvinylidene fluoride) fabricated by the non-solvent-induced phase separation method, single tubular and multichannel tubular ceramic membranes (commercial), as well as the defect size of supported Pd membranes prepared in-house by the electroless-plating method [26]. The geometrical details of these materials and operational characteristics during MLLDP analysis are depicted in Table 3.

Table 2. A comparison of operation pressures required for MBPM and MLLDP (20 °C, cos θ = 1).

Method	System	Interfacial Tension (mN/m)	Pressure (MPa) Corresponding to Different Pore Diameters (μm)			
			0.2	0.1	0.05	0.025
MBPM	water–nitrogen	72.8	1.45	2.91	5.82	11.64
	ethanol–nitrogen	22.39	0.44	0.89	1.79	3.58
MLLDP	n-amyl alcohol [a]/water	4.8	0.096	0.19	0.38	0.76
	isobutyl alcohol [a]/water	1.7	0.034	0.068	0.13	0.27
	oil–aqueous phase [b]/water	0.35	0.007	0.014	0.028	0.056

[a] n-amyl alcohol and isobutyl alcohol are saturated with water at a volumetric proportion of 1:1; [b] oil–aqueous phase is prepared using isobutyl alcohol, methanol, and water at a volumetric proportion of 15:7:25.

Table 3. Geometrical details of materials investigated and operational characteristics during MLLDP analysis.

Materials	Shape	Geometry	Nominal Pore Size	Liquid System (B/A)	Δ*p* (bar)	Test Rig
Anopore™ membrane	Planar	O.d.: 7 mm Thickness: 60 μm	100 nm	Isobutyl alcohol/water	0.4–1.0	Figure 3
PVDF membrane	Tubular	O.d.: 1.2 mm I.d.: 0.7 mm L: 40 cm	80 nm	Isobutyl alcohol/water	0.4–1.0	Figure 4
ZrO_2/γ-Al_2O_3/α-Al_2O_3 membrane	Single tubular	O.d.: 13 mm I.d.: 11 mm L: 5 cm	30 nm	Isobutyl alcohol/water	0.6–3.0	Figure 4
ZrO_2/γ-Al_2O_3/α-Al_2O_3 membrane	Multi-channel tubular	O.d.: 32 mm Channel i.d.: 4 mm L: 100 cm Channel no.: 19	100 nm	n-amyl alcohol/water	0.4–2.8	Figure 5
Defects of $Pd/ZrO_2/Al_2O_3$ membrane	Tubular	O.d.: 13 mm I.d.: 11 mm L: 5 cm	-	Isobutyl alcohol/water	0.32–3.0	Figure 4

2.3. Apparatus

Figure 3 shows the MLLDP apparatus designed for pore mouth size characterization of Anopore™ membranes while Figure 4 shows the apparatus for characterization of PVDF membranes, single tubular ceramic membranes, and dense Pd membranes (defects). The MLLDP apparatus shown in Figures 3 and 4 have different testing cells which can be used to fulfill the aim of characterizing samples with different shapes. For these membranes, a functional layer is deposited on the outer surface of porous substrates. For multichannel tubular ceramic membranes, the functional layer is deposited on the inner surface of tubes, and the pore mouth size distribution of the inner surface is what we are concerned with. Therefore, a distinct testing cell was designed for the pore mouth size distribution of multichannel membranes, as shown in Figure 5. Note that the apparatus in Figure 4 is also applicable for conventional BPM measurements and conventional LLDP measurements for pore throat characterization due to the similarity in the operations.

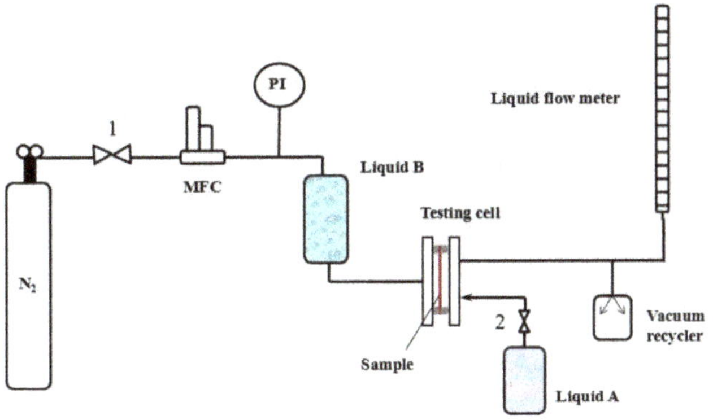

Figure 3. Schematic of apparatus for MLLDP measurements of an Anopore^{TM} membrane. MFC, mass flow controller. PI, pressure indicator.

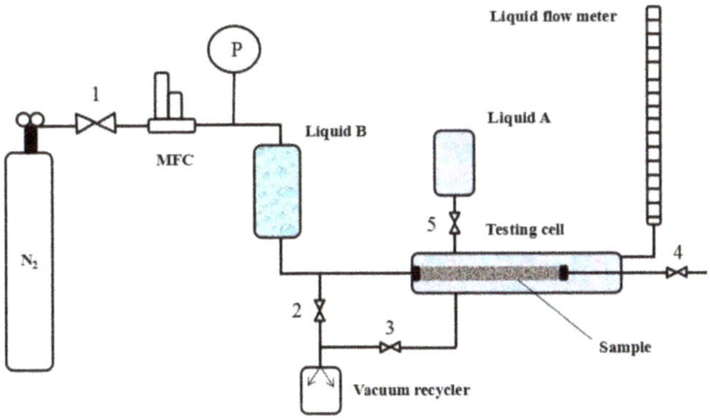

Figure 4. Schematic of apparatus for MLLDP measurements of a single-channel membrane. MFC, mass flow controller. PI, pressure indicator.

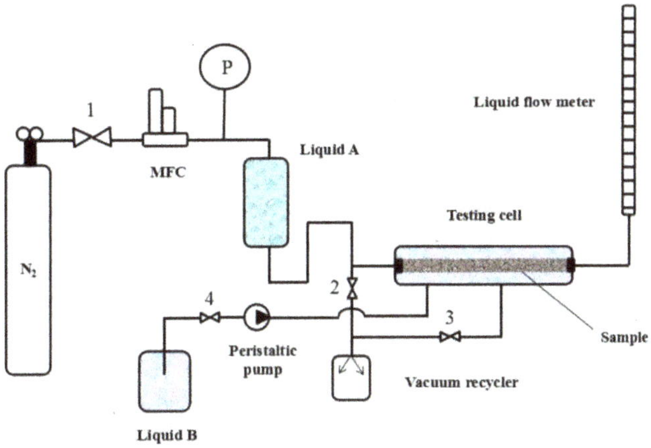

Figure 5. Schematic of apparatus for MLLDP measurements of a multichannel membrane. MFC, mass flow controller. PI, pressure indicator.

Take the testing of a single-channel ceramic membrane as an example (test rig in Figure 4); the operating procedure is as follows:

1. The dry sample is mounted in the testing cell (drying procedure and pretreatment are important for porous materials [18,27,28]; thus, samples were carefully dried without changing the pore structure);
2. Liquid B (acting as penetrating agent) from the reservoir fills the lumen of the sample by opening Valve 4 for a short time;
3. The pressure of Liquid B is increased to 3 bar by increasing the N_2 gas pressure while applying a negative pressure of -0.8 bar in the shell side by using a vacuum recycler (opening Valve 3) in order to achieve complete wetting of pore tunnels with Liquid B; this step lasts for at least 30 min;
4. Liquid A (acting as impregnating agent) from the reservoir is poured into the shell side by opening Valve 5 until the remaining Liquid B on the outer surface of the sample is completely replaced with Liquid A, i.e., until there is no Liquid B observed in the outflow (Liquids A and B can be easily distinguished by color), then Valve 5 is closed; Note that Liquid A is kept under ambient pressure and at a low flow rate below ca. 10 mL/min in order to avoid penetration of Liquid A into membrane pores;
5. The pressure of Liquid B increases step by step and a pressure–flux curve can be obtained by monitoring the N_2 gas pressure at a step of 0.05–0.1 bar and flux of Liquid B through the pore tunnels (measured by bubble flow meter). The flow rate is recorded until a steady-state flux is achieved. The testing cell is kept at RT during the operation.

For LLDP measurement, the operation procedure is as below:

1. The dry sample is mounted in the testing cell;
2. Liquid A (acting as impregnating agent) from the reservoir is poured into the shell side by opening Valve 5 for a short time;
3. The sample is impregnated in Liquid A for a period of 30 min to 1 h in order to achieve complete wetting of pore tunnels with Liquid A;
4. Liquid B (acting as penetrating agent) from the reservoir fills the lumen of the sample by opening Valve 4 for a short time;
5. The pressure of Liquid B increases step by step and a pressure–flux curve can be obtained by monitoring the N_2 gas pressure at a step of 0.05–0.1 bar and flux of Liquid B through the pore tunnels (measured by bubble flow meter). The flow rate is recorded until a steady-state flux is achieved. The testing cell is kept at RT during the operation.

It can be seen that Liquids A and B act as impregnating and penetrating agent, respectively, in cases of both MLLDP and LLDP measurement. In the former, Liquid B first fills the pore tunnels of the sample and then Liquid A occupies the outer surface of sample such that Liquids A and B contact each other at the position of the pore mouth. Liquid B penetrates through the pore tunnels when the pressure is high enough to overcome the interfacial tension at the pore mouth according to the Young–Laplace equation. In the latter, Liquid A first fills the pore tunnels of the sample and then Liquid B penetrates through the pore tunnels when the pressure is high enough to overcome the interfacial tension at the position of the pore throat according to the Young–Laplace equation.

2.4. Error Analysis

The error in the pore diameter for techniques based on the Young–Laplace Equation (1) is strongly dependent on the accuracy in determining pressure values. This also applies to BPM, MBPM, LLDP, and MLLDP methods.

According to the Young–Laplace Equation (1), the fractional error in the diameter *(Δd/d)* that results from the fractional error in the pressure analysis is given by

$$\frac{\Delta d}{d} = \frac{\Delta P}{P} \quad (4)$$

In should be emphasized that the pressure analysis error due to the difference in liquid levels between the reservoir of Liquid B and flow meter (Figure 4) has to be taken into account. The liquid level in the bubble flow meter may fluctuate frequently with the flow in and out during the measurement, and needs to be kept at a similar level to the reservoir of Liquid B in order to minimize corresponding pressure differentials. In this study, such a level difference is deliberately kept within ±20 cm during the operation process, which corresponds to a low pressure differential of ca. 0.02 bar. The pressure of Liquid B is then set above 0.32 bar in order to ensure a fraction error in the pressure analysis *(ΔP/P)* of less than ca. 6%. The fractional error in the diameter decreases with the increase of operational pressures, as shown in Figure 6.

As with other techniques, the accuracy of MLLDP decreases with increasing pore sizes. The analysis of large pores above ca. 100 μm requires measurements in a fairly low pressure range and thus degrades the accuracy.

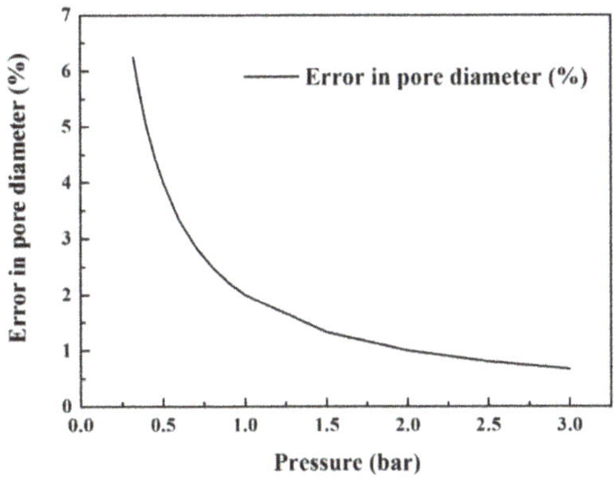

Figure 6. Percent error in the pore diameter as a function of the operation pressure.

3. Results and Discussions

3.1. Standard Membranes

3.1.1. Anopore^{TM} Membranes (100 nm)

In order to verify the feasibility of this novel MLLDP method in measuring the pore mouth size distribution of porous membranes, 100 nm AnoporeTM membranes and 80 nm PVDF membranes were used as standard membranes and characterized by MLLDP.

Figure 7a depicts the MLLDP analysis results for four replicate samples of the 100 nm Anopore^{TM} membrane, which indicates a good overlapping of the measurements with a fairly sharp pore mouth size distribution. This agrees well with the narrow pore size distribution of Anopore^{TM} membranes which exhibit regular columnar pore geometry. It should be mentioned that for the columnar pore geometry of Anopore^{TM} membranes, the pore mouth size is equal to the pore throat size. MLLDP analysis provides a number-averaged pore mouth size of 84.09 nm with $\sigma = \pm 3.52$ nm as listed in Table 4. The σ indicates the breadth of the pore mouth size distribution. The comparison in Table 5 shows a good agreement between MLLDP and other techniques, including evapoporometry [13], FESEM [13], SEM [29], AFM [30], and LLDP [21]. For example, FESEM [13] and SEM [21] analyses provide number-averaged pore mouth sizes of 87 nm and 83 nm, respectively, which are fairly close to that determined by MLLDP (84.09 nm). This has a strong relationship with the fact that MLLDP also provides a number-averaged pore mouth size. The pore diameter error with MLLDP analysis is kept within ca. 4%, which can be mainly ascribed to the proper selection of liquid pairs and minimization of errors in the pressure analysis. The slightly larger number-averaged pore size of 108 nm provided by AFM analysis [23] was attributed to a low lateral resolution resulting from the use of a 10 nm tip on AFM [4]. The mean-flow average pore size of 119 nm derived from LLDP measurement [21] is relatively larger than the number- or mass-averaged pore sizes determined by other techniques, which can be ascribed to the fact that the volume flow for cylindrical pores is proportional to d^4 instead of d^2 for pore mass.

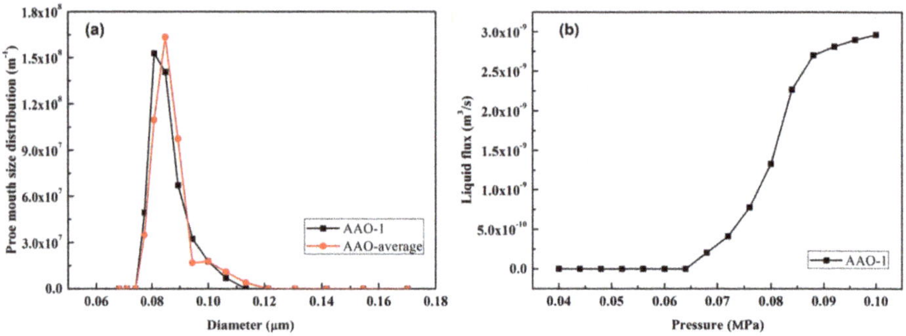

Figure 7. (**a**) Averaged pore mouth size distribution for Anopore™ membranes derived from four replicate samples; (**b**) liquid flux and pressure relationship of AAO-1 membrane.

Table 4. Summary of MLLDP results for four replicate samples of nominal 100 nm Anopore™ membrane.

Types	Membranes				Average	σ	Avg ± σ
	AAO-1	AAO-2	AAO-3	AAO-4			
Number-averaged pore mouth (nm)	80.95	80.95	89.47	85.00	84.09	3.52	84.09 ± 3.52

Table 5. Comparison of PSD (pore size distribution) measurements of nominal 100 nm Anopore™ membrane with different methods.

Methods	Average	Pore Mouth/Pore Throat/Mean Pore Size	Averaged Pore Size (nm)	Ref.
EP *	Mass-average	Mean pore size	94 ± 14	[13]
	Number-average	Mean pore size	69 ± 34	[13]
FESEM	Number-average	Pore mouth	87 ± 10	[13]
SEM	Number-average	Pore mouth	83 ± 26	[29]
LLDP	Mean-flow average	Pore throat	119	[29]
AFM	Number-average	Pore mouth	108 ± 26	[30]
MLLDP	Number-average	Pore mouth	84.09 ± 3.52	This work

* EP: Evapoporometry.

3.1.2. PVDF Membranes (80 nm)

A representative run was carried out with six different sections of the same membrane tube and the results are reported in Figure 8 and Table 6. The graph shows a good overlapping of these replicate measurements, obtaining an average pore mouth size of 75.6 nm with σ = ±3.18 nm. This is in fair agreement with the nominal pore size of 80 nm provided by the supplier. The representative FESEM image in Figure 3 corroborated MLLDP analysis results although some large pores approaching

100 nm can also be identified on the membrane surface. It has been indicated previously that the high operation pressures required in LLDP measurements may lead to considerable compression of polymeric membranes and, thus, the shifting of the pore size distribution towards smaller pore diameters [4]. In this study, the liquid system was selected with discretion, i.e., isobutyl alcohol/water with a low interfacial tension of 1.7 mN/m, which helps maintain the pressure differential below 1 bar and thus minimize the compaction of PVDF membranes. This approach can be universal for MLLDP analysis of polymeric membranes.

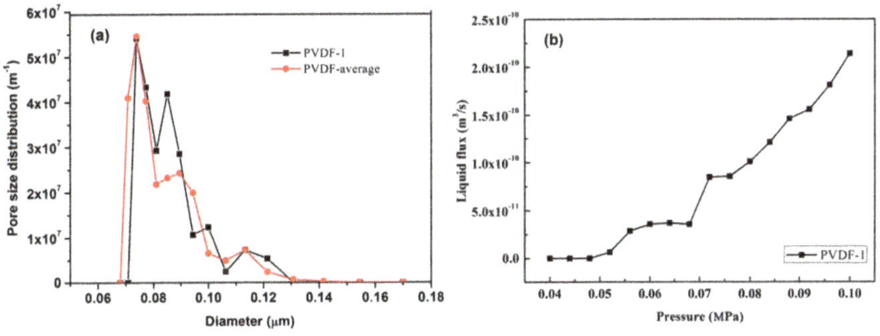

Figure 8. (**a**) Averaged pore mouth size distribution for PVDF membranes derived from six replicate samples; (**b**) liquid flux and pressure relationship of PVDF-1 membrane.

Table 6. Summary of MLLDP results for six replicate samples of nominal 80 nm PVDF membrane.

Types	Tubes						Average	σ	Avg ± σ
	P-1	P-2	P-3	P-4	P-5	P-6			
Number-averaged pore mouth size (nm)	73.9	70.8	73.9	80.9	77.0	77.0	75.6	3.18	75.6 ± 3.18

The pore mouth size information of those standard membranes obtained by the MLLDP method are well in agreement with data from the supplier and literature reports. Therefore, MLLDP as a novel characterization method is able to provide the pore mouth size information of porous membranes.

3.2. *Tubular Ceramic Substrates for Pd Composite Membranes*

3.2.1. Single Tubular $ZrO_2/\gamma\text{-}Al_2O_3/\alpha\text{-}Al_2O_3$ Membrane (30 nm)

The pore mouth size distribution of commercially fabricated single tubular $ZrO_2/\gamma\text{-}Al_2O_3/\alpha\text{-}Al_2O_3$ membranes was characterized by MLLDP, as reported in Figure 9 and Table 7. Figure 9a shows a good overlapping of the three replicate measurements in a representative run. The average pore mouth size was determined as 31.98 nm with $\sigma = \pm 1.06$ nm, which is slightly larger than the nominal pore throat size of 30.05 nm with $\sigma = \pm 7.02$ nm measured by conventional LLDP, as shown in Figure 9b. This agrees well with the fact that pore mouth size is slightly larger than pore throat size. It should be noted that, due to resistance in the sublayer [21], the measurement error of conventional liquid–liquid displacement porometry (LLDP) was $\sigma = \pm 7.02$ nm. However, "modified" liquid–liquid displacement porometry (MLLDP) gave $\sigma = \pm 1.06$ nm, indicating that the measurement errors due to resistance in the sublayer do not apply to MLLDP, which is an improvement over LLDP.

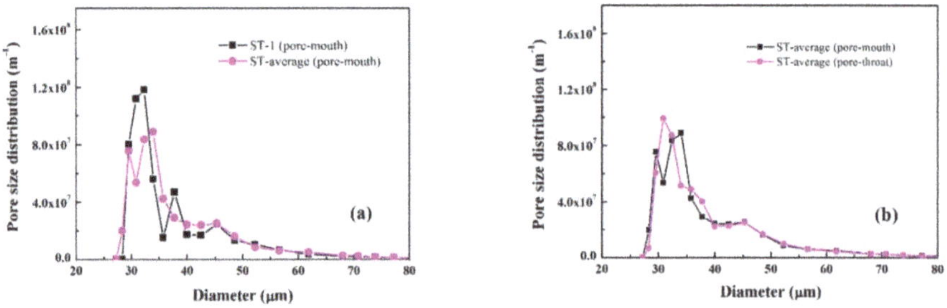

Figure 9. Characterization of single tubular ZrO_2/γ-Al_2O_3/α-Al_2O_3 membranes with 3 replicate samples: (**a**) Averaged pore mouth size distribution in comparison with that for Sample 1; (**b**) Comparison of averaged pore mouth size distribution and averaged pore throat size distribution.

Table 7. Summary of MLLDP results for three replicate samples of nominal 30 nm single tubular ZrO_2/γ-Al_2O_3/α-Al_2O_3 membrane.

Types	Repeat Times			Average	σ	avg ± σ
	ST-1	ST-2	ST-3			
Number-averaged pore mouth (nm)	32.38	34	29.56	31.98	1.06	31.98 ± 1.06
Number-averaged pore throat (nm)	30.91	30.91	28.33	30.05	7.02	30.05 ± 7.02

Inspection of the FESEM image in Figure 10c indicates a rather smooth surface with interconnected irregular-shaped pores. A wide range of pore sizes, particularly some large pores approaching up to 60 nm, can be discriminated on the membrane surface, which coincides well with the broad pore mouth size distribution determined by MLLDP as shown in Figure 9a. Note that the operation pressures applied during MLLDP or LLDP measurements will not affect the pore size distribution of ceramic membranes which exhibit strong mechanical strength.

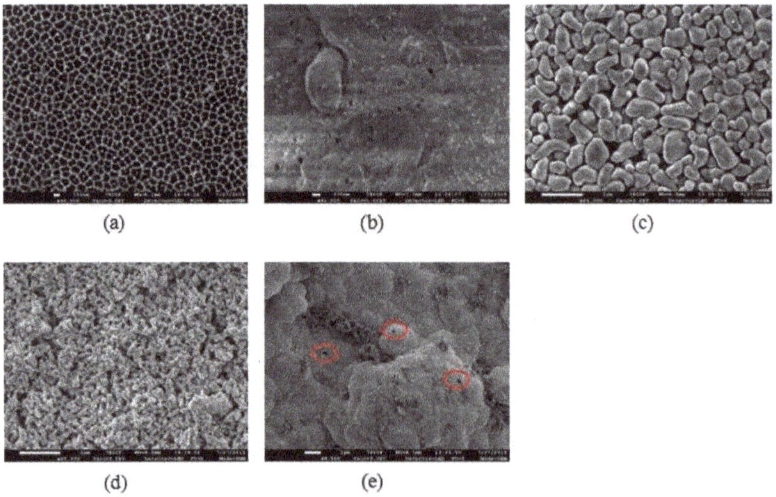

Figure 10. FESEM images: (**a**) 100 nm Anopore^{TM} membrane; (**b**) 80 nm PVDF membrane; (**c**) 30 nm single tubular ZrO_2/γ-Al_2O_3/α-Al_2O_3 membrane (outer surface); (**d**) 100 nm multichannel tubular ZrO_2/γ-Al_2O_3/α-Al_2O_3 membrane (inner surface); (**e**) defects of dense Pd composite membrane.

3.2.2. Multichannel Tubular ZrO_2/γ-Al_2O_3/α-Al_2O_3 Membrane (100 nm)

The pore mouth size distribution of multichannel tubular ZrO_2/γ-Al_2O_3/α-Al_2O_3 membranes (commercial) was also examined by MLLDP. Figure 11 shows that three repeated measurements of the same sample exhibit excellent reproducibility with similar patterns of distribution and an obtained average pore mouth size of 80.09 nm ($\sigma = \pm 2.72$ nm), the results were summarized in Table 8. This is in good agreement with the nominal pore size of 100 nm. The representative FESEM image in Figure 10d indicates a relatively narrow distribution of pore sizes on the membrane surface, which corresponds well with the MLLDP results in Figure 11. Despite the rough inner surface of the multichannel tubular membranes as shown in Figure 10d, the pores are assumed to be perpendicular to the local surface, which is apparently the case in a short range of ca. 80 nm. This assumption seemingly did not affect the pore mouth size analysis by MLLDP, which shows good agreement with nominal values. It should be emphasized that MLLDP exhibits a great advantage in terms of recyclability of the testing liquid, in comparison with MBPM measurement where gas such as air or N_2 acts as penetrant and, thus, cannot be easily recycled.

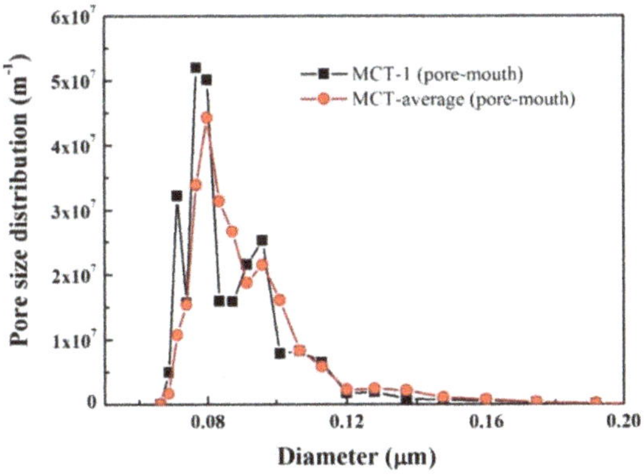

Figure 11. Averaged pore mouth size distribution for multichannel tubular ZrO_2/γ-Al_2O_3/α-Al_2O_3 membranes derived from six replicate samples.

Table 8. Summary of MLLDP results for three replicate samples of nominal 100 nm multichannel tubular ZrO_2/γ-Al_2O_3/α-Al_2O_3 membrane.

Types	Repeat Times			Average	σ	avg ± σ
	MCT-1	MCT-2	MCT-3			
Number-averaged pore mouth (nm)	76.80	83.48	80.00	80.09	2.72	80.09 ± 2.72

3.3. *Defects of a Dense Pd Composite Membrane*

Currently, there is lack of methods for defect size characterization of dense Pd composite membranes. Using the MLLDP analysis presented in this study, the defect size distribution was successfully obtained, as shown in Figure 12 and Table 9. The six repeated measurements of the same Pd composite membrane (prepared in-house by the electroless-plating method [26]) indicate excellent reproducibility with an average defect size around 24.44 nm (σ of ± 0.25 nm). The low σ suggests a relatively narrow distribution of defect sizes. This is in excellent agreement with the analysis of representative FESEM images in Figure 10e with defect sizes in a narrow range of 20–40 nm. Figure 10e also shows that the defects exhibit irregular-shaped geometries on rough surfaces. It can be claimed that the surface roughness has an inconsiderable effect on defect size analysis. In order to examine the defect size in a low range of ca. 20–40 nm, the liquid system of isobutyl alcohol/water with low interfacial tension was selected with discretion, which renders the operation in a proper pressure range (0.32–3.0 bar) and thus helps improve the accuracy of MLLDP measurements.

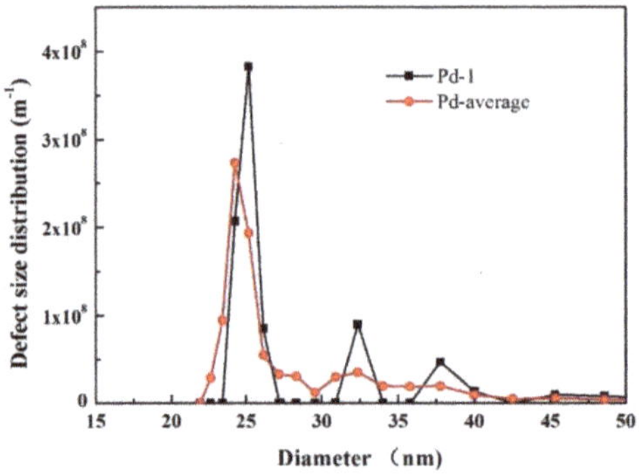

Figure 12. Averaged defect size distribution for Pd composite membranes derived from six replicate runs and four replicate samples.

Table 9. Summary of MLLDP results for six replicate runs of Pd composite membrane.

Types	Repeat Times						Average	σ	avg $\pm$ σ
	Pd-1	**Pd-2**	**Pd-3**	**Pd-4**	**Pd-5**	**Pd-6**			
Number-averaged defect size (nm)	24.29	24.29	25.11	24.29	24.29	24.29	24.44	0.25	24.44 $\pm$ 0.25

3.4. Advantages and Limitations of MLLDP

3.4.1. Advantages of MLLDP

MLLDP exhibits great advantages over direct observation methods including SEM, FESEM, and AFM, since it can provide a general picture of the whole membrane surface instead of local information on a small area (ca. 100 μm^2) which might not be representative. In addition, SEM, FESEM, and AFM require expensive dedicated equipment and broken pieces of samples for measurement. In comparison, MLLDP can be carried out with a simple setup at a much lower cost. The main instruments included are a mass flow controller (MFC) and a pressure gauge. In addition, the membrane sample can remain intact during the MLLDP measurement. Another important advantage of MLLDP over direct methods is that it can be used for characterization of irregular pore geometry instead of only well-defined pore structures.

Most indirect methods provide either the pore throat size distribution or average pore size, including LLDP, mercury porosimetry, BPM, GAD, permporometry, and EP. The recently developed MBPM can be used for the characterization of the pore mouth size distribution; however, it requires high operation pressures due to high surface tensions between gas and liquid. MLLDP presented in this study renders the possibility of characterizing a wide range of pore sizes at a much lower pressure as the interfacial tensions between liquid pairs is several orders of magnitude lower than surface tensions between gas and liquid. Thus, by selecting a liquid system with discretion, it is possible to reduce the operational pressure for thin and flexible membranes and determine the pore mouth size distribution at high accuracy, e.g., for polymeric membranes. Moreover, MLLDP can be extended to pore mouth size characterization of nanofiltration (NF) membranes and defect size analysis of dense membranes down to the nanometer scale. MLLDP analysis for defect characterization of Pd composite membranes can be regarded as a breakthrough technology in the relevant field.

Two other advantages of MLLDP are that it can be applied for characterization of both symmetric and asymmetric membranes, and it detects only continuous pores but not dead-end pores, thus providing more relevant information for flow analysis.

3.4.2. Limitations of MLLDP

As with other techniques, the accuracy of MLLDP decreases with increasing pore size. This has been suggested to be a common issue for other techniques including GAD and permporometry based on the Kelvin equation as well as LLDP based on the Young–Laplace equation. In MLLDP, the analysis of large pores above ca. 100 μm requires measurements in a fairly low pressure range and thus degrades the accuracy.

This technique employs a low pressure range; thus, small variations in the pressure can give higher relative errors in the measurement. The error in the pore diameter is highly sensitive to pressure analysis errors within Equation (1), which places urgency upon the accuracy of the pressure measurement. High-precision pressure analysis is thus a prerequisite for MLLDP measurement. In addition, the difference in liquid levels between the reservoir and liquid flow meter has to be minimized which may lead to some errors in the pressure analysis. Another limitation of MLLDP is that it assumes a noninterconnected, separated columnar pore structure, which may cause data interpretation problems for interconnecting pores on the surface. However, this is common for all other techniques mentioned above.

4. Conclusions

In the preparation of dense Pd composite membranes, the pore size distribution of the porous support immediately affects the quality of the membrane layer, including continuity and defect/pinhole formation. This work presents a novel method (MLLDP) to determine the pore mouth size distribution of porous supports and defect size distribution of Pd-based composite membranes by modifying the conventional liquid–liquid displacement method. MLLDP can be used to determine a wide spectrum of pore mouth sizes (defect sizes) under reasonably low pressures and thus achieve great accuracy and ease of operation, offering unique superiority compared with currently available technologies like FESEM, AFM, and the "modified" bubble-point method. Particularly, MLLDP exhibits great advantages for the characterization of multichannel tubular ceramic membranes in terms of recyclability of the testing liquid, and MLLDP analysis for defect characterization of dense membranes can be regarded as a breakthrough technology in the relevant field. It can thus be expected that this novel method will find wide applications in the surface analysis of porous materials as well as in defect analysis of dense membranes.

Author Contributions: Conceptualization, L.Z., H.L. and H.X.; Data curation, L.Z., H.Y. and T.X.; Formal analysis, L.Z., H.L., G.K., J.Y., X.L. and H.X.; Funding acquisition, H.L. and H.X.; Methodology, L.Z. and H.L.; Project administration, H.L. and H.X.; Supervision, H.L. and H.X.; Writing—original draft, L.Z.; Writing—review & editing, L.Z., H.L., H.Y., G.K., T.X., J.Y., X.L. and H.X.

Funding: We would like to thank the 100-Talent Project of CAS and National Natural Science Foundation of China (Grant No. 21676265; 51501177; 21306183) for their financial support.

Acknowledgments: The authors would like to gratefully acknowledge Shoufu Hou and Chunhua Tang for the technical support with respect to building up the setups.

Conflicts of Interest: The authors declare no conflict of interest.

References

1. Paglieri, S.N.; Way, J.D. Innovations in palladium membrane research. *Sep. Purif. Methods* **2002**, *31*, 1–169. [CrossRef]
2. Yun, S.; Oyama, S.T. Correlations in palladium membranes for hydrogen separation: A review. *J. Membr. Sci.* **2011**, *375*, 28–45. [CrossRef]

3. Yu, J.; Hu, X.J.; Huang, Y. A modification of the bubble-point method to determine the pore-mouth size distribution of porous materials. *Sep. Purif. Technol.* **2010**, *70*, 314–319. [CrossRef]
4. Ryi, S.K.; Ahn, H.S.; Park, J.S.; Kim, D.W. Pd-Cu alloy membrane deposited on CeO_2 modified porous nickel support for hydrogen separation. *Int. J. Hydrog. Energy* **2014**, *39*, 4698–4703. [CrossRef]
5. Zhao, H.B.; Pflanz, K.; Gu, J.H.; Li, A.W.; Stroh, N.; Brunner, H.; Xiong, G.X. Preparation of palladium composite membranes by modified electroless plating procedure. *J. Membr. Sci.* **1998**, *142*, 147–157. [CrossRef]
6. Chi, Y.H.; Yen, P.S.; Jeng, M.S.; Ko, S.T.; Lee, T.C. Preparation of thin Pd membrane on porous stainless steel tubes modified by a two-step method. *Int. J. Hydrog. Energy* **2010**, *35*, 6303–6310. [CrossRef]
7. Jabbari, A.; Ghasemzadeh, K.; Khajavi, P.; Assa, F.; Abdi, M.; Babaluo, A.; Basile, A. Surface modification of α-alumina support in synthesis of silica membrane for hydrogen purification. *Int. J. Hydrog. Energy* **2014**, *39*, 18585–18591. [CrossRef]
8. Tong, J.H.; Su, L.L.; Haraya, K.; Suda, H. Thin and defect-free Pd-based composite membrane without any interlayer and substrate penetration by a combined organic and inorganic process. *Chem. Commun.* **2006**, 1142–1144. [CrossRef] [PubMed]
9. Zheng, L.; Li, H.; Xu, H.Y. "Defect-free" interlayer with a smooth surface and controlled pore-mouth size for thin and thermally stable Pd composite membranes. *Int. J. Hydrog. Energy* **2016**, *41*, 1002–1009. [CrossRef]
10. Zheng, L.; Li, H.; Xu, T.Y.; Bao, F.; Xu, H.Y. Defect size analysis approach combined with silicate gel/ceramic particles for defect repair of Pd composite membranes. *Int. J. Hydrog. Energy* **2016**, *41*, 18522–18532. [CrossRef]
11. Ye, J.; Dan, G.; Yuan, Q. The preparation of ultrathin palladium membrane. In *Key Engineering Materials*; Trans Tech Publication: Zürich, Switzerland, 1992; pp. 437–442.
12. Calvo, J.I.; Bottino, A.; Capannelli, G.; Hernandez, A. Comparison of liquid-liquid displacement porosimetry and scanning electron microscopy image analysis to characterise ultrafiltration track-etched membranes. *J. Membr. Sci.* **2004**, *239*, 189–197. [CrossRef]
13. Krantz, W.B.; Greenberg, A.R.; Kujundzic, E.; Yeo, A.; Hosseini, S.S. Evapoporometry: A novel technique for determining the pore-size distribution of membranes. *J. Membr. Sci.* **2013**, *438*, 153–166. [CrossRef]
14. Reingruber, H.; Zankel, A.; Mayrhofer, C.; Poelt, P. A new in situ method for the characterization of membranes in a wet state in the environmental scanning electron microscope. *J. Membr. Sci.* **2012**, *399*, 86–94. [CrossRef]
15. Ochoa, N.; Pradanos, P.; Palacio, L.; Pagliero, C.; Marchese, J.; Hernandez, A. Pore size distributions based on AFM imaging and retention of multidisperse polymer solutes: Characterisation of polyethersulfone UF membranes with dopes containing different PVP. *J. Membr. Sci.* **2001**, *187*, 227–237. [CrossRef]
16. Calvo, J.; Hernandez, A.; Pradanos, P.; Martınez, L.; Bowen, W. Pore size distributions in microporous membranes ii. Bulk characterization of track-etched filters by air porometry and mercury porosimetry. *J. Colloid Interface Sci.* **1995**, *176*, 467–478. [CrossRef]
17. Cuperus, F.P.; Bargeman, D.; Smolders, C.A. Permporometry—The determination of the size distribution of active pores in UF membranes. *J. Membr. Sci.* **1992**, *71*, 57–67. [CrossRef]
18. Albo, J.; Hagiwara, H.; Yanagishita, H.; Ito, K.; Tsuru, T. Structural characterization of thin-film polyamide reverse osmosis membranes. *Ind. Eng. Chem. Res.* **2014**, *53*, 1442–1451. [CrossRef]
19. Cuperus, F.P.; Bargeman, D.; Smolders, C.A. Critical-points in the analysis of membrane pore structures by thermoporometry. *J. Membr. Sci.* **1992**, *66*, 45–53. [CrossRef]
20. Hernandez, A.; Calvo, J.I.; Pradanos, P.; Tejerina, F. Pore size distributions in microporous membranes. A critical analysis of the bubble point extended method. *J. Membr. Sci.* **1996**, *112*, 1–12. [CrossRef]
21. Gijsbertsen-Abrahamse, A.J.; Boom, R.M.; van der Padt, A. Why liquid displacement methods are sometimes wrong in estimating the pore-size distribution. *AIChE J.* **2004**, *50*, 1364–1371. [CrossRef]
22. Barrett, E.P.; Joyner, L.G.; Halenda, P.P. The determination of pore volume and area distributions in porous substances. I. Computations from nitrogen isotherms. *J. Am. Chem. Soc.* **1951**, *73*, 373–380. [CrossRef]
23. Grabar, P.; Nikitine, S. Sur le diamètre des pores des membranes en collodion utilisées en ultrafiltration. *J. Chim. Phys.* **1936**, *33*, 721–741. [CrossRef]
24. McGuire, K.S.; Lawson, K.W.; Lloyd, D.R. Pore-size distribution determination from liquid permeation through microporous membranes. *J. Membr. Sci.* **1995**, *99*, 127–137. [CrossRef]

25. Piatkiewicz, W.; Rosinski, S.; Lewinska, D.; Bukowski, J.; Judycki, W. Determination of pore size distribution in hollow fibre membranes. *J. Membr. Sci.* **1999**, *153*, 91–102. [CrossRef]
26. Li, H.; Goldbach, A.; Li, W.; Xu, H. PdC formation in ultra-thin Pd membranes during separation of H_2/CO mixtures. *J. Membr. Sci.* **2007**, *299*, 130–137. [CrossRef]
27. Albo, J.; Wang, J.; Tsuru, T. Application of interfacially polymerized polyamide composite membranes to isopropanol dehydration: Effect of membrane pre-treatment and temperature. *J. Membr. Sci.* **2014**, *453*, 384–393. [CrossRef]
28. Albo, J.; Wang, J.; Tsuru, T. Gas transport properties of interfacially polymerized polyamide composite membranes under different pre-treatments and temperatures. *J. Membr. Sci.* **2014**, *449*, 109–118. [CrossRef]
29. Hernandez, A.; Calvo, J.I.; Pradanos, P.; Palacio, L.; Rodriguez, M.L.; de Saja, J.A. Surface structure of microporous membranes by computerized sem image analysis applied to anopore filters. *J. Membr. Sci.* **1997**, *137*, 89–97. [CrossRef]
30. Bowen, W.R.; Hilal, N.; Lovitt, R.W.; Williams, P.M. Atomic force microscope studies of membranes: Surface pore structures of cyclopore and anopore membranes. *J. Membr. Sci.* **1996**, *110*, 233–238. [CrossRef]

membranes

MDPI

Review

Thermodynamic Aspects in Non-Ideal Metal Membranes for Hydrogen Purification

Stefano Bellini [1], Yu Sun [2,3], Fausto Gallucci [4,*] and Alessio Caravella [1,*]

1 Department of Environmental and Chemical Engineering (DIATIC), University of Calabria, Via P. Bucci, Cubo 44A, 87036 Rende, Italy; stefanobellini89@gmail.com
2 Institute for International Collaboration, Hokkaido University, Sapporo, Hokkaido 060-0815, Japan; sunyusunyu1983@gmail.com
3 Department of Chemistry, Faculty of Science, Hokkaido University, N10W8, Kita-ku, Sapporo, Hokkaido 060-0810, Japan
4 Inorganic Membranes and Membrane Reactors (SIR), Sustainable Process Engineering (SPE), Department of Chemical Engineering and Chemistry, Eindhoven University of Technology, P.O. Box 513, STW 1.45, 5600 MB Eindhoven, The Netherlands
* Correspondence: F.Gallucci@tue.nl (F.G.); alessio.caravella@unical.it (A.C.)

Received: 14 August 2018; Accepted: 10 September 2018; Published: 16 September 2018

Abstract: In this paper, an overview on thermodynamic aspects related to hydrogen-metal systems in non-ideal conditions is provided, aiming at systematically merging and analyzing information achieved from several different studies present in the open literature. In particular, the relationships among inner morphology, dissolved hydrogen and internal stresses are discussed in detail, putting in evidence the conformation complexity and the various types of dislocations induced by the presence of H-atoms in the lattice. Specifically, it is highlighted that the octahedral sites are preferentially occupied in the FCC metals (such as palladium), whereas tetrahedral sites are more energetically favored in BCC-structured ones (such as vanadium). These characteristics are shown to lead to a different macroscopic behavior of the two classes of metals, especially in terms of solubility and mechanical failure due to the consequent induced stresses. Furthermore, starting from the expression of the chemical potential generally presented in the literature, a new convenient expression of the activity of the H-atoms dissolved into the metal lattice as a function of the H-concentration is achieved. Such an activity expression is then used in the dissolution equilibrium relationship, which is shown to be the overall result of two different phenomena: (i) dissociative adsorption of molecular hydrogen onto the surface; and (ii) atomic hydrogen dissolution from the surface to the metal bulk. In this way, the obtained expression for equilibrium allows a method to calculate the equilibrium composition in non-ideal conditions (high pressure), which are of interest for real industrial applications.

Keywords: chemical potential; activity; hydrides; solubility; membranes

1. Introduction

Separation processes based on metal membranes with high hydrogen permeance and selectivity have been identified as a promising technology for efficiency improvement and cost reduction of hydrogen production. The virtually infinite selectivity of metal membranes towards hydrogen is related to the fact that the potential field of the (Pd) metal surface can dissociate only hydrogen among all the other light and heavier gas species. In particular, membranes composed of palladium and its alloys can play a fundamental role thanks to their unique affinity towards hydrogen, as they show high catalytic activity to the hydrogen molecule dissociation combined with the relatively high H-atoms diffusion [1].

The substantial difference between palladium and other metals is that the former shows a relatively small activation barrier (fast sorption kinetics) of the surface dissociation, while in the latter the activation energy of the surface dissociation is higher [2].

During the past two decades, the research on Pd-alloy membranes has led to a technology maturity that appears ready for up-scaling for applications at operating temperatures less than about 550 °C and in conditions where detrimental impurities are preventively eliminated by means of appropriate treatments (such as H_2S removal steps).

Based on the behavior of hydrogen dissolution into the metal lattice, metals can be classified into materials in which solubility is favored by temperature and materials showing the opposite trend. The former group, including palladium and vanadium, absorbs hydrogen by means of an exothermic process generating a metal-hydride phase, whereas the latter (e.g., Pt, Fe, Cu, Ni, Ag) absorbs hydrogen endothermically forming a covalent hydride phase [3].

In the fabrication of membranes and hydrogen-storage material, it is important to limit the hydrogen solubility to avoid embrittlement and mechanical failure. For this purpose, the main metal is alloyed with others such as Cr, Co and Ni [4]. Vanadium in particular has interesting characteristics, showing the highest diffusivity and lowest solubility compared with other BCC metals [5–7] along with a relatively low cost. Moreover, the insertion of Ni-atoms into the V-lattice has been demonstrated to enhance the alloy properties in terms of mechanical resistance to the internal stress induced by the presence of hydrogen [8].

Current research on metal membranes has been looking for the best combination of materials and manufacturing approaches. However, such a kind of research risks to turn into a trial-and-error activity if it is not driven by the knowledge of the phenomena involved in the hydrogen-metal system. For this purpose, the present paper aims at providing a critical review of thermodynamics related to metal-H systems, giving general guidelines to improve the performances of metal membranes.

2. Hydrogen Interactions in Transition Metals

2.1. Hydrogen Dissolution in Metal Lattices

Hydrogen absorbed in metals occupies interstitial positions in metal lattices, as demonstrated by neutron diffraction [9]. The FCC lattice (like that of palladium) features an atom at each corner of the cubic cell, and one atom at the center of each face. It shows one octahedral (O) and two tetrahedral (T) interstitial sites per metal atom (14 sites and 14 metal atoms in a single cell).

Differently, the BCC lattice (like that of vanadium, niobium and tantalum) features an atom at each corner of the cubic cell and one at the center. Such a configuration has three octahedral and six tetrahedral interstitial sites per metal atom. The structures of these two lattices are depicted in Figure 1. In particular, the octahedral sites of the FCC lattice have the largest free volume, whereas the tetrahedral sites are the largest in the BCC lattice [10]. This means that the occupation of tetrahedral interstices by hydrogen in BCC lattices of the groups IV and V is favored over the octahedral ones, whereas the opposite behavior is found for hydrogen in palladium [11].

In fact, Cser et al. [12] found by neutron holographic study that the H_2 molecule dissociates in H-atom in palladium and these atoms preferentially occupy the octahedral sites of the FCC lattice.

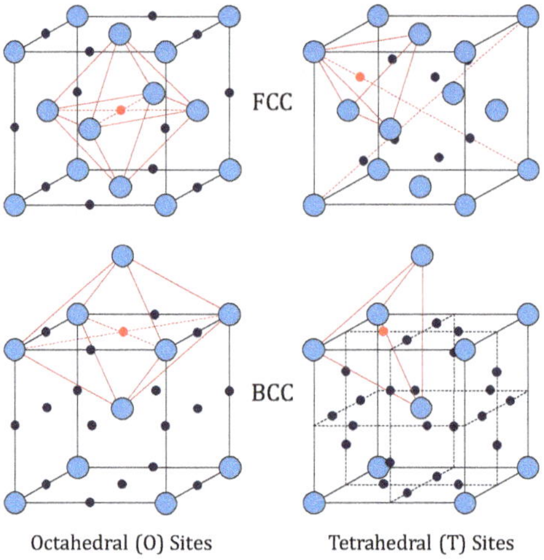

Figure 1. Octahedral (O) and Tetrahedral (T) sites in FCC lattice and BCC lattice [13].

As for the dissolution of hydrogen in vanadium, the V-H system presents a phase diagram consisting of several different phases: (i) the solid solution (α-phase), observed at high temperature, in which the H-atoms are randomly distributed in the tetrahedral sites of its BCC structure; (ii) the hydride V_2H β_1-phase, showing a monoclinic structure where the H-atoms are located in the octahedral sites; (iii) the hydride V_2H or VH β_2-phase, showing a monoclinic body-centered tetragonal (BCT) structure; (iv) the V_3H_2 phase with a monoclinic structure; (v) the VH_2 δ-phase, showing a CaF_2-like structure with the H-atoms occupying the tetrahedral sites; (vi) the γ-phase VH_x, observed at high H-concentration and temperature below 375 K ca. [13–19]. The overall result of such a complex situation is that neutron diffraction on vanadium-deuterium system at 50% of V-D atomic ratio showed that about 90% of the dissolved deuterium atoms occupies the tetrahedral BCC sites, with the rest being placed in the octahedral ones [20]. More recently, using a Density Functional Theory (DFT) analysis applied to pure vanadium, Lu et al. [21] estimated that the H_2 solution energy into the T-sites is −0.332 eV, which is higher than the value calculated into the O-ones (−0.149 eV). Since a more favorable solution energy has a more negative value, this result demonstrates that hydrogen is preferentially absorbed into the BCC T-sites with respect to the BCC O-ones.

In general, in a given circumstance, hydrogen prefers one type of interstitial site with respect to all the other available ones. In case of palladium, the preferred site are the octahedral ones. Understanding why a certain type of site is preferred over another one would allow getting crucial information on the diffusion mechanism of hydrogen into metals [13].

The interstitial sites represent a finite population, whose occupancy by the dissolved hydrogen can be described by means of a thermodynamic approach in terms of Fermi-Dirac statistics. Smirnov and Pronchenko provided a detailed expression of chemical potential of the Pd-H system, taking into account three different contributions: (i) conformational, (ii) oscillatory and (iii) electronic. Then, chemical potential can be obtained as the derivative of the Helmholtz free energy (F) with respect to the hydrogen molar fraction ξ in the metal lattice (Equation (1)), which is defined as the hydrogen/interstices concentration ratio [22].

$$\begin{aligned} \mu &= \frac{\partial F}{\partial \xi} = \mu_{conf} + \mu_{osc} + \mu_{el} \\ &= \frac{\partial F_{conf}}{\partial \xi} + \frac{\partial F_{osc}}{\partial \xi} + \frac{\partial F_{el}}{\partial \xi} \end{aligned} \quad (1)$$

The conformational chemical potential can be written as described in Equation (2):

$$\mu_{conf} = RT\ln\frac{\xi}{1-\xi} + U_{\mathrm{H}}^{conf} + U_{\mathrm{HH}}^{conf}\xi \tag{2}$$

where U_{H}^{conf} and U_{HH}^{conf} are the bond energy of atom H with the metal lattice and the H-H interaction energy within the interstices. The oscillatory contribution to the Helmholtz free energy is reported in Equation (3) [23]:

$$F_{osc} = 3RT\xi\ln\left(2\sinh\frac{\hbar\omega(\xi)}{2RT}\right) \tag{3}$$

from which the corresponding expression of chemical potential can be obtained as follows (Equation (4)):

$$\mu_{osc} = \frac{\partial F_{osc}}{\partial\xi} = 3RT\ln\left(2\sinh\frac{\hbar\omega(\xi)}{2RT}\right) - 3b_1\xi\coth\frac{\hbar\omega(\xi)}{2RT} \tag{4}$$

The parameter ω is the frequency of local oscillations of H-atoms, which depends on H-concentration. Smirnov et al. suggested the following linear approximation for ω (Equation (5)) [22]:

$$\hbar\omega(\xi) = \hbar\omega_0 - b_1\xi \Rightarrow \lim_{\xi\to 0}\hbar\omega(\xi) = \hbar\omega_0 \tag{5}$$

As for the electronic contribution to chemical potential, Smirnov and Pronchenko proposed the following two-parameter expression valid in the entire H-content range within an approximation of 5% (Equation (6)) [22]:

$$\mu_{el} = A\xi + B\xi^4,\ 0 \le \xi \le 1 \tag{6}$$

From the previous expressions, the following complete expression of chemical potential is obtained (Equation (7)):

$$\begin{aligned}\mu = {} & RT\ln\tfrac{\xi}{1-\xi} + U_{\mathrm{H}}^{conf} + U_{\mathrm{HH}}^{conf}\xi + 3RT\ln\left(2\sinh\tfrac{\hbar\omega(\xi)}{2RT}\right) \\ & -3b_1\xi\coth\tfrac{\hbar\omega(\xi)}{2RT} + A\xi + B\xi^4\end{aligned} \tag{7}$$

Once chosen an appropriate reference chemical potential μ_0, the corresponding activity of atomic hydrogen in the metal lattice is defined (Equation (8)):

$$\mu = \mu_0 + RT\ln a(\xi) \tag{8}$$

In the present paper, to obtain a complete expression of hydrogen activity in the metal bulk, the chemical potential in the infinite dilution conditions (i.e., hydrogen content tending to zero) is chosen as the reference chemical potential (Equation (9)):

$$\mu_0 = \mu(\xi \to 0) = 3RT\ln\left(2\sinh\frac{\hbar\omega_0}{2RT}\right) \tag{9}$$

From Equations (8) and (9), an explicit expression of the H-activity is finally obtained (Equation (10)):

$$\begin{aligned}a(\xi) = \exp\left(\tfrac{\mu-\mu_0}{RT}\right) = {} & \tfrac{\xi}{1-\xi}\exp\left(\tfrac{U_{\mathrm{H}}^{conf}}{RT} + \tfrac{U_{\mathrm{HH}}^{conf}}{RT}\xi\right) \times \left(-2\sinh\tfrac{b_1\xi}{2RT}\right)^3 \\ & \times\exp\left(-\tfrac{3b_1}{RT}\xi\coth\tfrac{\hbar\omega(\xi)}{2RT} + \tfrac{A}{RT}\xi + \tfrac{B}{RT}\xi^4\right)\end{aligned} \tag{10}$$

In the limit of infinite dilution system ($\xi \to 0$), the activity a coincides with the hydrogen composition ξ. Based on Equation (7), as ξ tends to the limit value of 1, the chemical potential μ tends to infinity. From a physical point of view, as the available sites are gradually more occupied, it becomes progressively more difficult to fill the remaining sites, which corresponds to a situation of a gradually higher chemical potential.

From the energetic point of view, the hydrogen atoms progressively occupy the interstitial sites from the lowest-energy class to the highest-energy one [20].

Indeed, despite a former theory, which expected the occupancy of just one class of sites, i.e., the octahedral ones in a Pd-H system, McLennan et al. confirmed through diffraction evidence that, for the palladium-deuterium system, the tetrahedral sites, which are the highest-energy class of sites, are partially occupied when the deuterides PdD_x are formed above the thermodynamic critical point. Moreover, the same authors found via Rietveld profile analysis that the maximum tetrahedral occupancy occurs at a D/Pd atomic ratio of around 0.6, at which one third of all D-atoms are placed in tetrahedral sites [24].

An explanation for these observations can be provided if considering that the interactions between the H-atoms dissolved in the metal bulk induce an energetic difference between occupied and empty sites belonging to the same class. In particular, it is proposed that the hydrogen atoms occupies tetrahedral sites on alternate planes, forming in this way a symmetric, ordered and more energetically favorable sub-lattice, as occurs in tantalum–hydrogen system [20].

The occupation of the interstitial sites induces a stress state [20,25,26]. In fact, the metal lattice expands upon H_2 absorption. To intuitively understand how the occupation of the interstices occurs, it is useful to consider the *ball-in-hole* model in continuum elasticity. In this model, the hydrogen atom is assumed to be an incompressible ball, while the metal cavity in which the ball fits is thought as an elastic, spherical *hole*. When the ball enters the hole, the hole size increases because of the ball-hole interactions, inducing an additional volume metal matrix. If ΔV_1 is the volume of the intruding ball, the ΔV_2 volume of the cavity is given by:

$$\Delta V_2 = \frac{3(1-\nu)}{1+\nu}\Delta V_1 \tag{11}$$

where ν is the Poisson ratio of the matrix. For a typical value of n_H/n_{Pd} of 0.3, it results that ΔV_2 is greater than ΔV_1 by 50%. Thus, the volume of the body increases by an extent larger than that required to accommodate an H-atom: this is the origin of the increase of the lattice parameters caused by the insertion of dissolved H-atoms [20].

Besides, Fukai [13] found that hydrogen-induced augmented volume is basically incompressible. The compression curves of pure vanadium and that of β-V_2H hydride, obtained by means of two different compression ways—i.e., through diamond anvil cell (open symbols) and through shock wave (closed symbols)—are shown in Figure 2. From this figure, one can notice that both vanadium and V_2H trend keep almost the same atomic volume gap. In particular, pure V and V_2H volume decrease of about 69% and 74% of its original value at 130 GPa, respectively, with the two values being relatively close to each other.

It must be remarked that, although the same amount of hydrogen could be included into the metal lattice, the method used to inject hydrogen could affect the lattice distortion and, thus, the internal thermodynamic state of the metal-solute system. Indeed, inserting hydrogen simply by applying a pressure difference contrasts the lattice expansion, whereas the expansion effect is not hindered when metal is subject, for example, to cathodic charging [20].

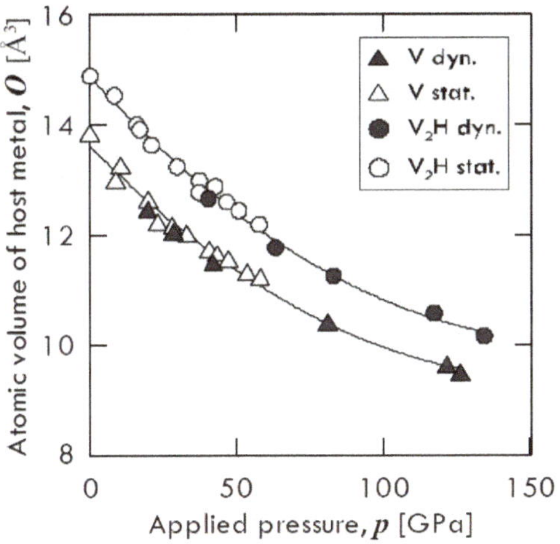

Figure 2. Compression curves of β-V_2H and pure vanadium. The open symbols refer to static compression data (diamond anvil cell), while the closed ones refer to dynamic compression data (shock wave). Adapted from Fukai [13].

2.2. Elastic and Electronic Effects of Alloying

The lattice constants are different for different systems as well as the volumes of the tetrahedrons. The larger the volume, the wider the free space available for the dissolution of the interstitial atoms, which thus results facilitated.

When an alloying element is inserted in the metal lattice, the lattice constants could change, affecting the solubility of hydrogen. Lu et al. [21] simulated the behavior of hydrogen in a V-based $2 \times 2 \times 2$ supercell made by 8 unit cells, in which an alloying M-atom (Al, Ti, Cr, Fe, Ni and Nb) replaces the central V-atom of such a supercell, thus forming a V-based alloy. This alloy is indicated as $V_{15}M$, because 15 V-atoms and a single M-atom form the unit cell lattice.

When an alloying element is added to vanadium, its atomic radius does not match the vanadium original matrix, leading to lattice distortions and change in the volume of tetrahedrons. These distortions produce an elastic stress field, which affects the H-atoms dissolution, resulting in the variation of the solution energies. Such variations were found to depend significantly on the lattice constants of $V_{15}M$, which in their turn depend on the size of alloyed atoms [21].

As the M-atom size increases, the cubic lattice tends to swell to a certain extent, thus causing the available space for the dissolution of an additional H-atom to increase as well. In this way, the M-atoms having a size smaller than the V-atom one (like Cr, Ni and Fe) reduce the solubility of hydrogen in vanadium by making the solution energy values less negative. The opposite occurrence is found for M-atoms with a bigger size (Al, Ti, Nb) [21]. Results are illustrated in Figure 3.

A remarkable exception to the previously explained general cases is found by observing the behavior of niobium. In fact, since it is the biggest atom in the analyzed series, it is expected to form an alloy with a solution energy more negative than that of titanium, which has a smaller atomic size. However, such an expectation is not confirmed by simulation. Those authors explain this apparently anomalous behavior by considering that the effective available space in the interstices of the V-based lattice is reduced by the size of the Nb-atom, which is too large compared with the V-atom one [21]. The addition of an alloying atom into the vanadium lattice affects the interstitial H-diffusion not only by producing elastic distortion, but also by modifying the original electronic structure around the interstitial solute. Investigating the local density of states, however, Lu and his co-workers found that the intensity values of the resonance peaks of the hydrogen in the pure vanadium system are similar to those of the hydrogen in $V_{15}M$ system. From this observation, it is withdrawn that the electronic effect due to the M-atom presence may be not the main influencing factor of hydrogen dissolution [21].

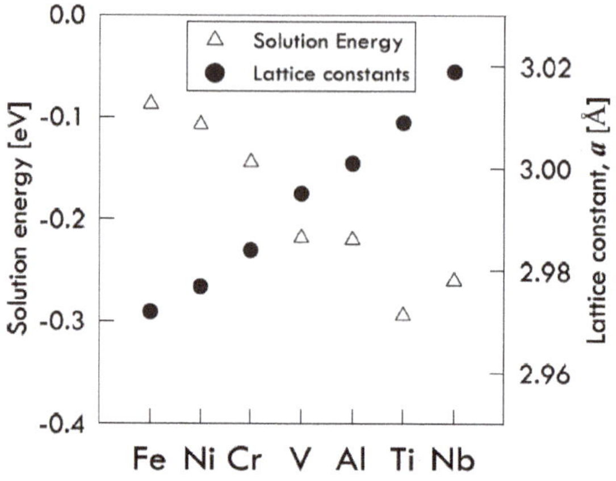

Figure 3. Solution energy of hydrogen in $V_{15}M$ alloys, compared with the lattice constants of each alloy. Adapted from Lu et al. [21].

2.3. *Effects of Dissolved Atoms on Global Electronic Structure*

The enthalpy of hydrogen dissolution $\Delta\overline{H}^0$ from the molecular state to the atomic one inside metal is given by the contribution of two terms. The first is related to the dissociation energy of the molecule and the second is related to the interaction energy between dissolved atoms as well as between atoms and metal. In particular, the metal–atoms interaction determines the metals classification into two classes of occluders: endothermic and exothermic ones. The former has a solubility increasing with increasing temperature, whereas the latter shows the opposite behavior [20].

Endothermic occluders, such as Mn, Fe, Co and Ni, form a metal–hydrogen solid solution in which the H-atom is located at random interstices [11]. However, the most interesting transition metals for hydrogen separation, e.g., Pd, Ta, V, Nb, are exothermic occluders, which show a tendency to form ordered hydrides at high H-concentrations [27].

The solubility isotherms of some metals and a V-alloy (vanadium, tantalum and $V_{85}Ni_{15}$ alloy at 400 °C, and palladium at 340 °C) are shown in Figure 4. As is possible to notice, vanadium clearly absorbs much more hydrogen than palladium. Under a hydrogen partial pressure of 7 bar, the H/V ratio at 400 °C would be around 0.6, i.e., 6 hydrogen atoms per 10 vanadium atoms, while, at the same pressure, the H/Pd ratio at 340 °C would settle at about 0.02 (ca. 2 ÷ 98). Since exothermic occluders show a solubility decreasing with increasing temperature, the difference between these two metals would become even more pronounced with palladium at 400 °C. Therefore, it is possible to state that the dominant contributor to permeability is solubility for BCC metals, whereas for FCC metals (Pd in particular), the dominant factor is represented by diffusivity [11].

Consequently, permeability of BCC metals decreases with increasing temperature, because this behavior is driven by solubility. On the contrary, permeability of FCC metals increases with increasing temperature, because it mainly driven by diffusivity. It is interesting to underline how alloying vanadium with a small percentage of nickel drastically reduces hydrogen solubility. At the same condition of pressure and temperature, the hydrogen/metal ratio would pass from 0.6 to 0.09. By reducing the hydrogen solubility, the embrittlement tendency reduces as well [11].

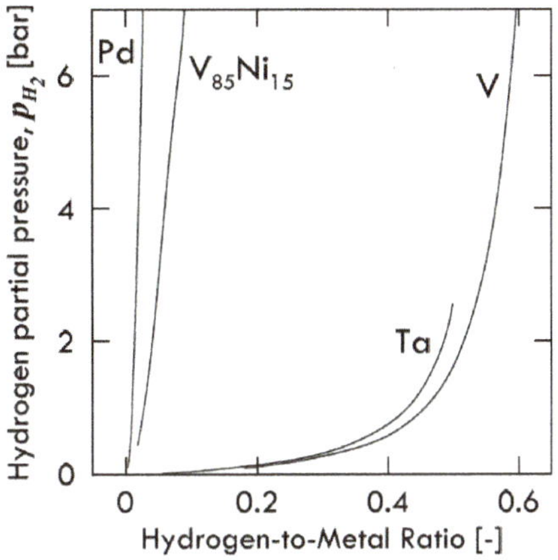

Figure 4. Isotherms for selected metals and alloys measured using Sieverts' technique, in which the metal sample is equilibrated in a hydrogen atmosphere. The test for Pd are carried out at 340 °C, whereas for all the others metals a temperature of 400 °C is used. Adapted from Dolan [11].

The H-content in the metal lattice affects also other physical properties, such as electrical resistance, magnetic susceptibility and thermoelectric power. For their extent, these variations cannot be simply explained by lattice expansion induced by solute atoms. Differently, it is clear that hydrogen consistently perturbs the electronic structure of the host metal [20]. Indeed, the *1s* electrons of the hydrogen atoms enters the *s*- and *d*- bands of the host metal, causing shifts in the energy bands. As a result, the Fermi electrons of the host metal start surrounding the positive H^+ nuclei, leading to the formation of a sort of *screened entity*, which could be thought as neutral atoms even if the electrons are not bound to the nuclei. Accordingly, it is possible to distinguish two kinds of effects. The first consists in the shift of metal bands, which has a *long-range* effect, as the induced interactions involve the entire metal domain. The second is related to the heaping of Fermi electrons placed on the hydrogen nuclei and is a *short-range* effect. An attractive interaction is observed between two hydrogens caused by the *d*-vacancies of the Pd-atom, which is the common neighbor of both H-atoms. This interaction could explain the elastic interactions as well as the hydrogen cluster formation in Pd-H systems. Since hydrogen remarkably affects the global electronic structure and, more importantly, the metal atoms inter-distance, it is not surprising that the cohesive forces of the host metal are unavoidably reduced. Such an occurrence is explained through molecular dynamic simulations, which show that hydrogen fills antibonding states in *4d* band, reducing the bond strength between Pd-atoms. The decrease of the electron density between metal atoms, resulting in the Pd-Pd bonds impairment in several transition metals, is confirmed through cluster calculation and embedded atom method [20].

2.4. Hydrogen-Lattice Defects Interactions

Hydrogen can also interact with structural defects. Metals usually exhibit defects of various size, which can be distinguished into three classes, so-called *zero-dimensional, one-dimensional,* and *two-dimensional* one.

The zero-dimensional defects, also known as *point defects*, are local defects, i.e., locations where an atom is missing (vacancy) or is placed into an interstitial void of the lattice structure (self-interstitial), as illustrated in Figure 5. The latter occurs only at low concentration because of the strain field induced in the tightly packed metal structure, whereas the former can be found more often, with a probability increasing with increasing temperature. In fact, the higher the temperature, the more frequent and random the position changes of the atom in the lattice. It can also happen that the vacancies population is in excess with respect to the equilibrium value, for example after a quenching from high temperature. Anyway, both defect types can also occur after plastic deformations.

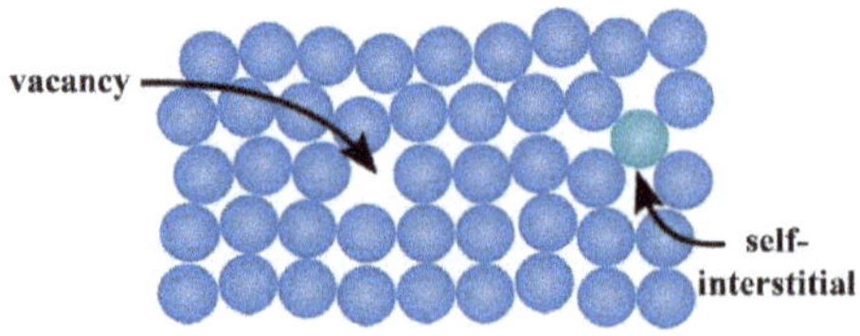

Figure 5. Point defects in a crystal structure: vacancy and self-interstitial/substitutional atom.

Regarding the interaction between these defects and H-atoms, a curious aspect was observed by Kirchheim [28], who prepared two samples of palladium: the first one, which was structurally defect-free, and the second one plastically stretched (up to 50%) through cold drawing and subsequent rolling, in order to enhance the number of defects. Afterwards, Kirchheim measured the partial molar volume $\widetilde{V}_H$ of hydrogen in both samples (Figure 6).

It can be observed that, for the virtually defect-free Pd-sample, $\widetilde{V}_H$ is constant within the whole range of H/Pd ratio, whereas the plastically deformed one shows a continuously increasing value tending to the same level as that of the former.

A peculiar aspect to be observed is that, for the defected sample, $\widetilde{V}_H$ is even negative below a certain value of hydrogen concentration (40 ppm). In other words, instead of expanding because of the hydrogen absorption, the lattice shrinks. Kirchheim explained this weird behavior considering the phenomena occurring when hydrogen fills an empty site or vacancy. In fact, if hydrogen is close enough to the nearest palladium atom at low concentration, attractive forces Pd-H come into play and, thus, the lattice contracts. From around 40 ppm on, the lattice starts growing in the positive values, as the effect of hydrogen in the substitutional vacancies is overcome by the expansion of the normal interstitial sites [28].

The one-dimensional defects, also known as linear defects, consist in lattice local dislocations, produced in the metal by plastic deformations and many other processes. One can basically distinguish two types of dislocations: the *edge* dislocation and the *screw* dislocation, even if most dislocations are probably a hybrid of these two limits.

The shear stress causes screw defects, which propagates along a direction normal to the shear stress, as depicted in Figure 7a. As well, the edge dislocations can be visualized as an extra half-planes inserted in the crystal homogeneous lattice, as sketched in Figure 7b. The interatomic bonds are significantly distorted but only in the immediate neighborhood. When a shear stress is applied, dislocations can not only originate but also move. It is important to go deeper into this phenomenon, to understand the hydrogen dragging induced by dislocation motion. To make an analogy, this type of movement can be compared to the motion of a caterpillar: first, it moves the rear part creating a hump, and then moves the front part, similar to how the defect slips one cell per time. In this way, only a small amount of bonds is broken at a given time, thus requiring a much smaller energy than that needed to break all the bonds simultaneously. For the same reason, dislocations move easily over dense planes, because the stress required to move the dislocations increases with increasing spacing between planes. As FCC and BCC metals have several dense planes, the dislocations move relatively easily, resulting in materials with a relatively high ductility.

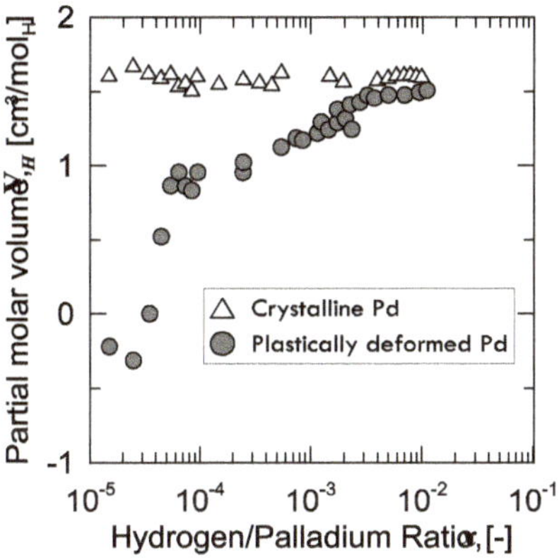

Figure 6. Partial molar volume of hydrogen measured in crystalline Pd (open triangles) and in plastically deformed Pd (closed circles) by cold drawing and rolling, as a function of hydrogen concentration in palladium (number of H-atoms per number of Pd-atoms). Adapted from Kirchheim [28].

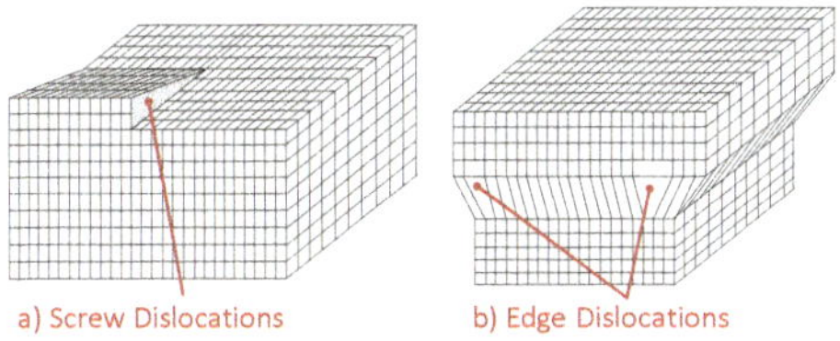

Figure 7. Illustration of linear defects: (**a**) Screw dislocation and (**b**) Edge dislocation.

Kirchheim [29] evaluated the interaction energy of hydrogen in palladium edge dislocations. He found that the closer the core of the edge defect, the larger the interaction energy, which includes a self H-H attractive interaction energy. Far from the cores, the H-H interactions do not provide an appreciable contribution and, in general, the attractive interaction energy decreases with the reciprocal of the distance. Anyway, the additional population of hydrogen can be measured only at very small quantities of dissolved hydrogen due to the attractive interaction with edge dislocations. When carbon embeds in α-Fe, it leads to tetrahedral distortions causing attractive interactions towards screw dislocations. On the contrary, the octahedral positions occupied by the H-atoms in palladium make it interact with this type of dislocation at negligible or null levels [29].

The two-dimensional or planar defects include several forms of dislocations. Overall, stacking faults, grain boundaries and internal surfaces of micro-voids and micro-cracks are mentioned. To explain the stacking fault phenomenon, one can consider the case of the crystalline structure of two common lattices, the FCC and the HCP structures, for which the former can be oriented in two different planes, i.e., square or triangular, whereas the latter can be seen in the triangular orientation. In particular, one can imagine the crystal structure as a sequence of stacks. If A indicates the triangle, B the hexagon and C the reverse triangle, one can describe the HCP structure as an AB sequence, whereas the FCC structure as an ABC sequence. With this in mind, a stacking fault occurs, for example, when an unpredicted C-layer is present in the AB sequence of the HCP structure or when a C-layer is missing, or ABC default sequence randomly changes in the FCC structure.

In their study, Whiteman and Troiano showed a hydrogen-charging operation in a stainless-steel sample (25%Cr–20%Ni) conducted through electrolytic way. In those conditions, they found that the activation energy required to create a stacking fault lowers. Such an energy variation can be associated with an attractive interaction between hydrogen dissolved in metal and stacking faults [30]. Another type of planar defect is represented by the grain boundary.

Generally, a solid consists of several crystallites or grains, whose size varies from nanometers to millimeters. Each grain has usually a different orientation with respect to neighboring grains. The border separating one grain from another one is known as grain boundary. A representation of the grain boundaries present in crystal is sketched in Figure 8.

Grain boundaries have been demonstrated to absorb an excess of H-atoms population in several metals other than palladium. Mütschele and Kirchheim [31] found that there is a Gaussian distribution of interaction energies between dissolved hydrogen and grain boundaries in palladium (Figure 9), denoting a variety of binding sites in grain boundaries.

The energy E_1 is equal to 9.2 kJ/mol$_H$ and represents the expected value (μ) of the interaction energies distribution around which site energies for the grain boundaries might vary, whereas E_0 is equal to 3.9 kJ/mol$_H$ and represents the interaction energy of the sites within the grains, which has the same value as that of the single crystals of Pd. The difference between these two energies (equal to 5.3 kJ/mol$_H$) gives the spectrum of the segregation energies, i.e., the energies of molecular partitioning from metal bulk to defects.

The standard deviation of the distribution (σ) is equal to 15 kJ/mol, so that it is possible that some grain boundary sites show even a negative interaction energy, which means that there can be also repulsive forces. However, grain boundaries in any metal exhibit in general attractive interactions towards dissolved hydrogen [31].

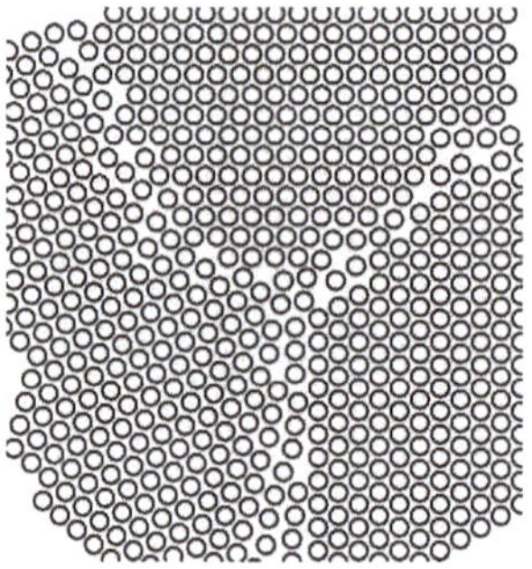

Figure 8. Illustration of grains and relative boundaries in a crystal.

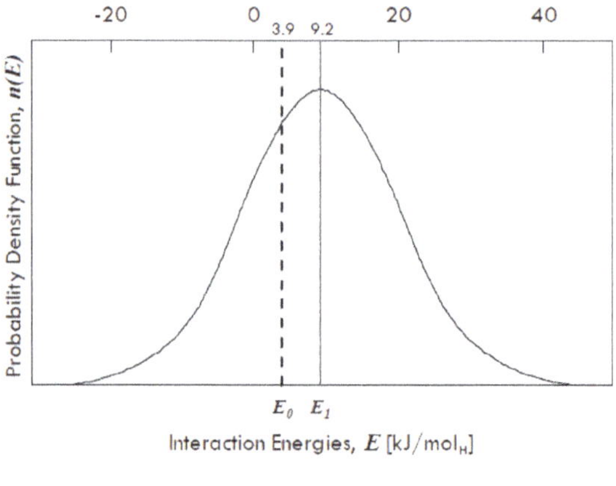

Figure 9. Gaussian distribution of site energies for hydrogen in nano-crystalline palladium. Adapted from Mütschele and Kirchheim [31].

In addition to stacking faults and grain boundaries, a third class of defects is represented by micro-voids, which can occur when a ductile material is pulled during a tensile strain. If the stress continues, a coalescence of some voids can occur, leading to the formation of micro-cracks. Hydrogen is absorbed upon these cracks in the same way as if it would be absorbed upon a generic external surface. In addition, the volume of the internal cavity (which can be referred to as a *three-dimensional* defect) would acquire a concentration of molecular hydrogen consistent with the input pressure of external environment at equilibrium [20]. It is possible to divide the interactions between dissolved hydrogen and lattice imperfections into two categories: chemical and elastic one. The former refers to positions where the atomic displacements are remarkable, e.g., at dislocation cores and internal interfaces, whereas the latter refers to positions where atomic displacements are small, which means sufficiently far from the imperfections. In this case, an approximated linear elasticity model can be applied [20].

3. Phase Diagrams and Phase Transitions

3.1. Thermodynamics of Hydrogen Dissolution in Metal Membranes

When the hydrogen mass transport is limited by the bulk diffusion (i.e., the diffusion in the metal lattice) as occurs in most cases, permeability is the best way to compare the intrinsic transport properties of different materials. This parameter can be seen as the overall result of the effect of two intrinsic properties: diffusivity and solubility, both crucially depending on the gas/solid system.

The temperature dependence of permeability of several transition metals is shown in Figure 10. We can observe that those metals with the highest permeability are early transition metals (like niobium, vanadium and tantalum) with a BCC structure. In terms of the sole permeability, those materials could easily overcome the performances of palladium. However, in practice, the major barrier preventing from achieving such a high flux is the embrittlement by hydrogen [11], which is the macroscopic result of all the above-discussed phenomena generating micro-cracks, dislocations and local strains eventually leading to mechanical failure of the metal lattice. Furthermore, an important phenomenon to understand is the permeability change with temperature. While permeability increases with increasing temperature for palladium, the BCC metals show the opposite behavior. This effect can be explained by considering the solubility influence. As solubility represents the strength of the H-metal interactions usually evaluated by measuring the amount of dissolved species, it generally decreases with increasing temperature, mainly because of the stronger lattice vibrations, which favor the release of the solute from the interstices.

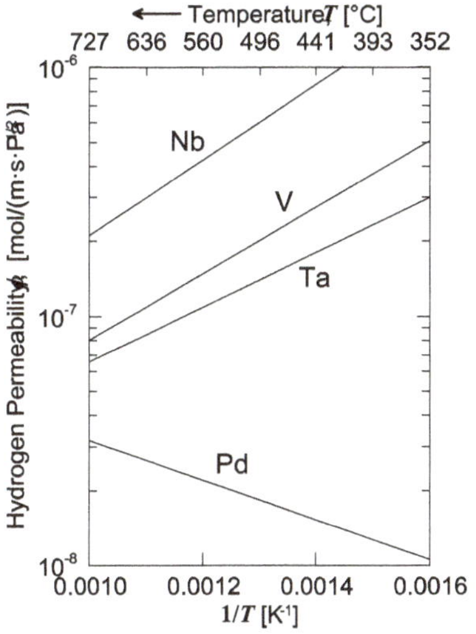

Figure 10. Permeability of selected metals as a function of temperature. Adapted from Buxbaum and Kinney [32].

In the hypotheses of the validity of the Langmuir adsorption model—i.e., mono-layer adsorption, energetically uniform adsorption sites, no lateral interactions among adsorbed atoms and ability of each adsorption sites to accommodate a single molecule/complex [33]—and in the absence of any inhibiting and/or poisoning species, the hydrogen surface coverage on feed and permeate side can be respectively expressed by Equation (12) [34]:

$$\theta_{\mathrm{H}} = \left. \frac{K_{\mathrm{H}_2}\sqrt{P_{\mathrm{H}_2}}}{1 + K_{\mathrm{H}_2}\sqrt{P_{\mathrm{H}_2}}} \right|_{\substack{\text{Feed} \\ \text{Perm}}} \tag{12}$$

From the equilibrium condition between surface and bulk Pd-H systems, which is always assured in the operating conditions of interests for hydrogen purification applications [35,36], we have Equation (13) [37]:

$$\xi = \left. \frac{a_1(T)\theta_{\mathrm{H}}}{a_1(T)\theta_{\mathrm{H}} + (1 - \theta_{\mathrm{H}})} \right|_{\substack{\text{Feed,} \\ \text{Perm}}} \tag{13}$$

Combining Equations (12) and (13), an explicit relationship expressing the H-concentration in the metal bulk as a function of the hydrogen partial pressure in the gas phase is obtained (Equation (14)).

$$\xi = \left. \frac{a_1(T)\frac{K_{\mathrm{H}_2}\sqrt{P_{\mathrm{H}_2}}}{1+K_{\mathrm{H}_2}\sqrt{P_{\mathrm{H}_2}}}}{a_1(T)\frac{K_{\mathrm{H}_2}\sqrt{P_{\mathrm{H}_2}}}{1+K_{\mathrm{H}_2}\sqrt{P_{\mathrm{H}_2}}} + \left(1 - \frac{K_{\mathrm{H}_2}\sqrt{P_{\mathrm{H}_2}}}{1+K_{\mathrm{H}_2}\sqrt{P_{\mathrm{H}_2}}}\right)} \right|_{\substack{\text{Feed,} \\ \text{Perm}}} \tag{14}$$

In the limit of Sieverts' hypotheses, the following approximation can be made (Equation (15)):

$$\xi \cong a_1(T)K_{\mathrm{H}_2}\sqrt{P_{\mathrm{H}_2}}\Big|_{\substack{\text{Feed,} \\ \text{Perm}}} \tag{15}$$

which leads to the well-known Sieverts' law if the constant K_{H_2}, which depends on temperature as well, is the same for both membrane side. This fact underlines that the use of an empirical Sieverts' law to get information on the non-ideal behavior of the hydrogen permeation through metal membranes should be used carefully to correctly interpret the experimental results at relatively high pressure and in the presence of other external phenomena [38,39].

Substituting Equation (14) into Equation (10), we have an explicit expression of the activity of the H-atoms dissolved in the lattice (H(*sol*)) as a function of the hydrogen partial pressure (Equation (16)):

$$a(\xi) \equiv a[\mathrm{H}(sol)] \Rightarrow a\left(P_{\mathrm{H}_2}\right) \tag{16}$$

The expression of the H-activity is needed to solve the equilibrium of the solution process of hydrogen in the metal lattice (r_{ov}), which is actually the overall result of two different processes, i.e., the dissociative adsorption of hydrogen onto the surface (r_1) and the solubilization (or dissolution) of the adsorbed H-atoms into the metal bulk (r_2).

$$\begin{array}{rl} r_1: & \mathrm{H}_2(g) = 2\mathrm{H}(ads) \\ r_2: & \underline{\mathrm{H}(ads) = \mathrm{H}(sol)} \\ r_{ov}: & \mathrm{H}_2(g) = 2\mathrm{H}(sol) \end{array} \tag{17}$$

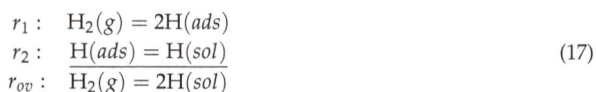

The equilibrium constant of the overall solubilization process can be expressed in terms of activities as follows (Equation (18)):

$$K_{e,ov} = \exp\left(-\frac{\Delta \overline{G}^0_{r_{ov}}(T)}{RT}\right) = \frac{a[\mathrm{H}(sol)]^2}{a[\mathrm{H_2}(g)]}$$
$$\Delta \overline{G}^0_{r_{ov}}(T) = \Delta \overline{H}^0_{r_{ov}}(T) - T\Delta \overline{S}^0_{r_{ov}}(T) \quad (18)$$

where the Gibbs free energy of the overall reaction as well as its enthalpy and entropy can be actually written as the appropriate linear combinations of the corresponding thermodynamic state functions of the reactions r_1 and r_2 (Equation (19)):

$$\Delta \overline{G}^0_{r_{ov}}(T) = \Delta \overline{G}^0_{r_1}(T) + 2\Delta \overline{G}^0_{r_2}(T)$$
$$\Delta \overline{H}^0_{r_{ov}}(T) = \Delta \overline{H}^0_{r_1}(T) + 2\Delta \overline{H}^0_{r_2}(T) \quad (19)$$
$$\Delta \overline{S}^0_{r_{ov}}(T) = \Delta \overline{S}^0_{r_1}(T) + 2\Delta \overline{S}^0_{r_2}(T)$$

In Equation (18), the activity of the dissolved hydrogen can be evaluated using Equation (16), whereas that of the gaseous hydrogen can be made explicit by using fugacity, which can be expressed in terms of hydrogen partial pressure and fugacity coefficient evaluated by an appropriate equation of state (Equation (20))

$$K_{e,ov} = \frac{a[\mathrm{H}(sol)]^2}{\frac{f[\mathrm{H_2}(g)]}{f^0=1\ \mathrm{bar}}} = \frac{a[\mathrm{H}(sol)]^2}{\frac{\phi[\mathrm{H_2}(g)]P_{\mathrm{H_2}}}{f^0=1\ \mathrm{bar}}} \quad (20)$$

In this way, in pure-hydrogen conditions, the right-hand side of Equation (20) can be expressed as a function of the hydrogen partial pressure only, whose complexity requires the implementation of an appropriate numerical method to obtain a solution.

3.2. PCT Diagrams

Since one of the major characteristics of the exothermic occluders is the formation of a hydride, it is possible to represent the equilibrium conditions of this type of metal–hydrogen system by means of a phase diagram, analogously to what done for the phase diagrams used to represent the metal alloy systems [27]. In particular, the formation of metal hydrides leads to the existence of different phases in the palladium–hydrogen system: α-phase and β-phase.

The α-phase is a solution phase and has lattice constants close to those of palladium. At room temperature, the H: Pd ratio for this phase is 0.03. As more hydrogen dissolves in metal, the lattice constants are observed to increase linearly with pressure until the β-phase (metal hydride) appears. However, the situation is actually more complex, since the transition from α- to β-phase is not sharp but passes through the existence of intermediate phases composed of a mix of both. Moreover, hysteresis is observed in increasing/decreasing cycles of hydrogen concentration [40]. The composition of the β-phase is approximately 0.6 at room temperature [41], which is considerably higher than the value of the α-phase. Both α- and β-phase show the same FCC lattice structure, with the same octahedrally coordinated H-atoms [42]. Please note that the same sites are occupied by hydrogen in α-palladium hydride at a considerably lower H: Pd ratio.

For the α-phase at room temperature (20 °C), the value of the H:Pd ratio is 0.008, while for the β-phase such a value is 0.607. For this reason, even if the α- and β-phase have the same lattice symmetry, the specific volumes are different: at room temperature, the values of the lattice constant a referred to the side of cubic unit cells for both phases are reported in Table 1. As is possible to notice, volume increases up to 10% in the α-to-β transition, determining a change of molar volume equal to 1.57 cm^3/mol_H [43]. Such a transition involves the formation of clusters of H-atoms such that 125 Pd-atoms have nearly 76 H-atoms in the β-phase and just one H-atom in the α-phase [20]. Therefore, for the performance of membrane-assisted purification processes, monitoring of the formation of the

β-phase is crucial, as it leads to a larger volume expansion as well as higher internal stresses that need be relaxed to avoid cracks and fractures [11].

In general, metal–hydrogen systems exhibit phase diagrams of various complexities. Phase transitions have been object of study since the first part of the 20th century, when Gillespie and Galstaun [44] first described the *Pressure-Composition-Temperature* (PCT) diagram for palladium (Figure 11). Specifically, they found a critical point at 295.3 °C and 19.8 atm, above which only a single phase exists. Therefore, such a critical point represents the highest temperature at which the β-phase can form. In the single-phase region, the composition varies smoothly, and the distortion phenomena are circumvented. Therefore, one method whereby the phase change can be avoided is to ensure that the palladium membrane is always operating within the single-phase region of the Pressure-Composition isothermal diagram [45].

Table 1. Lattice constant and corresponding unit cell volume evaluated at 20 °C.

System	a, Å	V_{cell}, Å^3	H:Pd Ratio
Palladium	3.890	58.86	-
α-phase	3.894	59.05	0.008
β-phase	4.025	65.21	0.607

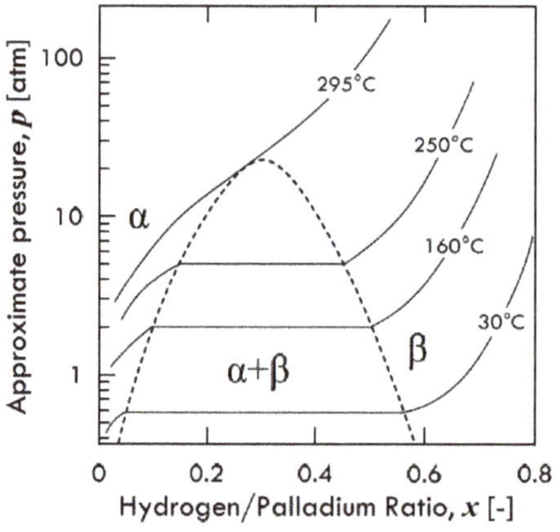

Figure 11. Approximate Pressure-Composition-Temperature phase diagram for the Pd-H system. Adapted from Knapton [45].

The bell-shape curve shown in Figure 11 takes the name of *coexistence curve*, since it borders a miscibility gap where the α- and β-phase can coexist. Such a PCT diagram for the PdH_x system could be simplified in a PT diagram, as shown in Figure 12. In this graph, the mixed ($\alpha + \beta$) area is not shown, and only the boundary between the two single phases is visible. Similar to palladium, also in the PCT diagrams of vanadium [46], niobium [47] and tantalum [48] isotherms exhibit the plateau typical of exothermic occluders. This plateau is due to the re-ordering of α-to-β transition in octahedral interstices [11]. The critical temperatures for each metal are reported in Table 2 along with the related H:M ratio.

Table 2. Critical temperatures T_c and corresponding H:M ratio for selected metals.

Metal	Ref.	T_c	H:M
Vanadium	[46]	170 °C	0.45
Niobium	[47]	140 °C	0.30
Tantalum	[48]	52 °C	0.42

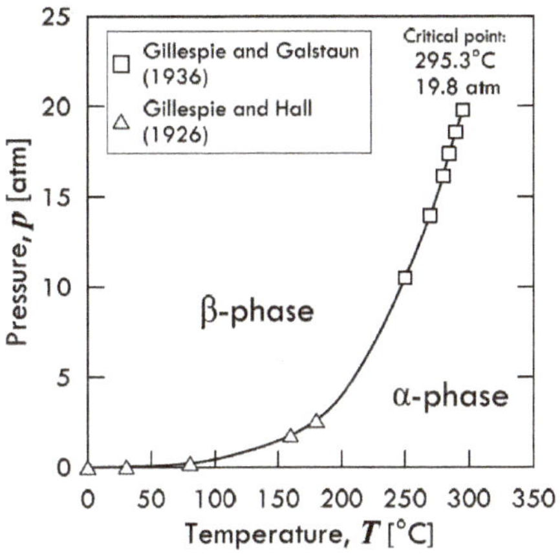

Figure 12. Pressure-Temperature phase diagram for the Pd-H system. Adapted from Jewell et al. [41].

3.3. Coherent and Incoherent States

The miscibility boundaries of the coexistence curve are influenced by the concentration of hydrogen in the metal. This effect is related to differences in zero-point energies, in lattice expansion and in phonon spectrum. Besides, a modification of the coexistence curve can be also related to the effects of lattice stresses. As is possible to observe in Figure 11, an increase of hydrogen concentration leads to β-phase that can coexists with the α-phase when operating below 300 °C and 20 atm, passing through an unalloyed phase (α + β), which enhances the gross distortion. In other words, in this miscibility gap, the Pd-clusters related to β-phase expands within the α-phase, generating mechanical stresses in analogy with the *ball-in-hole* model described above. To relax these stresses, the lattice can react by removing Pd-atoms from the surrounding α-phase to the external surface of the metal. Hence, two equilibrium states leading to corresponding equilibrium diagrams can be identified: the *coherent* state and the *incoherent* one. Specifically, coherent solid interfaces are interfaces across which the lattice planes are continuous although not straight. Indeed, in most solids, the lattice lines are elastically strained to keep continuity, and these elastic strains play a crucial role in determining the material properties.

Differently, incoherent solid interfaces are interfaces across which there is no continuity between lattice planes. This situation can be found, for example, in grain boundaries and inclusions in alloys. It must be remarked that a net distinction between both states is actually just ideal, since a distribution of coherent/incoherent phases, named *semi-coherent* state, is in practice achieved [20].

In an ideal coherent equilibrium between two α- and β-phases, in the α + β region shown in the bell-shaped diagram, one phase grows up within the parent phase without any diffusive motion of metal atoms to the surface and the lattice is only deformed elastically. As said, the lattice planes are curved but still continuous, and the interface between the phases is characterized by a change both in compositions and in lattice parameters.

The coherent state corresponds to a higher free energy condition than the incoherent state because of the elastic potential energy involved. This chemical potential accumulates until a shift-point from coherency to incoherency is reached. Once this happens, the nucleation of dislocations at the phase interfaces occurs. Ideally, these vacancies should carry metal atoms from the neighborhood to the free surface but, in practice, this transport is never complete, as it is stopped at the grain boundaries. Therefore, the excess free energy of coherent state collapses into plastic deformations characterizing the incoherent state, even though part of that excess energy is not completely removed [20]. The coherence-incoherence dualism is the reason for the well-known hysteresis observed during the definition of coexistence line in the bell-shaped diagram. Coherency stresses as well as their reduction caused by dislocation generation occur in charge/release cycles of hydrogen into/from the lattice whenever more than one phase is generated by the presence of hydrogen [20].

4. Conclusions

In the present paper, several thermodynamic aspects related to the hydrogen-metal systems are critically described and discussed. First, the influence of hydrogen-metal interactions on mechanical, structural and electronic properties of several metal alloys is discussed, putting in evidence the detrimental effects provided by a relatively high hydrogen content dissolved in metal lattices.

Within such an analysis, an explicit expression of activity as a function of the hydrogen content in the metal lattice is obtained starting from a complex expression of the hydrogen chemical potential already available in the open literature [22].

The so-obtained activity expression was then used to approach the solution of the solubilization process equilibrium in non-ideal conditions, showing that to be the overall result of two different processes: (a) the molecular hydrogen dissociative adsorption and (b) the solubilization process of the H-atoms adsorbed onto the surface into the metal bulk. For such an analysis, we used a Langmuir-based expression, giving an explicit relationship between surface coverage and hydrogen partial pressure in the gas phase. The obtained expression is shown to extend the original Sieverts' law, which, differently, is based on Henry's law. Concerning this aspect, it was highlighted that, in the limit of an infinite dilution system, the obtained non-ideal expression is coincident with Sieverts' law, as expected for low-concentration systems.

Acknowledgments: Alessio Caravella has received funding for this research through the "Programma Per Giovani Ricercatori *Rita Levi Montalcini*" granted by the "Ministero dell'Istruzione, dell'Università e della Ricerca, MIUR" (Grant no. PGR12BV33A), which is gratefully acknowledged.

Conflicts of Interest: The authors declare no conflict of interest.

List of Symbols

a	H-activity, -
a_1	Parameter depending on temperature only in Equation (13)
A	Parameter in Equation (6), J mol^{-1}
B	Parameter in Equation (6), J mol^{-1}
b_1	Parameter in Equation (4), mol^{-1} s^{-1}
E	Interaction energy, J mol^{-1}
F	Helmholtz free energy, J mol^{-1}
f	Fugacity, Pa
f^0	Reference fugacity, Pa
$\overline{G}$	Partial molar Gibbs free energy, J mol^{-1}
$\overline{H}$	Partial molar enthalpy, J mol^{-1}
$\hbar$	Reduced Planck constant, $h/2\pi$, J s
K_{H_2}	Langmuirian hydrogen dissociation constant, Pa$^{-0.5}$
K_e	Equilibrium constant, -
n	Number of moles, -
P	Pressure, Pa
r_1, r_2	Adsorption and dissolution reactions
R	Gas constant, J mol^{-1} K^{-1}
$\overline{S}$	Partial molar entropy, J mol^{-1} K^{-1}
U	Internal Energy, J mol^{-1}
T	Temperature, K
V	Volume, m^3

Greek Symbols	
ϕ	Fugacity coefficient, -
μ	Chemical potential, J mol^{-1}
μ_0	Chemical potential in the infinite dilution condition, J mol^{-1}
ν	Poisson ratio, -
θ	Surface coverage, -
ξ	H-concentration in the metal lattice, -
ω	H-atoms local oscillations frequency, mol^{-1} J^{-1} s^{-2}
ω_0	H-atoms local oscillations frequency at zero H-content, mol^{-1} J^{-1} s^{-2}
Subscripts, Superscripts and Abbreviations	
ads	Adsorbed phase (on the metal surface)
conf	Configurational
osc	Oscillatory
el	Electronic
g	gas phase
H	Referred to Metal–Hydrogen interactions
HH	Referred to Hydrogen–Hydrogen interactions
ov	Overall
Perm	Permeate side
sol	Solution state (dissolved in the lattice)
Acronyms	
BCC	Body-Centered Cubic
BCT	Body-Centered Triclinic
FCC	Face-Centered Cubic

References

1. Conde, J.J.; Maroño, M.; Sánchez-Hervás, J.M. Pd-Based Membranes for Hydrogen Separation: Review of Alloying Elements and Their Influence on Membrane Properties. *Sep. Purif. Rev.* **2017**, *46*, 152–177. [CrossRef]
2. Adams, B.D.; Chen, A. The role of palladium in a hydrogen economy. *Mater. Today* **2011**, *14*, 282–289. [CrossRef]
3. Mueller, W.M.; Bfackledge, J.P.; Libowitz, G.G. *Metal Hydrides*; Academic Press: New York, NY, USA; London, UK, 1968.
4. Phair, J.W.; Donelson, R. Developments and design of novel (non-palladium-based) metal membranes for hydrogen separation. *Ind. Eng. Chem. Res.* **2006**, *45*, 5657–5674. [CrossRef]
5. Völkl, J.; Alefeld, G. Diffusion of hydrogen in metals. In *Hydrogen in Metals I: Basic Properties*; Alefeld, G., Völkl, J., Eds.; Springer: Berlin/Heidelberg, Germany, 1978; pp. 321–348.
6. Fukai, Y. *Diffusion in the Metal–Hydrogen System: Basic Bulk Properties, 2nd Rev.*; Hull, R., Parisi, J., Osgood, R.M., Jr., Warlimont, H., Eds.; Springer: Berlin/Heidelberg, Germany, 2010; pp. 303–400.
7. Veleckis, E.; Edwards, R.K. Thermodynamic properties in the systems vanadium-hydrogen, niobium-hydrogen, and tantalum-hydrogen. *J. Phys. Chem.* **1969**, *73*, 683–692. [CrossRef]
8. Bellini, S.; Liang, X.; Li, X.; Gallucci, F.; Caravella, A. Non-Ideal Hydrogen Permeation through V-alloy Membranes. *J. Membr. Sci.* **2018**, *564*, 456–464. [CrossRef]
9. Alefeld, G.; Volkl, J. *Hydrogen in Metals*; Springer: Berlin, Germany, 1978; Volumes 1 and 2.
10. Bellini, S.; Azzato, G.; Grandinetti, M.; Stellato, V.; De Marco, G.; Sun, Y.; Caravella, A. A Novel Connectivity Factor for Morphological Characterization of Membranes and Porous Media: A Simulation Study on Structures of Mono-sized Spherical Particles. *Appl. Sci.* **2018**, *8*, 573–589. [CrossRef]
11. Dolan, M.D. Non-Pd BCC alloy membranes for industrial hydrogen separation. *J. Membr. Sci.* **2010**, *362*, 12–28. [CrossRef]
12. Cser, L.; Török, G.; Krexner, G.; Prem, M.; Sharkov, I. Neutron holographic study of palladium hydride. *Appl. Phys. Lett.* **2004**, *85*, 1149–1151. [CrossRef]
13. Fukai, Y. Site preference of interstitial hydrogen in metals. *J. Less-Common Metals* **1984**, *101*, 1–16. [CrossRef]

14. Maeland, A.J. Investigation of the Vanadium-Hydrogen System by X-Ray Diffraction Techniques. *J. Phys. Chem.* **1964**, *68*, 2197–2200. [CrossRef]
15. Fukai, Y.; Kazama, S. NMR studies of anomalous diffusion of hydrogen and phase transition in vanadium-hydrogen alloys. *Acta Metall.* **1977**, *25*, 59–70. [CrossRef]
16. Lasser, R.; Schober, T. The phase diagram of the vanadium-tritium system. *J. Less-Common Metals* **1987**, *130*, 453–458. [CrossRef]
17. Seo, C.Y.; Kim, J.H.; Lee, P.S.; Lee, J.Y. Hydrogen storage properties of vanadium-based b.c.c. solid solution metal hydrides. *J. Alloys Compd.* **2003**, *348*, 252–257. [CrossRef]
18. Ukita, S.; Ohtani, H.; Hasebe, M. Thermodynamic Analysis of the V-H Binary Phase Diagram. *Mater. Trans.* **2008**, *49*, 2528–2533. [CrossRef]
19. Kumar, S.; Jain, A.; Ichikawa, T.; Kojima, Y.; Dey, G.K. Development of vanadium based hydrogen storage material: A review. *Renew. Sustain. Energy Rev.* **2017**, *72*, 791–800. [CrossRef]
20. Oriani, R.A. The physical and metallurgical aspects of hydrogen in metals. *Fusion Technol.* **1994**, *26*, 235–266.
21. Lu, Y.; Gou, M.; Bai, R.; Zhang, Y.; Chen, Z. First-principles study of hydrogen behavior in vanadium-based binary alloy membranes for hydrogen separation. *Int. J. Hydrogen Energy* **2017**, *42*, 22925–22932. [CrossRef]
22. Smirnov, L.I.; Pronchenko, D.A. Chemical potential and phase diagrams of hydrogen in palladium. *Int. J. Hydrogen Energy* **2002**, *27*, 825–828. [CrossRef]
23. Smirnov, L.I.; Ruzin, E.V.; Goltsov, V.A. On the statistical theory of equilibrium isotherms in a system hydrogen–palladium. *Ukr. Fiz. Zh.* **1985**, *30*, 1392–1397.
24. McLennan, K.G.; Gray, E.M.A.; Dobson, J.F. Deuterium occupation of tetrahedral sites in palladium. *Phys. Rev. B Condens. Matter Mater. Phys.* **2008**, *78*, 1–9. [CrossRef]
25. Adrover, A.; Giona, M.; Capobianco, L.; Tripodi, P.; Violante, V. Stress-induced diffusion of hydrogen in metallic membranes: Cylindrical vs. planar formulation. I. *J. Alloys Compd.* **2003**, *358*, 268–280. [CrossRef]
26. Hara, S.; Caravella, A.; Ishitsuka, M.; Suda, H.; Mukaida, M.; Haraya, K.; Shimano, E.; Tsuji, T. Hydrogen diffusion coefficient and mobility in palladium as a function of equilibrium pressure evaluated by permeation measurement. *J. Membr. Sci.* **2012**, *421–422*, 355–360. [CrossRef]
27. Cotterill, P. The hydrogen embrittlement of metals. *Prog. Mater. Sci.* **1961**, *9*, 205–301. [CrossRef]
28. Kirchheim, R. Interaction of hydrogen with external stress fields. *Acta Metall.* **1986**, *34*, 37–42. [CrossRef]
29. Kirchheim, R. Interaction of hydrogen with dislocations in palladium-I. Activity and diffusivity, and their phenomenological interpretation. *Acta Metall.* **1981**, *29*, 835–843. [CrossRef]
30. Whiteman, M.B.; Troiano, A.R. The Influence of Hydrogen on the Stacking Fault Energy of an Austenitic Stainless Steel. *Phys. Status Solidi* **1964**, *7*, K109–K110. [CrossRef]
31. Mütschele, T.; Kirchheim, R. Segregation and diffusion of hydrogen in grain boundaries of palladium. *Scr. Metall.* **1987**, *21*, 135–140. [CrossRef]
32. Buxbaum, R.E.; Kinney, A.B. Hydrogen Transport through Tubular Membranes of Palladium-Coated Tantalum and Niobium. *Ind. Eng. Chem. Res.* **1996**, *35*, 530–537. [CrossRef]
33. Smith, J.M. *Chemical Engineering Kinetics*, 2nd ed.; McGraw-Hill, Inc.: New York City, NY, USA, 1970.
34. Mejdell, A.L.; Chen, D.; Peters, T.A.; Bredesen, R.; Venvik, H.J. The effect of heat treatment in air on CO inhibition of a ~3 μm Pd-Ag (23 wt. %) membrane. *J. Membr. Sci.* **2010**, *350*, 371–377. [CrossRef]
35. Ward, T.L.; Dao, T. Model of hydrogen permeation behavior in palladium membranes. *J. Membr. Sci.* **1999**, *153*, 211–231. [CrossRef]
36. Caravella, A.; Barbieri, G.; Drioli, E. Modelling and simulation of hydrogen permeation through supported Pd-alloy membranes with a multicomponent approach. *Chem. Eng. Sci.* **2008**, *63*, 2149–2160. [CrossRef]
37. Caravella, A.; Scura, F.; Barbieri, G.; Drioli, E. Sieverts Law Empirical Exponent for Pd-Based Membranes: Critical Analysis in Pure H_2 Permeation. *J. Phys. Chem. B* **2010**, *114*, 6033–6047. [CrossRef] [PubMed]
38. Caravella, A.; Hara, S.; Drioli, E.; Barbieri, G. Sieverts law pressure exponent for hydrogen permeation through Pd-based membranes: Coupled influence of non-ideal diffusion and multicomponent external mass transfer. *Int. J. Hydrogen Energy* **2013**, *38*, 16229–16244. [CrossRef]
39. Caravella, A.; Hara, S.; Sun, Y.; Drioli, E.; Barbieri, G. Coupled influence of non-ideal diffusion and multilayer asymmetric porous supports on Sieverts law pressure exponent for hydrogen permeation in composite Pd-based membranes. *Int. J. Hydrogen Energy* **2014**, *39*, 2201–2214. [CrossRef]
40. Manchester, F.D.; San-Martin, A.; Pitre, J.M. The H-Pd (hydrogen-palladium) System. *J. Phase Equilib.* **1994**, *15*, 62–83. [CrossRef]

41. Jewell, L.L.; Davis, B.H. Review of absorption and adsorption in the hydrogen-palladium system. *Appl. Catal. A Gen.* **2006**, *310*, 1–15. [CrossRef]
42. Mackay, K.M. *Hydrogen Compounds of the Metallic Elements*; Spon: Hamburg, Germany, 1966.
43. Wicke, E.; Brodowsky, H. Hydrogen in Palladium and Palladium alloys. In *Hydrogen in Metals 2*; Springer: Berlin, Germany, 1978; p. 73.
44. Gillespie, L.J.; Galstaun, L.S. The Palladium-Hydrogen Equilibrium and New Palladium Hydrides. *J. Am. Chem. Soc.* **1936**, *58*, 2565–2573. [CrossRef]
45. Knapton, A.G. Palladium Alloys for Hydrogen Diffusion Membranes. *Platin. Met. Rev.* **1977**, *21*, 44–50.
46. Griffiths, R.; Pryde, J.A.; Righini-Brand, A. Phase diagram and thermodynamic data for the hydrogen/vanadium system. *J. Chem. Soc. Faraday Trans. 1 Phys. Chem. Condens. Phases* **1972**, *68*, 2344–2349. [CrossRef]
47. Albrecht, W.M.; Goode, W.D.; Mallett, M.W. Reactions in the Niobium-Hydrogen System. *J. Electrochem. Soc.* **1959**, *106*, 981–986. [CrossRef]
48. Mallett, M.W.; Koehl, B.G. Thermodynamic Functions for the Tantalum-Hydrogen System. *J. Electrochem. Soc.* **1962**, *109*, 611–616. [CrossRef]

membranes

MDPI

Review

Progress in Methanol Steam Reforming Modelling via Membrane Reactors Technology

Adolfo Iulianelli [1,*], Kamran Ghasemzadeh [2] and Angelo Basile [1]

[1] ITM-CNR, Via P. Bucci Cubo 17/C, University of Calabria, 87036 Rende, CS, Italy; a.basile@itm.cnr.it
[2] Department of Chemical Engineering, Urmia University of Technology, Urmia 57166-93187, Iran; kamran.ghasemzadeh@uut.ac.ir
* Correspondence: a.iulianelli@itm.cnr.it; Tel.: +39-0984-492011; Fax: +39-0984-402103

Received: 9 July 2018; Accepted: 8 August 2018; Published: 17 August 2018

Abstract: Hydrogen has attracted growing attention for various uses, and, particularly, for polymer electrolyte membrane fuel cells (PEMFCs) supply. However, PEMFCs need high grade hydrogen, which is difficult in storing and transportation. To solve these issues, hydrogen generation from alcohols and hydrocarbons steam reforming reaction has gained great consideration. Among the various renewable fuels, methanol is an interesting hydrogen source because at room temperature it is liquid, and then, easy to handle and to store. Furthermore, it shows a relatively high H/C ratio and low reforming temperature, ranging from 200 to 300 °C. In the field of hydrogen generation from methanol steam reforming reaction, a consistent literature is noticeable. Despite various reviews that are more devoted to describe from an experimental point of view the state of the art about methanol steam reforming reaction carried in conventional and membrane reactors, this work describes the progress in the last two decades about the modelling studies on the same reaction in membrane reactors.

Keywords: methanol steam reforming; hydrogen production; modelling; membrane reactors

1. Introduction

In the last two decades, a growing interest towards the environment protection was noticed and more attention was gained by the research on renewable, alternative, and clean energy sources. In particular, fuel cells (FCs) were proposed as a viable option to the conventional processes for power production and to limit the greenhouse gases (GHGs) as well. Among various kinds of FCs, the polymer electrolyte membrane fuel cells (PEMFCs) are supplied by high grade hydrogen, which is industrially produced as a hydrogen-rich stream mainly via steam reforming of natural gas in fixed bed reformers (FBRs) [1]. Consequently, the hydrogen stream that is useful for PEMFCs supplying needs purification after the reforming process, which is conventionally realized in second stage processes, such as water gas shift (WGS) reactors (constituted of a high temperature reactor followed by a second low temperature reactor), partial oxidation (PROX) reactor, and pressure swing adsorption (PSA) [2]. Hence, the conventional process realizes the hydrogen purification with high costs and the need of various devices. As alternative candidates to the aforementioned conventional systems, inorganic membrane reactors (MRs) received much attention in the last 30 years because they combine the reforming reaction for hydrogen generation and its separation/purification in a single device [3,4]. In detail, among several inorganic membrane materials to be used in MR applications, a consistent literature was devoted on the development and utilization of palladium and/or palladium-alloyed membranes, which played a relevant role in this specific scientific field since palladium possesses the singular peculiarity of high hydrogen solubility and perm-selectivity with respect to all of the other gases [5–11].

In the meanwhile, according to the scientific literature, the methanol steam reforming (MSR) reaction has been studied a lot from an experimental point of view as attractive and promising process for generating hydrogen also in combination with inorganic MRs [12–29].

The whole MSR process can be described by the reactions reported in the following:

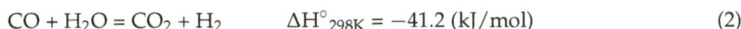

$$CH_3OH + H_2O = CO_2 + 3H_2 \qquad \Delta H^{\circ}{}_{8K} = +49.7\ (kJ/mol) \tag{1}$$

$$CO + H_2O = CO_2 + H_2 \qquad \Delta H^{\circ}{}_{298K} = -41.2\ (kJ/mol) \tag{2}$$

$$CH_3OH = CO + 2H_2 \qquad \Delta H^{\circ}{}_{298K} = +90.7\ (kJ/mol) \tag{3}$$

where reaction (1) represents MSR, reaction (2) is the WGS, and reaction (3) is the methanol decomposition. As stated above, hydrogen production from MSR process has been extensively studied in both FBRs and MRs, as reflected by the number of publications per year reported in Figure 1.

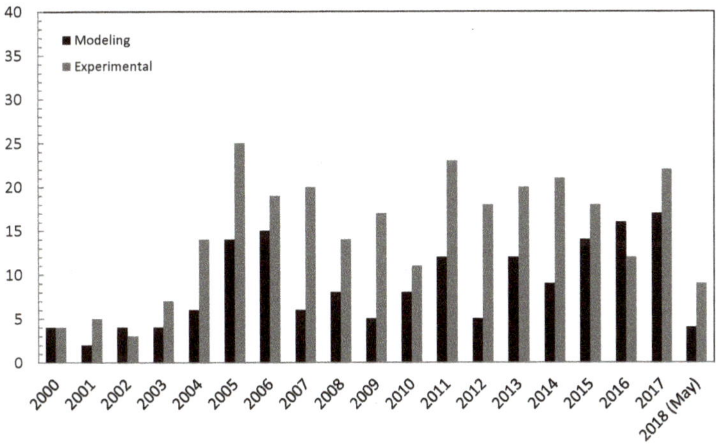

Figure 1. Number of scientific publication (modelling and experimental) versus year about methanol steam reforming (MSR) reaction in both fixed bed reformers (FBRs) and membrane reactors (MRs) (Scopus database: www.scopus.com).

Regarding the oscillation of the number of works per year for both modelling and experimental approaches, we believe that an experimental study may offer more opportunities for an original contribution due to new catalysts utilization and integration in a MR, new membranes tested, and so on. Otherwise, as also reported in Section 3, most of the simulations about the MSR reaction in MRs are one-dimensional (1-D) based, assuming full H_2 perm-selectivity for the palladium membranes. This—in our opinion—limited the number of original contributions in this field and only recently the development of simulations using two-dimensional (2-D) and three-dimensional (3-D) models made an increase of new and original works about modelling of MSR reaction in MRs possible, as confirmed by a higher number of modelling papers than the experimental ones in 2016.

Among the publications devoted to analyse MSR reaction from a modelling point of view, Figure 2 shows the percentage distribution between the studies involving FBRs and MRs.

This work is aimed to briefly describe the inorganic MRs technology from a modelling point of view, giving details regarding to the most recent findings in the MR assisted hydrogen generation from MSR reaction.

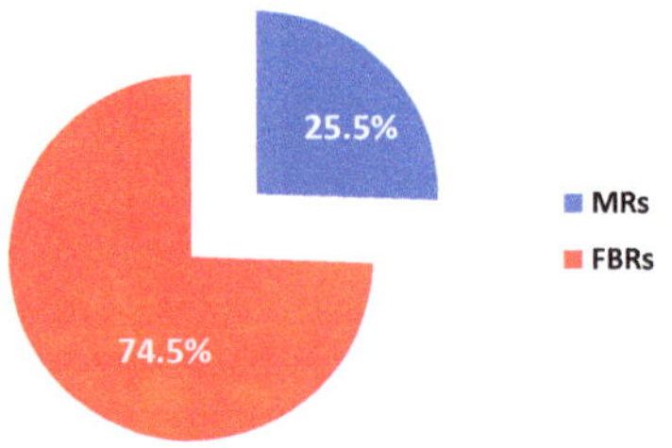

Figure 2. Percentage distribution about the modelling of MSR reaction involving FBRs and MRs (Scopus database: www.scopus.com).

2. Inorganic Membrane Reactor Technology

Today, MRs represent a mature technology and a relevant progress was done in the last years to propose them as a valid alternative technology to the conventional systems. This is clearly evident in the field of hydrogen generation where the application of MRs successfully follows the principles of Process Intensification Strategy (PIS), being more attractive than the conventional systems concerning modularity and reduced costs [3,4,30,31].

As an example of PIS pursued by MRs, the production of high grade hydrogen from natural gas steam reforming reaction (the first source of hydrogen production at industrial scale [30]) is conventionally realized by means of a multi-stages process, Figure 3A; on the contrary, the utilization of a MR housing a hydrogen perm-selective membrane makes the combination of the reaction process for generating hydrogen and its purification in a single device without needing further hydrogen separation stages possible, Figure 3B.

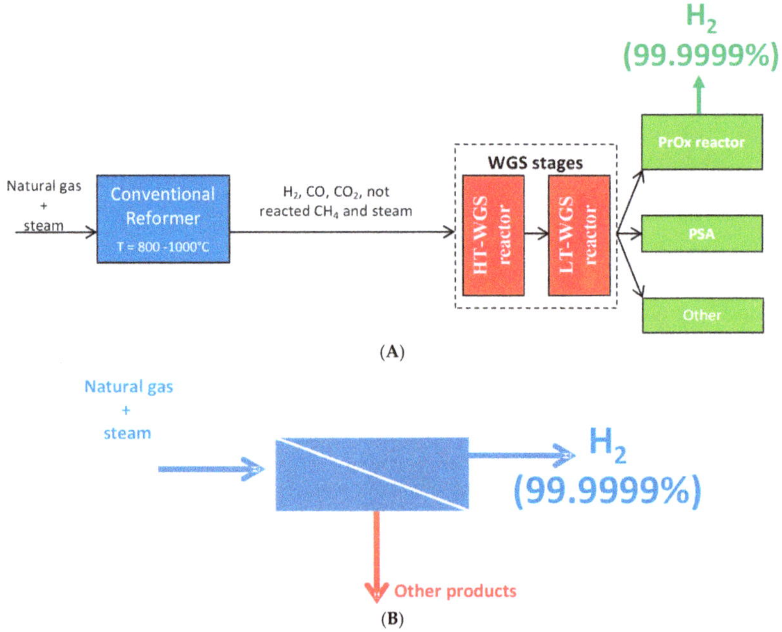

Figure 3. High grade hydrogen generation performed in: (**A**) a multi-stages conventional process; and (**B**) a MR housing a H_2 perm-selective membrane.

The MRs can be categorized in fluidized and packed bed, while two distinct design solutions are possible: planar or tubular configuration. Apart from the experimental investigations about MRs, the mathematical modelling of a MR plays an essential role in process engineering. Indeed, the reactor modelling is useful to predict the behaviour of the system in dynamic/steady state conditions, optimizing the operating parameters via a suitable optimizing algorithm. This approach may be realized prior to running such an experimental test to avoid a longer operating time and efforts in the experimental campaign.

Important parameters of a hydrogen perm-selective membrane are the permeability and selectivity, responsible of the performance enhancement of a catalytic reaction. It is worth of noting that, generally, a membrane may act as an extractor, facilitating the selective removal of one of the products (i.e., hydrogen) from the reaction side toward the permeate side, shifting the reaction to the chosen direction, according to the Le Chatelier principle. Furthermore, the MRs adoption allows for the enhancement of the reaction conversion, reducing the by-product formation and lowering the energy requirements, driving consequently to a much flexible process [4].

Commonly, a packed bed MR allocates the catalyst in the tube/shell side of the membrane, whereas part of the hydrogen produced in the reaction side permeates through the membrane and is recovered in the shell/tube side. In this case, the permeation driving force is represented by the hydrogen partial pressure difference across the membrane [4,32].

On the contrary, a fluidized bed MR utilization presents different benefits over the packed bed modality, such as: improved heat and mass transfer and elimination of external mass-transfer resistances [33]. Nevertheless, the fluidized bed MRs show difficulties in the reactor manufacturing and catalyst erosion [34].

2.1. Palladium Membranes

Many publications were addressed at the subject of hydrogen perm-selective membranes and much attention was received by metallic materials, such as Pd, Pd-alloy, Ti, Va, Mb, etc. [35–41]. Among them, palladium played a relevant role in this field and most of the literature is focused on this metal as a membrane material.

It is well known that the hydrogen permeation through the Pd-based membranes follows the so-called "solution/diffusion" mechanism, and, in the case of relatively low pressure, its rate-limiting step is assumed to be the diffusion [41]. The solution-diffusion mechanism may be expressed by six steps, as reported below:

1. hydrogen molecules adsorption from the membrane;
2. dissociation of hydrogen molecules on the membrane surface;
3. reversible dissociative chemisorption of atomic hydrogen;
4. reversible dissolution of atomic hydrogen in the metal lattice of the membrane;
5. diffusion into the metal of atomic hydrogen proceeds from the higher hydrogen pressure to the lower hydrogen membrane side;
6. desorption of re-combined atomic hydrogen into molecular form.

Hence, for a Pd-based membrane the hydrogen permeating flux may be expressed by the Equation (4):

$$J_{H_2} = Pe_{H_2} \left(p^n{}_{H2,retentate} - p^n{}_{H2,permeate}\right)/\delta \quad (4)$$

where J_{H_2} is the hydrogen flux permeating through the membrane, Pe_{H_2} the hydrogen permeability, δ the membrane thickness, $p_{H2,retentate}$ and $p_{H2,permeate}$ the hydrogen partial pressures in the retentate and permeate zones, respectively, and n (varying between 0.5–1) is the dependence factor of the hydrogen flux on the hydrogen partial pressure. In the case of dense Pd-based membranes having thickness higher than 5 μm, Equation (4) becomes Sieverts-Fick's law (5):

$$J_{H2,Sieverts\text{-}Fick,} = Pe_{H_2} \cdot \left(p^{0.5}{}_{H2,retentate} - p^{0.5}{}_{H2,permeate}\right)/\delta \quad (5)$$

Housing a Pd-based membrane in a MR, the enhancement of the reaction conversion with respect to the equivalent conventional reactor is promoted by the so-called "shift effect" [41]. Indeed, the hydrogen perm-selectivity characteristics of Pd-based membranes allow for the enhancement of conversion due to the hydrogen removal from the reaction toward the shell side.

2.2. MRs Modelling

Mathematical modelling of a MR represents an essential approach in membrane process engineering. Basically, MRs modelling is useful to predict the behaviour of the system in dynamic/steady state modality and optimize the operating conditions by a suitable optimizing algorithm. It can be realized prior to running the tests for avoiding longer operating time and efforts during the experimental campaign. Commonly, the models for realizing simulations on a MR are subdivided into three groups, which are described in brief.

2.2.1. White Box or Theoretical Models

They are developed by adopting physical and chemical principles that are based on the conservation laws of mass, energy, and momentum as well as kinetic and transport equations and taking into account the physical behaviours of the MR. These models are firstly validated by experimental data, and, then, are applicable within a wide range of operation, giving a full insight through the system. Meanwhile, they are able to develop and solve models requiring significant time, needing specific hardware possessing a high-level of computational capacity.

2.2.2. Black Box or Empirical Models

These tools work on the basis of the experimental data fitting and an example of them is represented by the Artificial Neural Network Models [42]. Among the characteristics of these models utilization, it is worth of noting that they are easy to derive but not helpful outside the conditions that are set in the experimental data.

2.2.3. Grey Box or Semi-Empirical Models

These theoretical models adopt some parameters, such as the coefficients for the reaction-kinetics rate, catalyst adsorption, etc., which are calculated using data fitting [43]. These tools provide a clear system understanding, combined to a good generalization over a wide range of data, needing lower computational efforts than other theoretical models.

2.2.4. Further Reactor Modelling Categorization

A further approach in categorizing the MRs modelling is based on the heterogeneous or pseudo-homogenous assumptions, which are responsible for the complexity and accuracy of the model itself. In a heterogeneous model, both fluid and the catalyst particles are seen as two different phases and the balance equations are described for both phases. On the contrary, in a pseudo-homogeneous model, they are considered as a single pseudo-phase and the balances equations are used for only one phase.

Generally, in modelling a conventional reactor using a white box, basically one should apply the mass, energy, and momentum balances, combined to the specific kinetics and transport equations. As a result, a set of differential equations (ODEs or PDEs) are obtained, and, through a suitable solving method, such results as the concentration profiles of each component, temperature and pressure in the reactor are simulated. By using the same approach for a MR, further parameters must be taken into account such as the hydrogen permeating flux and the heat transport inside and outside the reactor. Being a MR constituted of a reaction and a permeation zone, a hydrogen generation reaction is performed in the reaction side, whereas part of the produced hydrogen is removed from the reaction toward the permeate side for permeation through the membrane. Hence, four different streams may be considered: (1) feed: the inlet stream of the reaction side; (2) retentate: the outlet of the reaction side; (3) sweep: the inlet of the permeation side; and, (4) permeate: the outlet of the permeation side. Consequently, the balance equations are applied for both the reaction and the permeation sides, by using an accurate hydrogen permeating flux to correctly evaluate the removal of hydrogen from the reaction side. Furthermore, based on the type of reaction and the process needs, the catalyst may be packed in the tube or shell sides. Two configurations are then possible for the MRs, co-current and counter current. In the co-current mode, the feed in reaction zone and the sweep flow are introduced in the same direction, whereas in a counter current mode they are flowing in opposite directions.

2.2.5. Tubular MR Modelling

Most of the modelling approaches about MRs belong to the tubular ones, because they represent the majority of the scientific studies present in the open literature and the most abundant in industrial applications. The mass balance for each MR side is related to a differential volume of length dX (Figure 4). Commonly, it is assumed that the compositions, temperature, and pressure vary along the axial direction without axial diffusion. In these conditions, the plug flow assumption is hence acceptable. 1-D models are the most common type for simulating a reaction process in MRs.

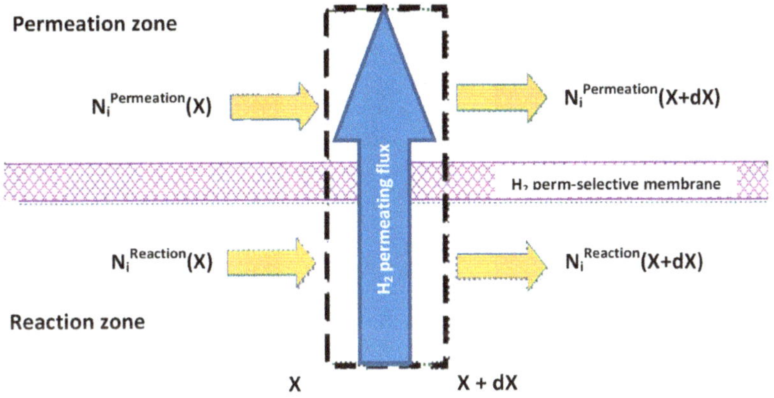

Figure 4. Schematic view of a mass transport in a MR.

The simplification made by considering the properties variations in only one dimension allows for a simple solution of their derivations.

Due to the hydrogen permeation and heat transfer through the membrane, 2-D models also take into account radial concentrations and temperature gradients as well, including radial profiles and applying a different differential volume for two-dimensional modelling. Consequently, the radial diffusion is included in the balances, avoiding considering the term of hydrogen permeation in the balance equations, which is inserted as a boundary condition for both the permeation and reaction zones.

3-D modelling represent the most complex for simulating a MR, since they include the whole geometry and take into account the profiles in angular directions as well. Therefore, their applications are limited unless a non-symmetrical reactor is adopted.

In summary, four different modelling strategies may be applied for simulating a reaction process to generate H_2 in MRs:

1. 1-D model, plug-flow;
2. 1-D model with axial diffusion;
3. 2-D model with axial and radial diffusion;
4. 3-D model with axial, radial and angular diffusion.

However, a deeper analysis about the main differences and peculiarities about the various models proposed in this work may be found in Alavi et al. [5].

Mass Balance

The mass balance is useful for calculating the concentration gradient in a MR. Generally, once a component balance is applied to a control volume of a dynamic reactor system, the principal terms of the mass balance equation are (1) the input and output flows of its component through the control volume, (2) the formation, permeation, and accumulation rates of its component in the control volume. The time dependent term is zero in steady state conditions and the reaction term only exists in the reaction zone modelling. The rate of hydrogen permeation depends on the type of membrane and on its permeation mechanism. It is positive for the permeation side and negative for the reaction side.

Concentration Polarization

The real hydrogen permeation driving force is represented by the Equation (4) and the exponent of the hydrogen partial pressure (n-value) may differ from 0.5, which is the typical coefficient describing the Sieverts-Fick law. As an example, during the permeation of hydrogen through a Pd-based membrane, a layer of non-permeating compounds accumulated on the membrane surface may involve a mass transfer resistance to the permeation process, representing the so-called concentration polarization effect, which may be responsible for a detrimental effect on the separation efficiency of the system. The fluidized bed MRs are able to minimize the concentration polarization effect [44]. Otherwise, the choice of smaller membrane diameter and bigger catalyst particles to avoid the pressure drop or the utilization of specific catalytic beds (structured catalysts) may facilitate the radial mixing properties, depressing the concentration polarization [45,46].

Energy Balance

For non-isothermal MRs, the energy balances are necessary to model their thermal performance. The prediction of the temperature variation inside a MR represents a crucial aspect because of the embrittlement phenomenon affecting the Pd-based membranes, which—as well known—takes place at relatively low temperatures (below 300 °C). Furthermore, in the case of composite Pd-based membranes applications, relatively high temperatures may determine different dilations of the two materials constituting the composite membrane, and then the membrane failure (especially in the case of ceramic substrates). Not less important is the control of the maximum operating temperature of the catalyst for ensuring that its activity cannot be affected.

Momentum Balance

In the case of the hydrogen permeation through the membrane and non-ideal flow pattern (not plug), the flow rate may be different and in small diameters the bed porosity that are close to the wall may be responsible for a non-uniform radial velocity profile, so that the momentum balance has to be included in the equations. Nevertheless, the inclusion of the momentum balance is not so common in the specialized literature, despite its advantages [5].

3. Modelling of MSR Reaction in MRs

Nowadays, to the best of our knowledge, there is not a consistent literature about numerical simulations of MSR reaction in MRs, also taking into account that this review analyzed the papers published after 2000 to illustrate the state of the art about modeling of MSR reaction in MRs in the last 20 years. Furthermore, most of the modeling papers about MSR in the aforementioned period were related to conventional reactors (please see Figure 2, in which it is indicated that ~75% of the papers on modeling of MSR reaction refers to conventional reactors), not in the scope of this review. Nevertheless, various studies focused on both conventional reactors and MRs [47–62] and most of them are based on 1-D model. As an exception, Fu and Wu [55] modeled MSR reaction in MRs while considering transient conditions and a couple of them used 2-D models to take into account the concentration polarization effect [47,53]. Most of the above referenced manuscripts refer to isothermal conditions, which are achieved only with an imposed cooling profile. Only a few of them deal with the study of non-isothermal conditions during MSR reaction [47,54,55].

Ideal gas behavior and plug flow are often assumed on both retentate and permeate sides, and heat gradients are neglected. Pressure drops are described by Ergun equation and palladium membranes are assumed to be defect free, hence showing a full hydrogen perm-selectivity, whereas mass and heat diffusions along the flow direction are usually neglected. Physical properties, such as heat capacity, gas density, and heat transfer coefficient are assumed constant with respect to the temperature variation.

In most of the aforementioned references, the pseudo-homogeneous formulation is adopted, and consequently, chemical reactions on the palladium surface are ignored. However, hydrogen removal from the reaction to the shell side is realistically considered by using experimental permeabilities. The finite difference method is used to solve the governing equations.

The prediction and optimization of the MRs performance during MSR reaction in terms of hydrogen yield and recovery and methanol conversion, as well constitute the final objective of the theoretical modelling by varying parameters such as H_2O/CH_3OH feed ratio, reaction temperature and pressure, sweep gas flow rate, etc.

Gallucci and Basile [52] compared the performance of a MR with respect to the equivalent conventional reactor, while adopting a 1-D numerical model. A Runge-Kutta solution method was developed, including as reaction rate equations those of Peppley et al. [57], meanwhile comparing two different flow patterns (co-current and counter-current modality). These authors theoretically demonstrated that such parameters as reaction temperature, pressure, and feed molar ratio strongly affect positively the hydrogen production, while the adoption of the counter-current configuration is responsible of higher methanol conversions, Figure 5.

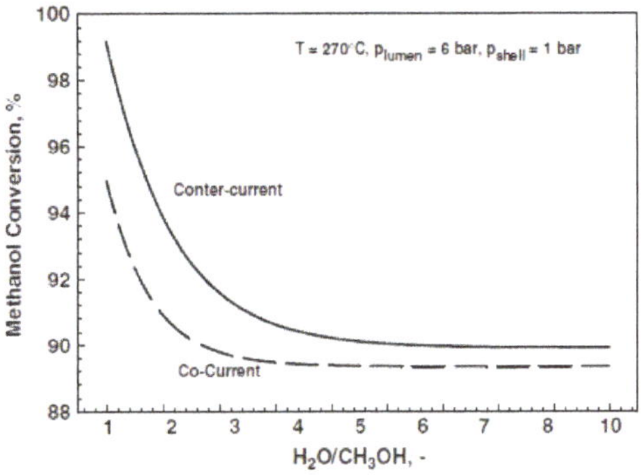

Figure 5. One-dimensional (1-D) modelling of MSR reaction in a Pd-based MR: methanol conversion versus feed molar ratio. With permission of reprint of Elsevier from Gallucci and Basile [52].

In detail, the model predicted that, with a $H_2O/MeOH$ ratio equal to 1/1 and at 270 °C and 6 bar, 99.5% methanol conversion may be reached in counter-current modality against 95% obtained with co-current mode adoption. Furthermore, the counter-current configuration acts positively on the hydrogen recovery, which is almost 100% in the whole range of feed ratio investigated. On the contrary, the co-current configuration is responsible for the decreasing trend of the hydrogen recovered in the permeate stream by increasing the feed ratio, Figure 6.

Nair and Harold [53] modeled MSR reaction in a Pd-Ag MR, in which the simulations were carried out in order to determine the conditions that are useful for maximizing the reactor productivity and the corresponding hydrogen utilization. In the simulations, the dusty gas model (DGM) was taken into account for pellet-scale diffusion and reaction.

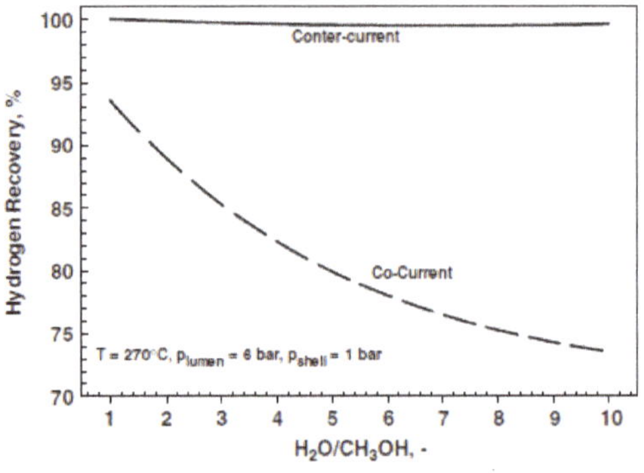

Figure 6. 1-D modelling of MSR reaction in a Pd-based MR: hydrogen recovery versus feed molar ratio. With permission of reprint of Elsevier from Gallucci and Basile [52].

Furthermore, the size of the catalyst particle was strictly considered for predicting the productivity. Indeed, Nair and Harold estimated that the maximum achievable productivity might be decreased by increasing the particle size. Furthermore, these authors identified both membrane thickness and surface to volume ratio as key design parameters. Consequently, they concluded that, for a given particle size, an optimum value of membrane surface/volume ratio allows for the maximum productivity to be reached. This conclusion is supported by the consideration that, at high membrane surface/volume ratio, insufficient catalyst is available to generate hydrogen and reduce the volumetric productivity.. Otherwise, the effect of particle size on productivity becomes negligible. Concerning the role of membrane thickness, Nair and Harold stated that this is in strict correlation with the surface/volume ratio. Adopting thicker membranes a higher surface/volume ratio is achievable and smaller is the catalyst particles size. For instance, they simulated that a hydrogen productivity higher than 50 mol/m^3·s is achievable at 260 °C and 10 bar, while using catalyst particles size smaller than 2 mm, membranes thickness lower than 10 µm, and membrane surface/volume ratio around 500 m^2/m^3.

Fu and Wu [54,55] modeled the MSR reaction by using a double-jacketed Pd-based MR, while also considering a non-isothermal numerical model. The schematic diagram of the MR used by Wu and Wu [54] is shown in Figure 7, which illustrates that, from the external to the internal layer, three components are placed: the catalytic combustor, the reformer, and the permeator.

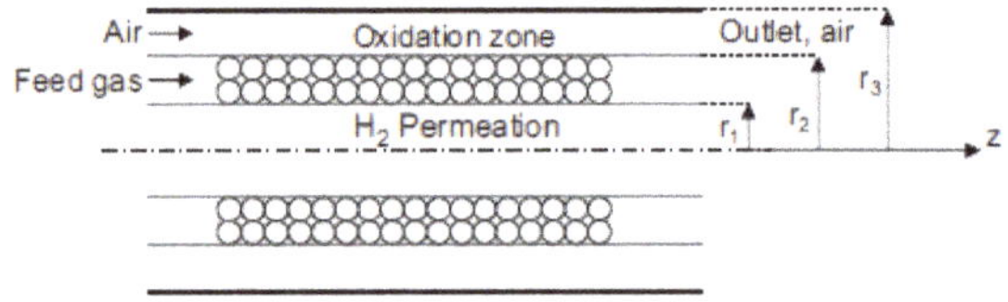

Figure 7. Conceptual scheme of a double-jacketed MR. With permission of reprint of Elsevier from Wu and Fu [54].

The feed gas flows into the annular side (reformer) packed with a catalyst and the hydrogen permeates through the membrane. The unreacted gases flow into the oxidized zone and are mixed with air. As far as the modelling aspects are concerned, both mass and energy balances were evaluated simultaneously in annular and oxidation zones, as well as in the permeation side. The simulations pointed out that an increase of hydrogen volumetric flow rate in the permeation side produces an enhancement of hydrogen permeation rate across the membrane. Therefore, Fu and Wu established that an optimum ratio between radial (permeation) and axial (annulus) velocity is close to 10. Furthermore, for a specific Damköhler number, the higher the temperature the higher the hydrogen production, whereas methanol conversion and hydrogen recovery yield (defined as the percentage of pure hydrogen collected from the total hydrogen generation) are decreased. These authors also compared the performance of the aforementioned double-jacketed MR with an auto-thermal conventional reactor in terms of methanol conversion, hydrogen recovery, yield, and production rate. The simulations showed that, at the same operating conditions and for a definite reactor volume as well, a higher methanol conversion is reached by using the double-jacketed MR, Figure 8.

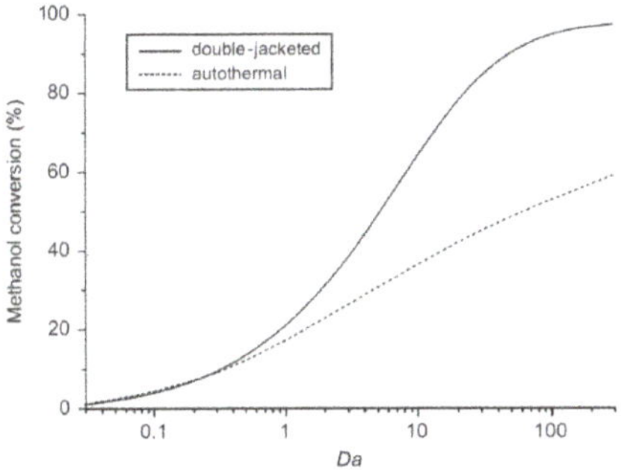

Figure 8. Methanol conversion *vs* Damköhler number: comparison between the double-jacketed MR and an equivalent autothermal conventional reactor. With permission of reprint of Elsevier from Wu and Fu [54].

The numerical results evidenced this result for a Damköhler number higher than 1, while for a Damköhler number equal to 100, 95% methanol conversion is attained in the double jacketed MR with respect to 55% achievable adopting the auto-thermal conventional reactor. The double-jacketed MR was theoretically studied by the same authors also developing a transient model [55]. At the start-up, the temperatures of gases and catalysts as well as the consumption and species production were investigated in two MR conditions: (1) feed temperature higher than catalyst temperature, and (2) the reverse of condition (1). The simulations showed that condition (1) allows for obtaining higher methanol conversion and reactor temperature than condition (2). Moreover, during start-up the instability of species can be reduced with condition (1). The model analyzed also the MR response when a temporary extra hydrogen demand occurs under steady-state conditions. The latter could be satisfied by increasing the MR temperature from additional methanol oxidation or by increasing the inlet methanol.

MSR reaction was also theoretically studied by Mendes and co-workers [48], who compared the performance of Pd-based palladium and carbon molecular sieve (CMS) MRs in terms of methanol conversion and hydrogen recovery. Based on the experimental H_2 permeabilities of CMS and Pd-based membranes from the open literature, the model predicted a similar methanol conversion being reached in both membrane reactors. This theoretical result could indicate that the permeation behaviors of the two membranes do not have any effect on the methanol consumption. On the contrary, the simulations evidenced that higher hydrogen recovery is achievable adopting a CMS-MR than a Pd-based MR, even though the latter may produce a pure hydrogen stream, Figure 9.

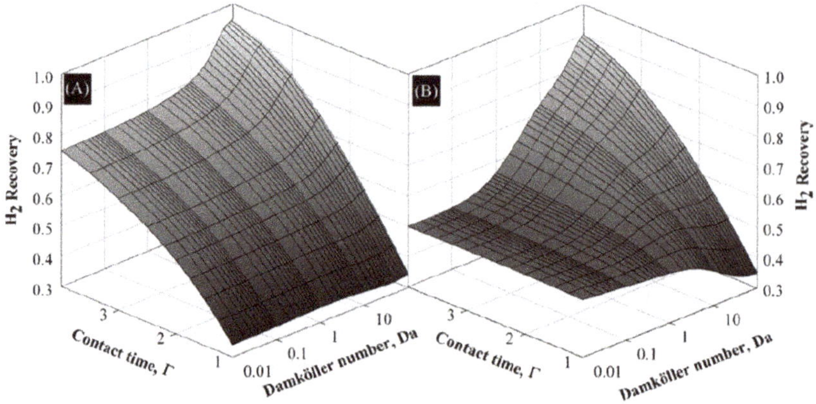

Figure 9. Simulation results of MSR reaction: Hydrogen recovery vs. contact time and Damköhler number: (**A**) carbon molecular sieve (CMS) MR, (**B**) Pd MR. With permission of reprint of Elsevier from Sà et al. [48].

The hydrogen recovery is enhanced in both the MRs by increasing the Damköhler number. However, the Sà et al. [48] confirmed that the adoption of a Pd-based MR is more adequate if high grade hydrogen has to be produced because the simulations demonstrated that the H_2/CO reaction selectivity is increased by using this reactor with respect to the CMS-MR.

As a further investigation, these authors analyzed a hybrid MR configuration consisting of a CMS membrane being positioned in series with a Pd-based membrane. This new configuration revealed some benefits, such as higher hydrogen recovery values when compared to MRs equipped with single membranes, Figure 10.

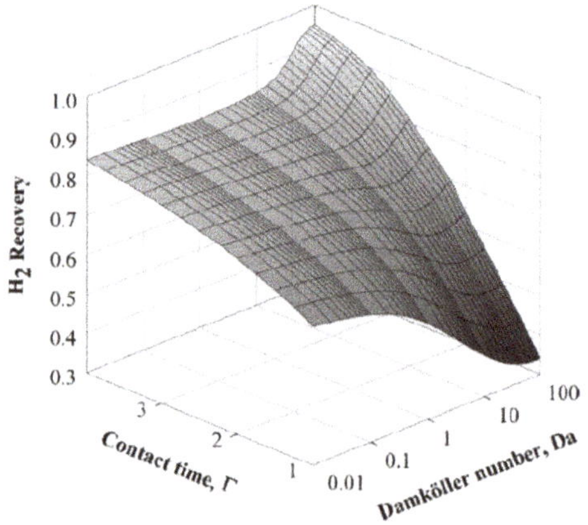

Figure 10. Simulation results of MSR reaction: Hydrogen recovery vs contact time and Damköhler number for a hybrid MR solution. With permission of reprint of Elsevier from Sà et al. [48].

Consequently, this hybrid MR solution allows for a reduced membrane area and higher feed flow rates than the Pd-MR, without a significant decrease in the performance. In another work, Sà et al. [14] modeled two different MRs for carrying out MSR reaction with the intent of obtaining high grade hydrogen for PEMFC supplying.

The first system was based on a MR setup, whereas the second system was constituted by a PROX reactor in addition to a MR. In both cases, the governing equations were discretized with a finite volume method and the simulations showed the advantages due to the adoption of a PROX reactor. Indeed, the solution of the MR combined to the PROX reformer allowed for converting into CO_2 (with a percentage below 20%) most of the CO contained in the MR permeate stream, showing a final CO content below 2 ppm, Figure 11. The simulations about this hybrid configuration showed that, at higher Damköhler number and contact time, both methanol conversion and hydrogen recovery are enhanced, Figure 12A,B, reaching the optimum at contact time values close to 2, where high methanol conversion (up to 85%) and hydrogen recovery (up to 75%) are >85% and 75%, respectively, meanwhile lowering the amount of CO_2 to 10%.

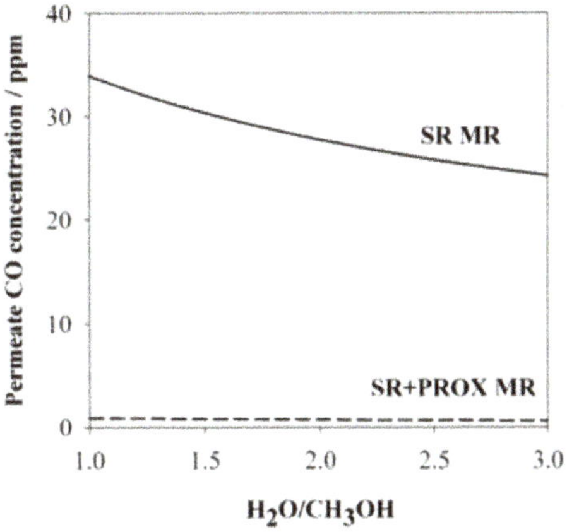

Figure 11. Simulation results of MSR reaction carried out in a MR and in a MR followed by a PROX reactor: Permeate CO concentration vs H_2O/CH_3OH feed molar ratio. With permission of reprint of Elsevier from Sà et al. [14].

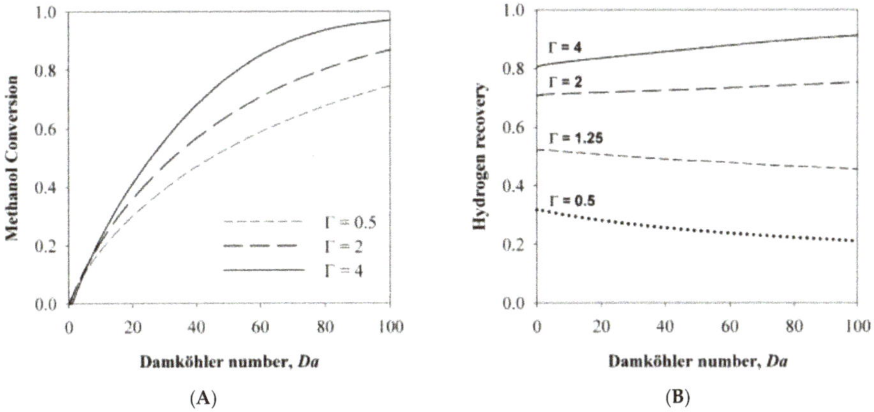

Figure 12. Simulation results of MSR reaction carried out in a MR followed by a PROX reactor: Methanol conversion (**A**) and hydrogen recovery (**B**) vs. Damköhler number at various contact time; With permission of reprint of Elsevier from Sà et al. [14].

Israni and Harold [47] proposed a 2-D and non-isothermal model to simulate MSR reaction in a MR housing a composite ~4 μm thick Pd-Ag membrane, when comparing the results with an equivalent conventional reactor. The authors included in the model the Maxwell equations for considering the molecular diffusion phenomena in the MR. Furthermore, a model of hydrogen flux inhibition due to the competitive adsorption of the primary MSR species was introduced, revealing a good agreement between the experimental and predicted concentrations. Figures 13 and 14 show both experimental and modelling results concerning the influence of space velocity at different reaction temperature and pressure, on methanol conversion and hydrogen productivity, respectively. In both figures, the symbols represent the experimental results and the solid lines indicate the numerical results for the conventional reactor, while the dashed lines the numerical MR results.

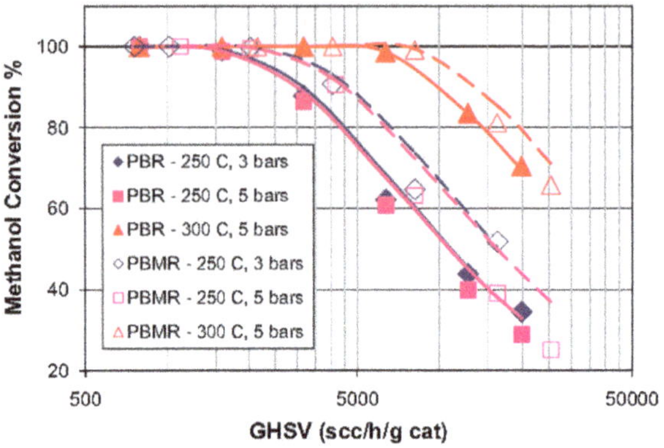

Figure 13. Simulation (lines) and experimental (symbols) results of MSR reaction carried out in a supported Pd-Ag MR and in a conventional reactor: Methanol conversions vs. gas hourly space velocity (GHSV) at different reaction temperature and pressure. With permission of reprint of Elsevier from Israni and Harold [47].

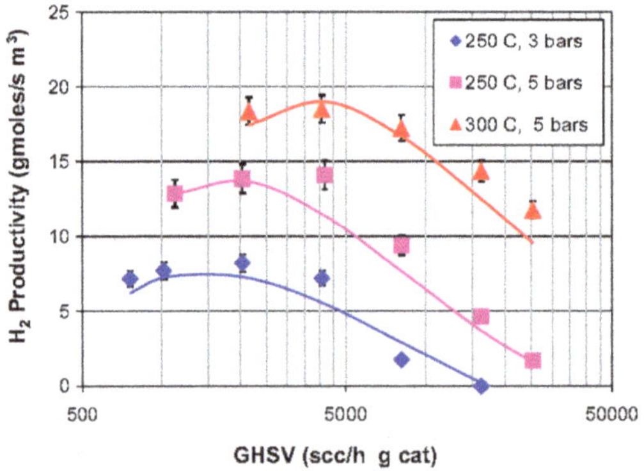

Figure 14. Simulation (lines) and experimental (symbols) results of MSR reaction carried out in a supported Pd-Ag MR and in a conventional reactor: H_2 productivities vs GHSV at different reaction temperature and pressure. With permission of reprint of Elsevier from Israni and Harold [47].

More recently, Ribeirinha et al. [63] simulated an integrated process in which a MSR catalyst was packed into the anodic compartment of a high temperature polymer electrolyte fuel cell (HT-PEMFC), Figure 15. Here, both reforming and electrochemical reactions took place simultaneously.

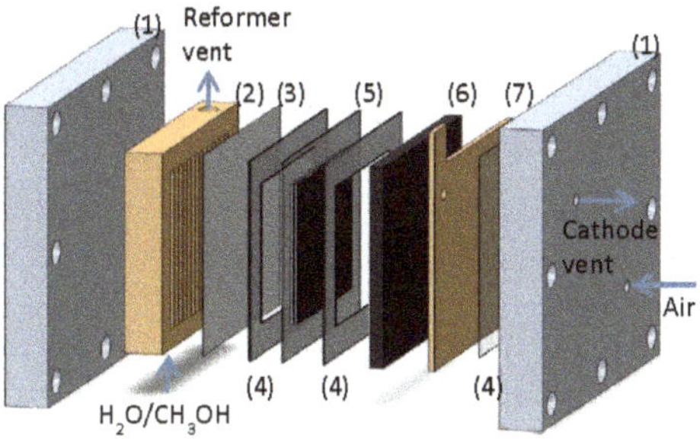

Figure 15. Schematic representation of the integrated system (MR + high temperature polymer electrolyte fuel cell (HT-PEMFC)): two metal endplates frames (1), gold coated reformer (2), Pd-Ag membrane (3), gasket (4), MEA (5), graphite composite bipolar plate (6) and current collector (7). With permission of reprint of Elsevier from Ribeirinha et al. [63].

In addition, a Pd-Ag membrane, with a thickness of a few micrometers, was placed between the reforming catalyst and the membrane electrode assembly in order to avoid the contamination of the anode electro-catalyst by methanol. Hence, they used a 3-D non-isothermal model that was developed in Fluent (Ansys™) to simulate a packed bed MR combined to a HT-PEMFC in a single unit. The permeation characteristics of the membrane were taken by the experimental tests on a self-supported Pd-Ag membrane having a layer of 4 μm, which was produced by the magnetron sputtering technique and without pin-holes. At 200 °C, the H_2/N_2 perm-selectivity of the membrane was around 5800 and the hydrogen permeability around 2.9×10^{-6} mol·m·s^{-1}·m^{-2}·bar$^{-0.8}$.

After proper validation of the model starting from the permeation behaviors of the membrane, the simulations pointed out high performance for the integrated system similar to the one that was obtained with a HT-PEMFC supplied by hydrogen, allowing for efficient heat integration between electrochemical and MSR reaction. In the MR simulations, Ribeirinha et al. [63] demonstrated the enhancement in the methanol conversion as the permeate pressure decreases, Figure 16. In this figure, methanol conversion is plotted against the space-time ratio in a Pd-Ag MR for different permeate pressures. To simulate a conventional reactor, the membrane was considered impermeable and—as shown—methanol conversion was always lower than the MR. In the combined configuration of the MR integrated with a HT-PEMFC, the hydrogen that was consumed by the electrochemical reaction may be responsible of the permeate pressure reduction below 1 bar, causing a faster hydrogen permeation through the membrane, acting as electrochemical hydrogen pumping. The success of the integrated system was confirmed by other simulations. These authors simulated the polarization curves by using a HT-PEMFC fed with pure hydrogen and the integrated system constituted of a HT-PEMFC coupled with the Pd-Ag MR-C at 200 °C, Figure 17.

This figure shows how the results of the integrated system completely match those that were obtained with the HT-PEMFC supplied by pure hydrogen, indirectly confirming that the high methanol conversion and a very high hydrogen permeability of the Pd-Ag membrane cover the hydrogen consumption request for the HT-PEMFC supplying. Saidi [64] developed a comprehensive 2-D non-isothermal model to simulate the performance of MSR reaction in a supported Pd-Ag MR, evaluating the influence of different operating parameters, such as temperature, pressure, sweep ratio, and steam ratio on methanol conversion and hydrogen recovery, Figure 18.

The simulations evidenced that a temperature increase improves the kinetic catalytic properties and the hydrogen permeance through the supported membrane, resulting in higher methanol conversion and hydrogen production, Figure 19. This figure also shows that an increase of reaction pressure from 2 to 16 bar enhances methanol conversion more than 30%. This result points out that the role of the Pd-Ag membrane is crucial, because, by increasing the operating pressure, the hydrogen permeation driving force is intensified, determining a higher hydrogen removal with a consequent shift effect on MSR reaction toward further products formation and methanol conversion improvement. According to several literature data, it is evident that the permeation effect overcomes the thermodynamic effect, which represents the detrimental influence that is operated by higher pressures on MSR reaction, which proceeds from reaction to products with an increase of the moles number.

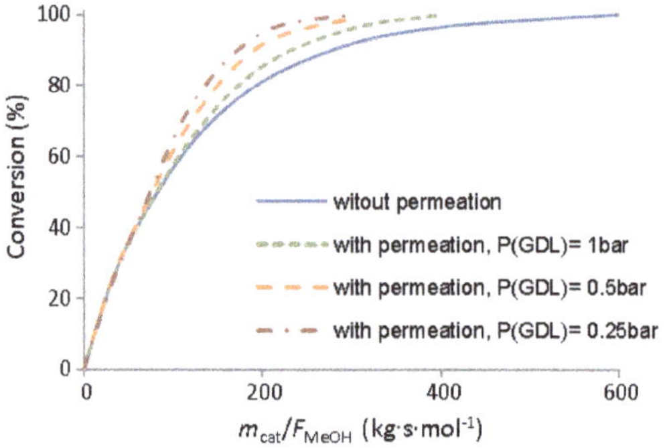

Figure 16. Simulated methanol conversion vs space-time in a Pd-Ag MR, at 200 °C, total retentate pressure of 3.0 bar. With permission of reprint of Elsevier from Ribeirinha et al. [63].

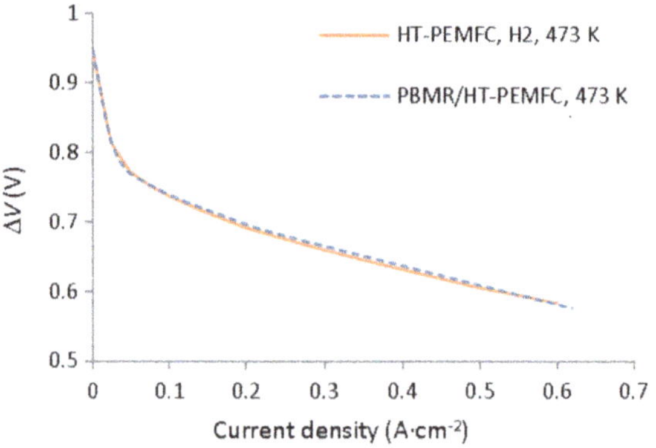

Figure 17. Simulations about the electric potential difference vs current density for a HT-PEMFC supplied by pure hydrogen and the integrated system combining a HT-PEMFC with a Pd-Ag MR operated at 200 °C. With permission of reprint of Elsevier from Ribeirinha et al. [63].

In conclusion, among several theoretical works [65–67], an original modelling study was realized by Ghasemzadeh et al. [65], who theoretically studied MSR reaction in a silica-based MR. These authors presented both a qualitative safety and a quantitative operating analysis of a silica-based MR adoption to perform MSR reaction for hydrogen generation. The safety analysis on the system was realized by using the HAZOP method (Figure 20). HAZOP (Hazard and Operability) analysis was developed in the early 1970s from a tentative approach to hazard identification for process plants to an almost universally accepted approach today, and a central technique of safety engineering. More details about HAZOP analysis may be found in Taylor [68].

Based on the HAZOP analysis approach, Ghasemzadeh et al. [65] made a comprehensive investigation about the most important operating parameters affecting the silica-based MR performance (quantitative analysis), which was realized by developing a 1-D isothermal model. The simulations evidenced that the reaction pressure and feed molar ratio have dual effects on the silica-based MR performance. On one hand, methanol conversion is decreased by increasing the reaction pressure from 1.5 to 4 bar, while over 4 bar, it is improved. On the other hand, the hydrogen recovery decreases by increasing the feed molar ratio from 1 to 5, while over 5 bar, it assumes a constant trend. Afterwards, the HAZOP analysis was carried out by using the analyzed operating variables as key parameters.

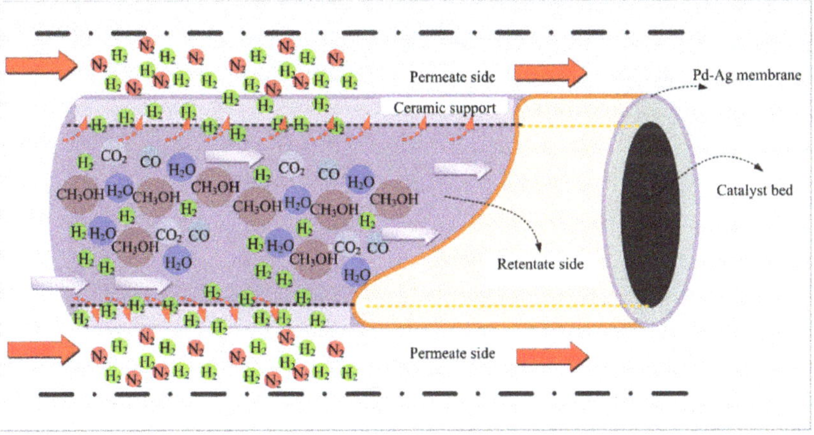

Figure 18. Scheme of the supported Pd-Ag MR. With permission of reprint of Elsevier from Saidi [64].

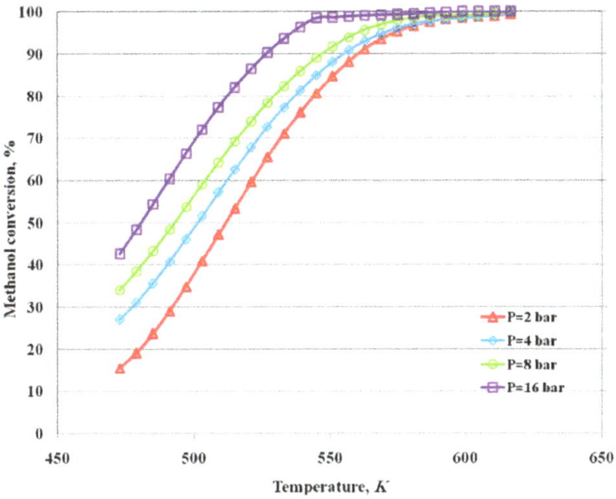

Figure 19. Simulation of MSR reaction in a supported Pd-Ag MR: methanol conversion vs temperature. With permission of reprint of Elsevier from Saidi [64].

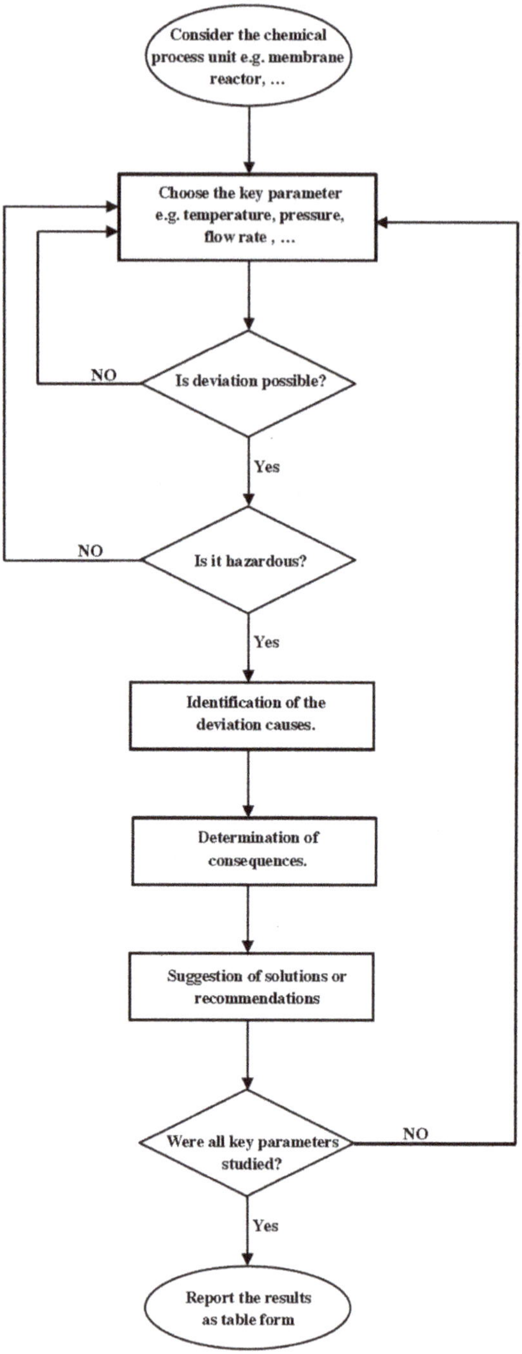

Figure 20. Scheme of the HAZOP approach. With permission of reprint of Elsevier from Ghasezmadeh et al. [65].

The safety assessment of MSR reaction performed in a silica-based MR was hence contained in different tables as check list. As an example, Table 1 reports the input-output table for deviations of the reaction temperature in the silica-based MR. However, the main goal of this modelling work was the nature itself of the recommendations resulted by the HAZOP analysis, constituting the solutions to avoid the economic and safety loss during MSR reaction in a silica-based MR.

Table 1. Safety assessment of MSR reaction performed in a silica-based MR via HAZOP analysis: input-output table for deviations of the reaction temperature in the MR. With permission of reprint of Elsevier from Ghasezmadeh et al. [65].

Guide Words	Causes	Consequences	Recommensations
Less	1. Heater controller fails	• Lower reaction rate	1. Check of the heater controller
	2. Lower temperature of feed reactants or HPLC pump fails	• Lower hydrogen permeation	2. Check values and lines or HPLC pump
	3. Lower temperature of sweep gas	• Lower hydrogen productivity	3. Check the sweep gas cylinder
	4. Isolation of MR set up fails	• Lower hydrogen selectivity	4. Check of the isolation system
		• Lower conversion	
		• Lower hydrogen recovery	
		Condensation of vapors	
More	1. Heater controller fails	• Thermal stress for the silica membrane	1. Check control system of heater
	2. Higher temperature of feed reactants or HPLC pump fails	• Catalyst sintering	2. Check values and lines or HPLC pump
	3. Isolation of MR set up fails	• Sealing of module fails	3. Check of the isolation system
		• Formation hot spot in MR	
Tolerance	1. Heater controller fails	• Defect formation on silica membrane	1. Check control system of heater

4. Conclusions

The modelling approach represents a useful tool for simulating the physical and chemical phenomena in a MR, acquiring the adequate prior knowledge on the system by a reliable model. Indeed, the prior knowledge makes hence unnecessary the consumption of time and energy in the experiments. In this review, different modelling types were described when applied to MSR reaction carried out in MRs, giving a panoramic view on the recent advancements present in the open literature. Particular role was reserved to palladium membranes and much attention was dedicated from a theoretical point of view to the intensified process allowed by housing a Pd-based membrane in a MR. Most of the modelling studies present in literature involving MSR reaction and MRs include 1 or 2D models due to their easier derivation, even though more recently more complex 3D-modelling analyses were dedicated to a more accurate understanding toward the systems. Last but not least, this review also reported a new approach by combining the HAZOP analysis, useful for a safety evaluation of a silica-based MR used for generating hydrogen from MSR reaction, and a modelling analysis, useful for evaluating the main variables constituting the principal parameters of HAZOP algorithm. However, some issues still need to be better developed in the future of modeling approach about MSR reaction in MRs such as the membrane deactivation or the membrane performance variation in long-term uses, as main responsible effects on the global MR efficiency.

Funding: This research received no external funding.

Conflicts of Interest: The authors declare no conflict of interest.

List of Symbols and Acronyms

J_{H_2}	Hydrogen flux permeating through the membrane
n	Dependence factor of the hydrogen flux on the hydrogen partial pressure
Pe_{H_2}	Hydrogen permeability
$p_{H_{2,permeate}}$	Hydrogen partial pressures in the permeate zone
$p_{H_{2,retentate}}$	Hydrogen partial pressures in the retentate zone
δ	Membrane thickness
CMS	Carbon molecular sieve
FBR	Fixed bed reactor
FC	Fuel cell
GHG	Greenhouse gas
MR	Membrane reactor
MSR	Methanol steam reforming
ODE	Ordinary differential equation
PDE	Partial differential equation
PEMFC	Proton exchange membrane fuel cell
PIS	Process intensification strategy
PROX	Preferential oxidation
PSA	Pressure swing adsorption
WGS	Water gas shift

References

1. Heinzel, A.; Vogel, B.; Hübner, P. Reforming of natural gas hydrogen generation for small scale stationary fuel cell systems. *J. Power Sources* **2002**, *105*, 202–207. [CrossRef]
2. Pettersson, L.J.; Westerholm, R. State of the art of multi-fuel reformers for fuel cell vehicles: Problem identification and research needs. *Int. J. Hydrogen Energy* **2001**, *26*, 243–264. [CrossRef]
3. Brunetti, A.; Sun, Y.; Caravella, A.; Drioli, E.; Barbieri, G. Process intensification for greenhouse gas separation from biogas: More efficient process schemes based on membrane-integrated systems. *Int. J. Greenhouse Gas Control* **2015**, *35*, 18–29. [CrossRef]
4. Iulianelli, A.; Basile, A. Advances on inorganic membrane reactors for production of hydrogen. In *Encyclopedia of Sustainability Science and Technology*; Meyers, R.A., Ed.; Springer: Berlin, Germany, 2018; pp. 1–11.
5. Alavi, M.; Iulianelli, A.; Rahimpour, M.R.; Eslamloueyan, R.; De Falco, M.; Bagnato, G.; Basile, A. Chapter. 8. Fixed bed membrane reactors for ultrapure hydrogen production: Modelling approach. In *Hydrogen Production, Separation and Purification for Energy*; Basile, A., Dalena, F., Tong, J., Nejat Veziroğlu, T., Eds.; Institution Engineering and Technology: Michael Faraday House, Stevenage, England, 2017; pp. 231–257, ISBN 978-1-78561-100-1.
6. Basile, A. Hydrogen production using Pd-based membrane reactors for fuel cells. *Top. Catal.* **2008**, *51*, 107–122. [CrossRef]
7. Damle, A.S. Hydrogen production by reforming of liquid hydrocarbons in a membrane reactor for portable power generation—Experimental studies. *J. Power Sources* **2009**, *186*, 167–177. [CrossRef]
8. Meénendez, M. Inorganic membrane reactors for energy applications. In *Nanoporous Materials for Energy and the Environment*; Rios, G., Centi, G., Kanellopoulos, N., Eds.; Pan Stanford Publishing Pte. Ltd.: Singapore, 2011; pp. 283–297.
9. Li, H.; Caravella, A.; Xu, H.Y. Recent progress in Pd-based composite membranes. *J. Mater. Chem. A* **2016**, *4*, 14069–14094. [CrossRef]
10. Dolan, M.D.; Dave, N.C.; Ilyushechkin, A.Y.; Morpeth, L.D.; McLennan, K.G. Composition and operation of hydrogen-selective amorphous alloy membranes. *J. Membr. Sci.*, **2006**, *285*, 30–55. [CrossRef]
11. Al-Mufachi, N.A.; Rees, N.V.; Steinberger-Wilkens, R. Hydrogen selective membranes: A review of palladium-based dense metal membranes. *Renew. Sustain. Energy Rev.* **2015**, *47*, 540–551. [CrossRef]

12. Chein, R.Y.; Chen, Y.C.; Lin, Y.S.; Chung, J.N. Hydrogen production using integrated methanol-steam reforming reactor with various reformer designs for PEM fuel cells. *Int. J. Energy Res.* **2012**, *36*, 466–476. [CrossRef]
13. Lindström, B.; Pettersson, L.J. Development of a methanol fuelled reformer for fuel cell applications. *J. Power Sources* **2003**, *118*, 71–78. [CrossRef]
14. Sá, S.; Sousa, J.M.; Mendes, A. Methanol steam reforming in a dual-bed membrane reactor for producing PEMFC grade hydrogen. *Catal. Today* **2010**, *156*, 254–260. [CrossRef]
15. Iulianelli, A.; Ribeirinha, P.; Mendes, A.; Basile, A. Methanol steam reforming for hydrogen generation via membrane reactors: A review. *Renew. Sustain. Energy Rev.* **2014**, *29*, 355–368. [CrossRef]
16. Rei, M.H.; Yeh, G.T.; Tsai, Y.H.; Kao, Y.L.; Shiau, L.D. Catalysis-spillover-membrane. III: The effect of hydrogen spillover on the palladium membrane reactor in the steam reforming reactions. *J. Membr. Sci.*, **2011**, *369*, 299–307. [CrossRef]
17. Sá, S.; Sousa, J.M.; Mendes, A. Steam reforming of methanol over a $CuO/ZnO/Al_2O_3$ catalyst part II: A carbon membrane reactor. *Chem. Eng. Sci.* **2011**, *66*, 5523–5530. [CrossRef]
18. Lee, D.W.; Nam, S.E.; Sea, B.; Ihm, S.K.; Lee, K.H. Preparation of Pt-loaded hydrogen selective membranes for methanol reforming. *Catal. Today* **2006**, *118*, 198–204. [CrossRef]
19. Lin, Y.M.; Rei, M.H. Study on hydrogen production from methanol steam reforming in supported palladium membrane reactor. *Catal. Today* **2001**, *67*, 77–84. [CrossRef]
20. Lee, D.W.; Park, S.J.; Yu, C.Y.; Ihm, S.K.; Lee, K.H. Study on methanol reforming-inorganic membrane reactors combined with water-gas shift reaction and relationship between membrane performance and methanol conversion. *J. Membr. Sci.* **2008**, *316*, 63–72. [CrossRef]
21. Briceño, K.; Iulianelli, A.; Montanè, D.; Garcia-Valls, R.; Basile, A. Carbon molecular sieve membranes supported on non-modified ceramic tubes for hydrogen separation in membrane reactors. *Int. J. Hydrogen Energy* **2012**, *37*, 13536–13544. [CrossRef]
22. Zhang, X.; Hu, H.; Zhu, Y.; Zhu, S. Methanol steam reforming to hydrogen in a carbon membrane reactor system. *Ind. Eng. Chem. Res.* **2006**, *45*, 7997–8001. [CrossRef]
23. Iulianelli, A.; Longo, T.; Basile, A. Methanol steam reforming in a dense Pd-Ag membrane reactor: The pressure and WHSV effects on CO-free H_2 production. *J. Membr. Sci.* **2008**, *323*, 235–240. [CrossRef]
24. Itoh, N.; Kaneko, Y.; Igarashi, A. Efficient hydrogen production via methanol steam reforming by preventing back-permeation of hydrogen in a palladium membrane reactor. *Ind. Eng. Chem. Res.* **2002**, *41*, 4702–4706. [CrossRef]
25. Lin, Y.M.; Lee, G.L.; Rei, M.H. An integrated purification and production of hydrogen with a palladium membrane-catalytic reactor. *Catal. Today* **1998**, *44*, 343–349. [CrossRef]
26. Iulianelli, A.; Longo, T.; Basile, A. Methanol steam reforming reaction in a Pd-Ag membrane reactor for CO-free hydrogen production. *Int. J. Hydrogen Energy* **2008**, *33*, 5583–5588. [CrossRef]
27. Ghasemzadeh, K.; Liguori, S.; Morrone, P.; Iulianelli, A.; Piemonte, V.; Babaluo, A.A.; Basile, A. H_2 production by low pressure methanol steam reforming in a dense Pd-Ag membrane reactor in co-current flow configuration: Experimental and modelling analysis. *Int. J. Hydrogen Energy* **2013**, *38*, 16685–16697. [CrossRef]
28. Liguori, S.; Iulianelli, A.; Dalena, F.; Piemonte, V.; Huang, Y.; Basile, A. Methanol steam reforming in an Al_2O_3 supported thin Pd-layer membrane reactor over $Cu/ZnO/Al_2O_3$ catalyst. *Int. J. Hydrogen Energy* **2014**, *39*, 18702–18710. [CrossRef]
29. Mateos-Pedrero, C.; Silva, H.; Pacheco Tanaka, D.A.; Liguori, S.; Iulianelli, A.; Basile, A.; Mendes, A. CuO/ZnO catalysts for methanol steam reforming: The role of the support polarity ratio and surface area. *Appl. Catal. B Environ.* **2015**, *174*, 67–76. [CrossRef]
30. Iulianelli, A.; Liguori, S.; Wilcox, J.; Basile, A. Advances on methane steam reforming to produce hydrogen through membrane reactors technology: A review. *Catal. Rev.* **2016**, *58*, 1–35. [CrossRef]
31. Brunetti, A.; Caravella, A.; Drioli, E.; Barbieri, G. Process intensification by membrane reactors: High-temperature water gas shift reaction as single stage for syngas upgrading. *Chem. Eng. Technol.* **2012**, *35*, 1238–1248. [CrossRef]
32. Rahimpour, M.R.; Samimi, F.; Babapoor, A.; Tohidian, T.; Mohebi, S. Palladium membranes applications in reaction systems for hydrogen separation and purification: A review. *Chem. Eng. Process. Proc. Intensif.* **2017**, *121*, 24–49. [CrossRef]

33. Gallucci, F.; Medrano, J.; Fernandez, E.; Melendez, J.; van Sint Annaland, M.; Pacheco, A. Advances on high temperature Pd-based membranes and membrane reactors for hydrogen purification and production. *J. Membr. Sci. Res.* **2017**, *3*, 142–156.
34. Rahimpour, M.R.; Mirvakili, A.; Paymooni, K.; Moghtaderi, B. A comparative study between a fluidized-bed and a fixed-bed water perm-selective membrane reactor with in situ H_2O removal for Fischer–Tropsch synthesis of GTL technology. *J. Nat. Gas Sci. Eng.* **2011**, *3*, 484–495. [CrossRef]
35. Tereschenko, G.F.; Ermilova, M.M.; Mordovin, V.P.; Orekhova, N.V.; Gryaznov, V.M.; Iulianelli, A.; Gallucci, F.; Basile, A. New Ti-Ni dense membranes with low palladium content. *Int. J. Hydrogen Energy* **2007**, *32*, 4016–4022. [CrossRef]
36. Basile, A.; Gallucci, F.; Iulianelli, A.; Tereschenko, G.F.; Ermilova, M.M.; Orekhova, N.V. Ti-Ni-Pd dense membranes—The effect of the gas mixtures on the hydrogen permeation. *J. Membr. Sci.* **2008**, *310*, 44–50. [CrossRef]
37. Cardoso, S.P.; Azenha, I.S.; Lin, Z.; Portugal, I.; Rodrigues, A.E.; Silva, C.M. Inorganic membranes for hydrogen separation. *Sep. Purif. Rev.* **2018**, *47*, 229–266. [CrossRef]
38. Conde, J.J.; Maroño, M.; Sánchez-Hervás, J.M. Pd-Based membranes for hydrogen separation: Review of alloying elements and their influence on membrane properties. *Sep. Purif. Rev.* **2017**, *46*, 152–177. [CrossRef]
39. Jiang, P.; Song, G.; Liang, D.; Wu, W.; Hua, T. Vanadium-based alloy membranes for hydrogen purification. *Rare Met. Mater. Eng.* **2017**, *46*, 857–863.
40. Arratibel Plazaola, A.; Pacheco Tanaka, D.A.; Van Sint Annaland, M.; Gallucci, F. Recent advances in Pd-based membranes for membrane reactors. *Molecules* **2017**, *22*, 51. [CrossRef] [PubMed]
41. Basile, A.; Iulianelli, A.; Longo, T.; Liguori, S.; De Falco, M. Chapter. 2. Pd-based selective membrane state-of-the-art. In *Membrane Reactors for Hydrogen Production Processes*; Marrelli, L., De Falco, M., Iaquaniello, G., Eds.; Springer London: Dordrecht Heidelberg, NY, USA, 2011; pp. 21–55.
42. Basile, A.; Curcio, S.; Bagnato, G.; Liguori, S.; Jokar, S.M.; Iulianelli, A. Water gas shift reaction in membrane reactors: Theoretical investigation by Artificial Neural Networks model and experimental validation. *Int. J. Hydrogen Energy* **2015**, *40*, 5897–5906. [CrossRef]
43. Agarwal, M. Combining neural and conventional paradigms for modelling, prediction and control. *Int. J. Syst. Sci.* **1997**, *28*, 65–81. [CrossRef]
44. Dang, N.T.; Gallucci, F.; van Sint Annaland, M. Micro-structured fluidized bed membrane reactors: Solids circulation and densified zones distribution. *Chem. Eng. J.* **2014**, *239*, 42–52. [CrossRef]
45. Tiemersma, T.P.; Patil, C.S.; van Sint Annaland, M.; Kuipers, J.A.M. Modelling of packed bed membrane reactors for autothermal production of ultrapure hydrogen. *Chem. Eng. Sci.* **2006**, *61*, 1602–1616. [CrossRef]
46. Shigarov, A.B.; Meshcheryakov, V.D.; Kirillov, V.A. Use of Pd membranes in catalytic reactors for steam methane reforming for pure hydrogen production. *Theor. Found. Chem. Eng.* **2011**, *45*, 595–609. [CrossRef]
47. Israni, S.H.; Harold, M.P. Methanol steam reforming in single-fiber bed Pd-Ag membrane reactor: Experiments and modelling. *J. Membr. Sci.* **2011**, *369*, 375–387. [CrossRef]
48. Sá, S.; Silva, H.; Sousa, J.M.; Mendes, A. Hydrogen production by methanol steam reforming in a membrane reactor: Palladium vs carbon molecular sieve membranes. *J. Membr. Sci.* **2009**, *339*, 160–170. [CrossRef]
49. Ouzounidou, M.; Ipsakis, D.; Voutetakis, S.; Papadopoulou, S.; Seferlis, P. A combined methanol autothermal steam reforming and PEM fuel cell pilot plant unit: Experimental and simulation studies. *Energy* **2009**, *34*, 1733–1743. [CrossRef]
50. Patel, S.; Pant, K.K. Experimental study and mechanistic kinetic modelling for selective production of hydrogen via catalytic steam reforming of methanol. *Chem. Eng. Sci.* **2007**, *62*, 5425–5435. [CrossRef]
51. Islam, M.A.; Ilias, S. Steam reforming of methanol in a Pd-composite membrane reactor. *Sep. Sci. Technol.* **2012**, *47*, 2177–2185.
52. Gallucci, F.; Basile, A. Co-current and counter-current modes for methanol steam reforming membrane reactor. *Int. J. Hydrogen Energy* **2006**, *31*, 2243–2249. [CrossRef]
53. Nair, B.K.R.; Harold, M.P. Hydrogen generation in a Pd membrane fuel processor: Productivity effects during methanol steam reforming. *Chem. Eng. Sci.* **2006**, *61*, 6616–6636. [CrossRef]
54. Fu, C.H.; Wu, J.C.S. Mathematical simulation of hydrogen production via methanol steam reforming using double-jacketed membrane reactor. *Int. J. Hydrogen Energy* **2007**, *32*, 4830–4839. [CrossRef]
55. Fu, C.H.; Wu, J.C.S. A transient study of double-jacketed membrane reactor via methanol steam reforming. *Int. J. Hydrogen Energy* **2008**, *33*, 7435–7443. [CrossRef]

56. Suh, J.S.; Lee, M.T.; Greif, R.; Grigoropoulos, C.P. Transport phenomena in a steam-methanol reforming microreactor with internal heating. *Int. J. Hydrogen Energy* **2009**, *34*, 314–322. [CrossRef]
57. Peppley, B.A.; Amphlett, J.C.; Kearns, L.M.; Mann, R.F. Methanol steam reforming on $Cu/ZnO/Al_2O_3$ catalysts. Part 2. A comprehensive kinetic model. *Appl. Catal. A* **1999**, *179*, 31–49. [CrossRef]
58. Hao, Y.; Du, X.; Yang, L.; Shen, Y.; Yang, Y. Numerical simulation of configuration and catalyst-layer effects on micro-channel steam reforming of methanol. *Int. J. Hydrogen Energy* **2011**, *36*, 15611–15621. [CrossRef]
59. Arzamendi, G.; Die´guez, P.M.; Montes, M.; Centeno, M.A.; Odriozola, J.A.; Gandì, L.M. Integration of methanol steam reforming and combustion in a microchannel reactor for H_2 production: A CFD simulation study. *Catal. Today* **2009**, *143*, 25–31. [CrossRef]
60. Chein, R.; Chen, Y.C.; Chung, J.N. Numerical study of methanol–steam reforming and methanol–air catalytic combustion in annulus reactors for hydrogen production. *Appl. Energy* **2013**, *102*, 1022–1034. [CrossRef]
61. Katiyar, N.; Kumar, S.; Kumar, S. Comparative thermodynamic analysis of adsorption, membrane and adsorption-membrane hybrid reactor systems for methanol steam reforming. *Int. J. Hydrogen Energy* **2013**, *38*, 1363–1375. [CrossRef]
62. Perng, S.-W.; Horng, R.-F.; Ku, H.-W. Numerical predictions of design and operating parameters of reformer on the fuel conversion and CO production for the steam reforming of methanol. *Int. J. Hydrogen Energy* **2013**, *38*, 840–852. [CrossRef]
63. Ribeirinha, P.; Abdollahzadeh, M.; Pereira, A.; Relvas, F.; Boaventura, M.; Mendes, A. High temperature PEM fuel cell integrated with a cellular membrane methanol steam reformer: Experimental and modelling. *Appl. Energy* **2018**, *215*, 659–669. [CrossRef]
64. Saidi, M. Performance assessment and evaluation of catalytic membrane reactor for pure hydrogen production via steam reforming of methanol. *Int. J. Hydrogen Energy* **2017**, *42*, 16170–16185. [CrossRef]
65. Ghasemzadeh, K.; Morrone, P.; Iulianelli, A.; Liguori, S.; Babaluo, A.A.; Basile, A. H_2 production in silica membrane reactor via methanol steam reforming: Modelling and HAZOP analysis. *Int. J. Hydrogen Energy* **2013**, *38*, 10315–10326. [CrossRef]
66. Ghasemzadeh, K.; Morrone, P.; Babalou, A.A.; Basile, A. A simulation study on methanol steam reforming in the silica membrane reactor for hydrogen production. *Int. J. Hydrogen Energy* **2015**, *40*, 3909–3918. [CrossRef]
67. Ghasemzadeh, K.; Andalib, E.; Basile, A. Evaluation of dense Pd-Ag membrane reactor performance during methanol steam reforming in comparison with autothermal reforming using CFD analysis. *Int. J. Hydrogen Energy* **2016**, *41*, 8745–8754. [CrossRef]
68. Taylor, J.R. Automated HAZOP revisited. *Process Saf. Environ. Prot.* **2017**, *111*, 635–651. [CrossRef]

membranes

MDPI

Review

Review of Supported Pd-Based Membranes Preparation by Electroless Plating for Ultra-Pure Hydrogen Production

David Alique, David Martinez-Diaz, Raul Sanz and Jose A. Calles *

Department of Chemical and Energy Technology, Rey Juan Carlos University, C/Tulipán s/n, Móstoles, 28933 Madrid, Spain; david.alique@urjc.es (D.A.); david.martinez.diaz@urjc.es (D.M.-D.); raul.sanz@urjc.es (R.S.)

* Correspondence: joseantonio.calles@urjc.es; Tel.: +34-91-488-7378; Fax: +34-91-488-7068

Received: 11 December 2017; Accepted: 15 January 2018; Published: 23 January 2018

Abstract: In the last years, hydrogen has been considered as a promising energy vector for the oncoming modification of the current energy sector, mainly based on fossil fuels. Hydrogen can be produced from water with no significant pollutant emissions but in the nearest future its production from different hydrocarbon raw materials by thermochemical processes seems to be more feasible. In any case, a mixture of gaseous compounds containing hydrogen is produced, so a further purification step is needed to purify the hydrogen up to required levels accordingly to the final application, i.e., PEM fuel cells. In this mean, membrane technology is one of the available separation options, providing an efficient solution at reasonable cost. Particularly, dense palladium-based membranes have been proposed as an ideal chance in hydrogen purification due to the nearly complete hydrogen selectivity (ideally 100%), high thermal stability and mechanical resistance. Moreover, these membranes can be used in a membrane reactor, offering the possibility to combine both the chemical reaction for hydrogen production and the purification step in a unique device. There are many papers in the literature regarding the preparation of Pd-based membranes, trying to improve the properties of these materials in terms of permeability, thermal and mechanical resistance, poisoning and cost-efficiency. In this review, the most relevant advances in the preparation of supported Pd-based membranes for hydrogen production in recent years are presented. The work is mainly focused in the incorporation of the hydrogen selective layer (palladium or palladium-based alloy) by the electroless plating, since it is one of the most promising alternatives for a real industrial application of these membranes. The information is organized in different sections including: (i) a general introduction; (ii) raw commercial and modified membrane supports; (iii) metal deposition insights by electroless-plating; (iv) trends in preparation of Pd-based alloys, and, finally; (v) some essential concluding remarks in addition to futures perspectives.

Keywords: review; palladium; membrane; Pd alloy; electroless plating; membrane reactor; hydrogen separation; hydrogen production

1. Introduction

The continuous population growth and economy intensification imply an increase of global energy demand. Up to date, this increasing demand has been usually covered by massive use of fossil fuels, causing global warming due to the large emission of anthropogenic greenhouse gases, as well as other combustion pollutants [1]. This situation is even more problematic because the fossil resources worldwide depletion and thus, it represents a clearly unsustainable scenario for the future. In the last years, it has been suggested a wide set of alternatives for the progressive replacement of fossil fuels as primary energy resource [2,3]. In this situation, the implementation of the so-called hydrogen-based economy is considered as a real choice and it is receiving great attention in the last years [4–8].

Hydrogen is advised as a very promising energy carrier due to its long-term viability, high energy density (14 $J \cdot kg^{-1} \cdot {}^{\circ}C^{-1}$), environmentally welcoming combustion emissions and high resources to be produced from. Indeed, hydrogen is the most abundant element in the Earth, although it is usually found combined with other elements, mainly in water and hydrocarbon molecules. The idea is to transfer the energy obtained for different primary energy sources, preferentially renewables (i.e., wind, solar or biomass, among others), to hydrogen, which can be stored, transported and eventually used in different energy applications. Ideally, hydrogen will be obtained from water by using these renewable energies, thus minimizing the environmental impact while the energy demand is covered [9,10]. However, the hydrogen generation by thermochemical processes seems to be a more realistic option in the near future for cost-cutting [11–14]. Hence, hydrogen can be generated from a wide variety of raw materials containing hydrocarbons for both centralized and distributed production systems by using relatively mature technologies [14], being the use of biomass and waste materials especially attractive [15–19]. In these cases, a mixture of gaseous compounds is frequently produced, being necessary to purify the hydrogen up to required levels accordingly to the final application, i.e., PEM fuel cells, turbines or combustion engines [20]. Indeed, the hydrogen purification step is a crucial process in the successful implementation of the hydrogen energy system from both technical and economical point of view.

Among readily available alternatives for hydrogen purification, the use of membranes for hydrogen separation/production applications has been proposed and used in practice. This technology shows relevant advantages such as low energy consumption, environmentally good properties and also additional potential to be combined with a reaction unit in a multifunctional membrane reactor [21,22]. The combination of simultaneous chemical reaction and hydrogen separation in one unique step results in additional benefits in terms of conversion increase by shifting the reaction equilibrium as one of the products, hydrogen, is selectively separated from the reaction media [23,24]. Particularly, dense metallic based membranes have been proposed for years due to their potential to transport hydrogen in a dissociative form with a theoretically complete perm-selectivity [25,26]. Thus, the structure of metals belonged groups III-V, such as Pd, Ni and Pt (in pure form and alloyed), has the ability to allow the hydrogen diffusion through the metal lattice, while avoiding the permeation of other molecules [27,28]. In this way, the solution-diffusion mechanism, depicted in Figure 1, is used to describe the hydrogen permeation process in these H_2-selective membranes.

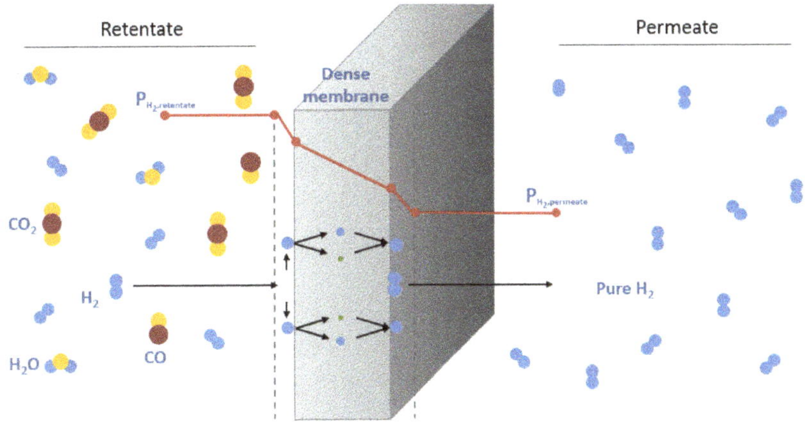

Figure 1. Solution-diffusion mechanism for hydrogen permeation through the metal lattice of a dense membrane.

Up to date, palladium is the most studied metal for preparing these H_2-selective membranes. The earliest studies date from the XIX century, when Deville and Troost discovered the capability of hydrogen to penetrate into bulk palladium [29,30] and Graham determined that this metal was able to absorb hundred times its own volume in hydrogen [31]. However, the use of palladium membranes for hydrogen separation/production applications does not appear until the fifties years. From this decade, as the related research publications evidence, these membranes have been gaining increasing interest. This trend can be observed in Figure 2, where it is shown the number of scientific documents published by year and region considering hydrogen, palladium and membrane or membrane reactor as keywords. It has to be pointed out the increase of related publications during the most recent years, mainly due to a greater awareness on environmental protection and renewable energies development, where hydrogen emerges as a very promising clean energy vector that, as mentioned before, requires to be purified [20,25].

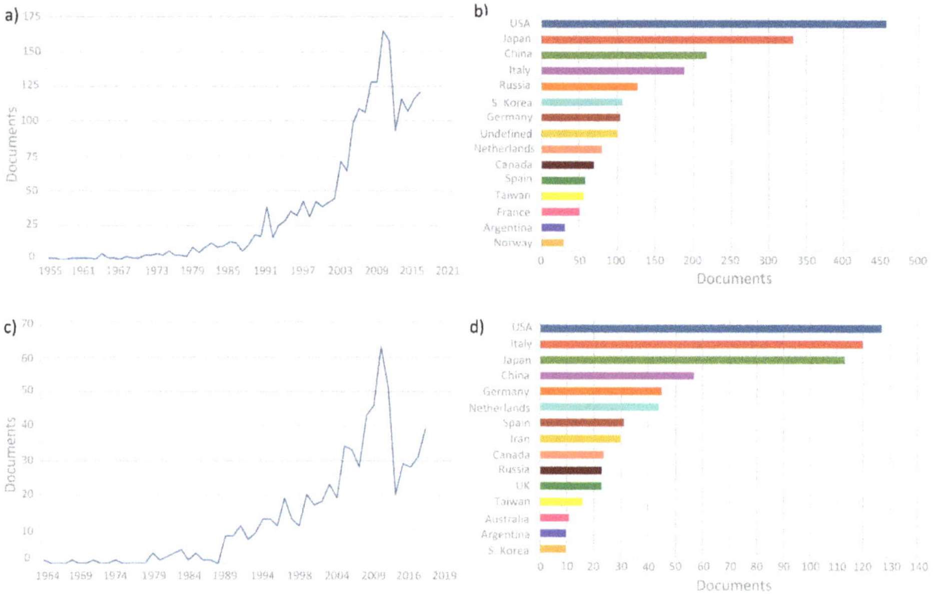

Figure 2. Citation analysis report by Scopus for keywords: palladium + membrane + hydrogen (**a**,**b**) and palladium + membrane reactor + hydrogen (**c**,**d**).

Analysing the number of publications by region, it is evident that this topic is investigated widely around the word, topping the list the United States of America with very ambitious policies but closely followed by different countries of Asia (Japan and China) and Europe (mainly Italy, Germany and The Netherlands).

Currently, main efforts are focused to reduce the cost of these membranes and to increase its mechanical resistance, lifespan and fabrication reproducibility [32,33]. Palladium is expensive and scarce and the growing demand for its use in large-scale applications is expected to keep driving up its price [34]. Two of the most studied strategies to reduce the cost of the membranes are: (a) minimizing the amount of palladium required to achieve a fully dense layer [35–38] and (b) increasing the use life-span since these membranes can suffer deactivation by poisoning and cracking by thermal or mechanical stress [39–43]. Taking into account the typical equation used to describe the hydrogen permeation flux (J_{H_2}) through a Pd-based membrane (Richardson equation, Equation (1)) as function of hydrogen permeability (k), metal thickness (t) and pressure driving force $(P^n_{H_2,ret} - P^n_{H_2,per})$, it is obvious that a decrease in the metal thickness provokes an increase of the permeation capability [28,44].

$$J_{H_2} = \frac{k}{t}(P^n_{H_2,ret} - P^n_{H_2,per}) \tag{1}$$

In case of the Pd-based membrane is totally free of defects, the hydrogen permeation is determined by the solution-diffusion in the bulk metal and the exponential factor takes the value $n = 0.5$, denoting the equation as Sieverts' law.

However, the preparation of ultrathin palladium layers entails two main problems: (i) limitation of membrane mechanical resistance and (ii) difficulty to obtain films free of defects. The use of porous supports tries to overcome these problems and thus, to maintain adequate mechanical properties saving palladium [45–49].

On the other hand, many authors focus their efforts in developing new fabrication processes for ensuring a better reproducibility and reducing the number of rejected membranes [49–51] or modifying the selective layer (Pd-based alloys) in order to improve some particular properties, such as resistance to hydrogen embrittlement or deactivation by sulphur compounds [52–56].

Several technologies can be used to incorporate a thin film of the hydrogen selective metal, preferentially Pd or Pd-based alloy, onto a porous support. Cold-rolling [57–59], physical vapour deposition [60–63], chemical vapour deposition [64–66], electrochemical plating [67–69] and electroless plating can be mentioned [33,62,70,71]. The last option (electroless plating, or its acronym ELP) provides important advantages in terms of adherence and uniformity of deposits on both conducting and non-conducting surfaces with complex geometries. Additionally, it has manufacturing low cost, becoming very popular for most of the studies carried out in the literature [72–75].

Considering all these facts, this review expects to provide a general overview of the most recent and relevant advances for preparation of dense Pd-based membranes for hydrogen production in membrane reactors, particularly focused on supported membranes obtained by electroless plating technology onto inorganic porous supports. The manuscript is divided into different sections focused on: (i) materials, pre-treatments and surface modifications of raw membrane supports, (ii) palladium deposition by electroless plating and (iii) development of new metal Pd-based alloys. Finally, some essential concluding remarks and brief comments about trendy futures perspectives have been also included.

2. Membrane Supports

Dense Pd-based membranes can be classified in two main groups, unsupported and supported ones, in which a thin selective film is deposited onto a porous substrate. The first type is usually prepared from relatively thick palladium (or Pd-based alloys) foils that, as Tosti et al. indicate in case of requiring tubular geometry, are cold-rolled and welded [17,76]. Typical thicknesses are ranged from 50 to 150 µm. However, as previously mentioned, a thick Pd layer strongly hinders both hydrogen permeate rate and membrane cost. Thus, development of new ultrathin membranes without jeopardizing mechanical resistance and presence of defects is the main objective of many researchers in this field [59,60,77]. This goal is usually achieved by incorporating a thin Pd layer on the surface of a porous material that provides the required mechanical resistance to the supported membrane [71,78–80]. This complex task is subject of numerous studies since many factors must be considered, i.e., the compatibility between support and selective layer, which strongly determines the mechanical resistance of the membrane due to cracks can be formed at high temperatures because of different expansion coefficients, as it will be discussed in detail later.

Numerous porous materials, such as Vycor glass [81,82], sintered metals [71,78,83], a wide variety of ceramics [53,71,84,85] and even polymers [86–88], can be used as supporting materials for the H_2-selective layer. The most relevant attributes of supports to be selected include porosity properties (mainly average porosity and pore sizes distribution), surface roughness and mechanical, chemical and thermal stabilities [89]. In this context, it is expected great porosity with a narrow distribution of small pore sizes, high mechanical strength and chemical resistance and similar thermal expansion coefficient to that of Pd [90]. In regards of textural properties, the support porosity needs to be open and interconnected enough to ensure a non-limiting gas transport through the support, besides the critical sizes of pore-mouth and of pore-throat [91]. It is accepted that usually both pore size and roughness strongly determine morphology and continuity of the selective layer. In this mean, Mardilovich et al. [92] indicated that the minimum thickness necessary to prepare an electroless plated Pd film onto a porous support is around three times the average size of the greatest pores. Despite Vycor glass was one of the first porous supports used to incorporate Pd by electroless plating [81,82], currently is more frequent the use of sintered porous metals [71,78,93,94] or ceramic materials [53,71,84,95], making a clean sweep on the majority of scientific publications in this field.

The use of porous polymers as substrates for membranes reactors, which usually operates a high temperature, is currently scarce due to the low thermal resistance of these materials [96]. For this particular application, the metallic supports are the preferred ones, such as stainless-steel 316L [71,83], Hastelloy [97,98], Inconel [94], nickel [99] or, in some particular cases, Ti-based alloys such as Ti-Al [100] or Ni-Ti [101]. They usually ensure good mechanical properties, hardness and adequate thermal expansion coefficient, similar to that of palladium, in the range of 10.5–12.5 $\times$ 10^{-6} °C^{-1}. Moreover, these materials are easily sealed and coupled to membrane reactor modules, conventionally made of stainless-steel [96]. However, these supports present relatively large pores with a wide pore sizes distribution that makes the generation of a thin and free-defect Pd layer difficult. In fact, it is usual that manufactures do not provide the concrete value of pore sizes in these supports, giving an average related value, known as media grade, that represents the particle size that is rejected in a 95% for a filtration process with this support [62]. Moreover, it is also possible that metal inter-diffusion between support and Pd-based selective layer takes place after operating the membrane at high temperatures for long times. This phenomenon causes a marked decrease in the permeation capacity [89]. To overcome both drawbacks, the original support has to be modified prior to the incorporation of the H_2 selective layer, as we detail in following sections [89,93,99].

On the other hand, ceramic supports provide a smoother surface with accurate control on porosity and narrow pore sizes distributions up to a few nanometres [38]. These properties facilitate the deposition of defect-free palladium layers with really low thickness and many researchers choose to use them as support for membranes [41,53,84]. Among some possibilities, the use of alumina, Al_2O_3 [102,103], is predominant, usually combining both α-Al_2O_3 and γ-Al_2O_3 particles in order to prepare asymmetric supports with big pores in the core to ensure greater permeabilities and smaller ones on the top layer to facilitate the palladium incorporation [71]. However, this material presents a thermal expansion coefficient noticeable different to that of palladium, besides a weak mechanical resistance, jeopardizing the integrity of the supported membrane, which is quite important on membrane reactors [91]. Other alternative less frequent is the use of yttria-stabilized zirconia (ZrO_2-YSZ), with closer thermal expansion coefficient to that of palladium (10.0 $\times$ 10^{-6} °C^{-1}), to prepare ceramic supports [85,96]. Anyway, different metallic or ceramic supports can be used to prepare totally dense supported Pd-based membranes, although a prevalent solution is still not reached. The advantages provided for ceramic ones are problems when using metal supports and *vice versa*, so different trends can be observed in literature. Some authors lean towards ceramic supports, mainly formed by alumina, in order to ensure the incorporation of an ultrathin Pd-based layer without defects, focusing on the membrane preparation, while other ones prefer to use the metallic supports thinking on real application of membranes in stainless-steel industrial devices.

Independently of constituent material of supports, the geometry is also important and quite a few configurations can be found in the literature, distinguishing mainly planar [100,101], tubular [71,78] and hollow fibre geometries [104,105]. In general, tubular geometries of both ceramics and metallic materials are prevalent in case of considering the use in a membrane reactor, while porous metals with planar geometry are most frequent in case of studying the membrane preparation with only purification purposes [106–108]. However, this situation has changed in the last years with the appearing of plate-type geometries and hollow fibres in design of trending reactor systems with micro-channels [109].

Table 1 summarizes the most frequent inorganic supports present in the literature for Pd-based membrane preparation, indicating important parameters such as material, geometry, average porosity and illustrative pore sizes. Several relevant manufacturers around the world have been considered, such as Mott Metallurgical Corp. (USA), Pall Corp. (USA), GKN Sinter Metal (UK), Inopor GmbH (Germany), TAMI industries (France) or NGK Insulators Ltd. (Japan). Currently, lower prices can be achieved for ceramic supports, despite they present smaller pore sizes than metallic ones, although their reutilization is not easy due to the frequent breaking during operation.

Table 1. Usual inorganic commercial supports for Pd-based membrane preparation.

Company	Material	Geometry	Thickness (mm)	Porosity (%)	Pore Size (nm)
Mott	Stainless steel: 304 L, 316 L, 310, 347, 430	Disc, sheet, cup, tube	1–3		$0.1–100 \times 10^3$ (media grade)
	Hastelloy: C-22, C-276, X, N, B, B2 Inconel: 600, 625, 690				
GKN	Stainless steel: 304 L, 316 L, 904 L, 310	Disc, tube	1.5–3		$0.1–200 \times 10^3$ (media grade)
	Hastelloy: C-22, C-276, X Inconel: 600, 625 Monel: 400 Bronze Titanium				
Pall	Stainless steel: 304 L, 316 L, 310 SC	Cup, tube	- [(a)]		$>0.1 \times 10^3$ [(a)] (media grade)
	Hastelloy: X Inconel: 600 Monel: 400 SiC/Al_2O_3 Mullite				
Inopor	α-Al_2O_3	Tube, multichannel tube	-	40–55	70–800
	TiO_2			40–55	100–800
				30–55	5–30
				30–40	1
	ZrO_2			40–55	110
				30–55	3
	γ-Al_2O_3			30–55	5–10
	SiO_2			30–40	1
Tami	TiO_2/ZrO_2	Tube, multichannel tube	2		4.5×10^3 [(b)]

[(a)] On request, [(b)] Ultrafiltration grade with ZrO_2 active layer (15 kg/mol).

As previously mentioned, it is not common the direct use of commercial supports to prepare supported membranes, especially in case of metallic substrates. On the opposite, it is usual to carry out some pre-treatments and surface modifications of the support to improve the final quality of the membrane. In addition to conventional initial cleaning procedures, most of these modifications are focused on the improvement of layers adherence and/or the reduction of average pore sizes and roughness in the support surface to achieve thinner hydrogen selective layers. These treatments can be classified in three general categories: (i) chemical treatment, (ii) physical treatment and (iii) incorporation of an intermediate layer. Considering the great importance of these steps on the final properties of the membrane and its costs, some of the most relevant advances and extended practices are summarized in the next paragraphs. We have focused on metallic supports since, as mentioned above, they are the most suitable to use in membrane reactors to hydrogen production, which is the aim of the review. Moreover, it should be noted that the external surface of ceramic supports is not usually modified prior to deposit the selective layer due to the good original properties in terms of average pore diameter and surface roughness.

2.1. Chemical Treatment

The use of chemicals to modify the surface of supports is known as etching. It is commonly applied for polymers but it can also be used to modify some original properties of inorganic materials. These treatments consist of dipping the support in a corrosive solution, traditionally a strong acid and maintaining it at controlled temperature for a short period of time. The main effect of these treatments is to dissolve oxides thin films formed on the top of the supports, being also possible to remove part of the support bulk material. This action is primary determined by support composition, acid concentration, temperature and time of the treatment. Mardilovich et al. [110] used a solution of hydrochloric acid to treat a commercial stainless-steel support, achieving a noticeably increase of roughness on the surface of stainless-steel particles that form the support after only 5 min. of immersion. Moreover, the new treated surface evidenced better properties for the subsequent palladium incorporation as a pre-activated surface, increasing the plating rate and improving the adherence. A similar treatment was reported by Li et al. [111] but mixing the hydrochloric acid with some amount of nitric acid and Kim et al. [101] for preparing a supported Pd membrane over a porous nickel support. In this way, the etching pre-treatment of an inorganic support can provide benefits for the pre-activation of support surface and adherence of the selective layer at a relatively low cost, independently of using any other additional treatment such as mechanical modification or the incorporation of an additional layer.

2.2. Mechanical Treatment

A different alternative to modify original supports, mainly the metallic ones, can be carried out by the polishing of the external surface. The plasticity of the metal particles that form the support is used to reduce both external pore size and roughness through a mechanical treatment with an abrasive material. One of the first references about the use of this alternative to prepare Pd supported membranes, published by Jayaraman et al. in the nineties, utilized commercial sandpapers with different grit numbers to smooth the original surface of the support [112]. Particularly, they used commercial sandpapers with grades #320, #500 and #800. Later, Mardilovich et al. [110] used a similar polishing process with commercial sandpapers to modify the surface of porous stainless-steel supports. They indicated that it was possible to reduce both external average pore size and roughness, although most porosity was lost, decreasing the permeation capacity of the modified support up to 20% of the untreated one. Most recently, a similar technique based on the use of an abrasive sandpaper has been reported in the literature, as evidence the works published by Li et al. [111], Ryi et al. [113] or Pinacci et al. [114]. This polishing technique has not been only proposed for modifying the surface properties of supports, being also possible the reparation of defects in palladium thin films of supported membranes by mechanical treatments [115]. Despite this type of mechanical treatment is the predominant one, it is also possible to find some work in which high velocity shot peening with ion particles is used to achieve the plastic deformation of the metal particles of the support. However, the high cost of this alternative makes the traditional abrasive ones prevalent [116].

However, some researchers have critical opinions about these mechanical treatments due to the reduction of the permeation capability and adhesion properties of the thin selective layer, which in turn constrain the performance of the supported membrane. In this context, it is accepted that the adhesion between a support and a thin selective layer depends on the mechanical binding and anchoring effects. Consequently, it is necessary a minimal support roughness for ensuring a good adhesion of the top coatings [117,118]. This is clearly indicated in works published by Collins [119] and Huang [120], where larger pores and a certain external roughness in supports improve the adhesion of the thin coating layer. In this manner, it can be stated that it is necessary to achieve a compromise solution between the original surface modification and maintaining certain anchoring points to guarantee a suitable adherence of the Pd selective layer.

2.3. Incorporation of Intermediate Layers

In spite of using chemical and/or mechanical treatments, the incorporation of an intermediate layer between the commercial support and the top selective layer is the most preferred alternative to improve the external surface of the support. This option can be used simultaneously for different objectives such as modification of the original morphology, metal inter-diffusion mitigation, adhesion improvement of the Pd selective layer, corrosion prevention of support or even incorporation of first metal nuclei as pre-activated surface. The last one is usually the main reason to incorporate intermediate layers on ceramic supports due to these materials usually present a very smooth surface with very narrow pore sizes distribution, up to 3 nm [71,121] without need of additional modifications. However, metallic supports display a typical rough surface and wide pore sizes [62,71], being the incorporation of an intermediate layer a critical issue to achieve a really thin palladium layer. Considering the final target of the intermediate layer, its composition and thickness need to be adjusted at a reasonable cost and, up to now, a unique solution is not reached.

Anyway, one of the most important factor to be considered is the compatibility between the different components of the supported membrane. Figure 3 shows the thermal expansion coefficient for some of the materials most frequently used as intermediate layer, besides common metal support raw materials (316L stainless-steel or Hastelloy X) and selective layer constituents (mainly palladium, silver, copper and gold).

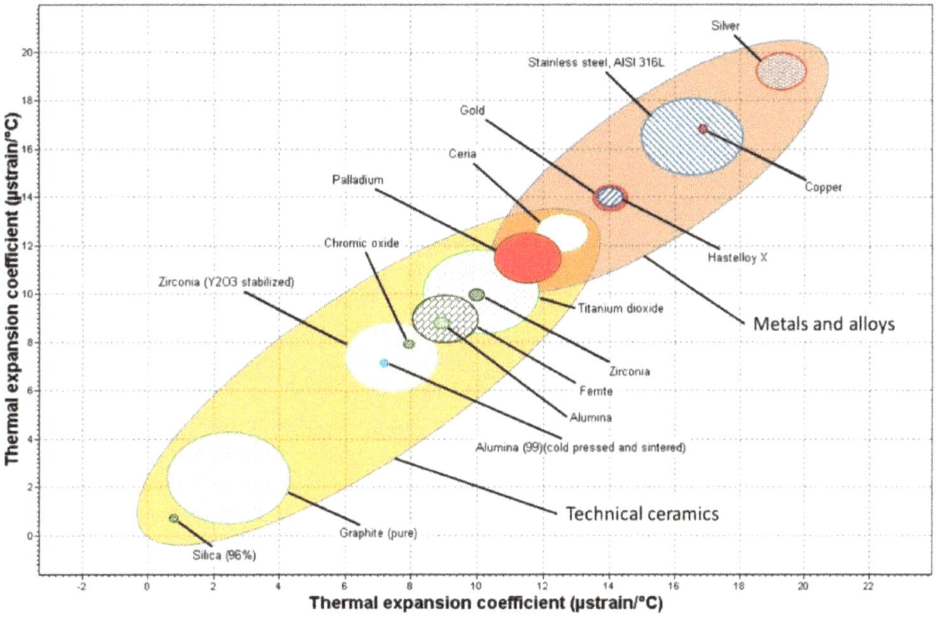

Figure 3. Thermal expansion coefficients for typical constituents of supported membranes for hydrogen separation.

At large, small differences between the thermal expansion coefficients of the supported membrane elements are recommended to ensure enough mechanical resistance at operating conditions, usually at moderate or high temperatures. According to the data shown in Figure 3, cerium oxide appears as a very attractive alternative, with a thermal expansion coefficient between of palladium and common metallic supports (i.e., AISI 316L stainless-steel or Hastelloy X). This material was employed by Tong et al. [122] to modify a macro porous stainless-steel tubular support to prepare a supported Pd membrane with a selective layer of around 13 μm thickness. They evidenced a really good stability of the supported system after long-term experiments, obtaining almost equal hydrogen permeability to the theoretical value for a pure Pd membrane. A similar intermediate layer of CeO_2 was prepared by Qiao et al. [123] to prevent intermetallic diffusion between a PSS support and a PdCu alloy selective layer. This intermediate layer was prepared through a sol-gel method and the modification of the original support also improved the adherence between the metallic support and the selective layer.

In addition to CeO_2, many other materials have been also successfully incorporated as an effective intermediate layer, even though they present different thermal expansion coefficient to that of the selective layer or the support. A first relevant group is formed by zirconium oxide and related materials. Some authors, such as Wang et al. [124] or Gao et al. [125], modify commercial PSS supports by the incorporation of ZrO_2 particles to reduce the thickness of the hydrogen selective layer up to around 10 μm. A similar thickness was achieved by Tarditi et al. [93] by the incorporation of the ZrO_2 particles through a vacuum-assisted method, while Lee at al. [126] reduced this thickness up to 3.5 μm for a better permeability. Other researchers added small amounts of yttria to the based-zirconia material in order to increase the structure stability of the material, obtaining an yttria-stabilized zirconia (YSZ) [50,117]. References in literature present the use of YSZ as an effective intermediate layer with the double aim of reducing the palladium thickness and preventing the intermetallic diffusion between support and selective layer, indicating sol-gel methods or atmospheric plasma spraying as successful techniques for the material incorporation [70,78,127].

Considering the relatively good surface properties of alumina as support, the use of this material as intermediate layer to modify the metallic supports has been also proposed by different authors. In this way, Yepes et al. [128] and Li et al. [129] decreased the original pore size of the metallic support by incorporating an alumina top layer that prevents possible inter-diffusion processes between the original support and the selective layer. Broglia et al. [130] reported the incorporation process of γ-Al_2O_3 particles by dip-coating onto a PSS support to achieve a totally defect-free Pd layer of around 11 μm. Chi et al. [131] detailed the use of different graded alumina particles for a better modification of commercial PSS tubes. They used particles with a size close to 10 μm to fill the widest pores and smaller particles (size around 1 μm) for a final smooth of the surface. Thus, they eventually achieved a thin free-defects Pd layer with less than 5 μm in thick and good thermal stability. Lee et al. [126] compared the effect of using Al_2O_3 and ZrO_2 technical ceramics with similar thickness as support modifiers and they indicated that both materials act effectively as diffusion barrier, although the use alumina yield a lower membrane permeability.

Other conventional material used as intermediate layer is the SiO_2, being possible to accomplish different functions such as surface support modifier, intermetallic diffusion limiter, perm-selectivity booster or even catalyst for some chemical processes. For instance, Nam et al. [132] modify a commercial 316L stainless-steel substrate by the incorporation of amorphous silica. In this way, they reduced the selective layer, constituted by a PdCu alloy, up to 2 μm but maintaining an excellent separation behaviour with hydrogen permeance of 8.37×10^{-7} $mol \cdot m^{-2} \cdot s^{-1} \cdot Pa^{-1}$ and H_2/N_2 selectivity of around 70,000 at 450 °C. Calles et al. [62] published the use of three different siliceous materials as intermediate layer for preparing supported Pd-PSS membranes: amorphous disordered silica, amorphous ordered silica (HMS) and crystalline silica (silicalite-1). In all cases, both roughness and pore size of the original supports were reduced and, consequently, the minimum Pd thickness required to obtain a defect-free selective membrane. The best results were obtained for the silicalite-1 material, reducing the Pd thickness up to 5 μm and yielding a hydrogen permeance of 1.423×10^{-4} $mol \cdot m^{-2} \cdot s^{-1} \cdot Pa^{-0.5}$ with a complete hydrogen selectivity at 400 °C. Similar modifications of metallic supports with microporous silica layers can be found for increasing the H_2 perm-selectivity of the composite without any other additional layer [133] or even combined with palladium in a mixed-matrix structures [134]. Recently, these materials have been also applied on the top of finished supported Pd membranes in order to repair small defects and pinholes, significantly increasing the H_2 selectivity with a very low cost [135].

Materials that combine silica and alumina are the well-known zeolites, crystalline materials with controlled pore sizes distribution and additional catalytic properties. Among the wide variety of possible structures, the use of zeolites NaA [136], NaX [137], Z-21 [138], FAU-type [139] and TS-1 [140,141] as effective intermediate or protective layers in membrane preparation can be found in the literature. On the whole, the higher cost of these materials limits their use to very specific processes, mainly for membrane reactors in which undesirable products are presented and the zeolite plays the role of both support modifier and catalyst.

Other method quite simple to modify the support, with high reproducibility and reasonable cost, is the direct oxidation of 316L PSS supports in air atmosphere at high temperatures. This process yields a top coating of mixed Fe_2O_3-Cr_2O_3, which is able to prevent the inter-diffusion process [78]. Ma et al. [142] patented a controlled in-situ oxidation method to prepare Pd composite membranes over porous stainless-steel supports and thus, they achieve effective inter-diffusion barriers with thermal treatments upper than 600 °C. Following this pioneering work, other researchers such as Guazzone et al. [143] or Mateos-Pedrero et al. [144] modified PSS supports by the incorporation of metal oxides derived from an oxidation process at temperatures higher than 400 °C. Mostly, only slight modifications on the support surface can be observed after the thermal treatments due to the very limited thickness of the new oxide layer and, consequently, the Pd thickness is not reduced as much as when other alternatives are used. In case of using really high temperatures for the treatment (>700 °C), more oxides are generated, although in that cases the original porosity of the support drastically drops.

In the last years, some other materials have been investigated to develop more efficient intermediate layers and achieve better supported membranes. Some of these new materials are thin TiN thin layers obtained by sputtering [145], a combination of silver as diffusion barrier and aluminium hydroxide gel for filling in the biggest pores of the support [146], bi-metal multi-layers formed by staked layers of Pd and Ag [147], nickel [148] or even tungsten powders [47]. However, despite these promising results, a definitive solution has not yet been found.

One original alternative consists of using a temporary material to make the incorporation of the selective Pd layer easier. For instance, Tong et al. reported this methodology for the first time, employing an aluminium hydroxide gel or a polymer to modify the top surface of a PSS support. Then, they deposited the Pd layer over the modified surface and, finally, the temporary intermediate layer was removed in order to recover the original pores of the support [149,150]. Figure 4 collects the main steps carried out during this attractive method. Following this procedure, the authors prepare membranes with around 5 μm of palladium thickness that exhibited a maximum hydrogen permeation flux of 0.82 $mol \cdot m^{-2} \cdot s^{-1}$ with infinite hydrogen selectivity at 600 °C and ΔP 200 kPa.

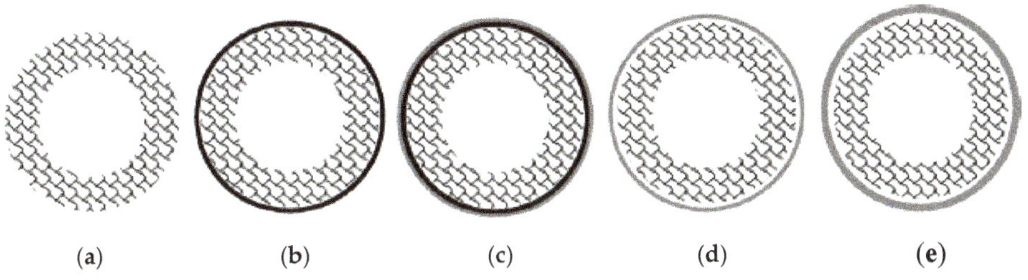

Figure 4. Use of a temporary intermediate layer for the preparation of a Pd-composite membrane: (**a**) original support; (**b**) polymer + support; (**c**) Pd layer + polymer + support; (**d**) Pd layer + small gap + support; and (**e**) defect-free Pd layer + small gap + support [150], with permission from © Elsevier.

Finally, despite the presence of an intermediate layer in ceramic substrates is less common, some examples can also be found in the literature. For instance, the work published by Hu et al. [151], in which a low-cost macroporous Al_2O_3 support is modified with graphite and clay from a conventional

2B-pencil. With this method, they achieved a totally defect-free supported membrane with a palladium thickness of 5 μm. In spite of the incorporation of intermediate layers onto the ceramic supports prior to incorporate the final selective coating is scarce, is possible to found some works that use this alternative to improve the surface activation, as published by Zhao et al. [152]. They used a Pd(II)-modified boehmite sol for modifying the original surface and achieved a thickness of the selective layer of only 1 μm. A very particular application of this methodology is the synthesis of pore-filled membranes, in which YSZ particles are used to modify the original surface of ceramic supports in a double layer. The aim is to get a good adhesion and uniform coating of the membrane film onto the support, as well as create a barrier that plays as protection of the Pd-selective layer [50]. More details about this alternative can be found in the next section, talking about recent developments for improving the metal deposition processes via Electroless Plating.

The morphology of the external surface of a typical commercial metal support and its modification after the incorporation of some of the previously described materials as intermediate layer are collected in Figure 5. As it can be seen, the original PSS surface is practically covered after the incorporation of the different materials, obtaining a very homogeneous external surface while surface roughness and original pore sizes are significantly decreased.

The most relevant information about the wide alternatives included in this section to modify commercial raw supports has been also summarized in Table 2. Support nature and modification alternatives are collected, as well as other relevant parameters such as composition and thickness of H_2 selective layer and permeation properties of the final supported membrane.

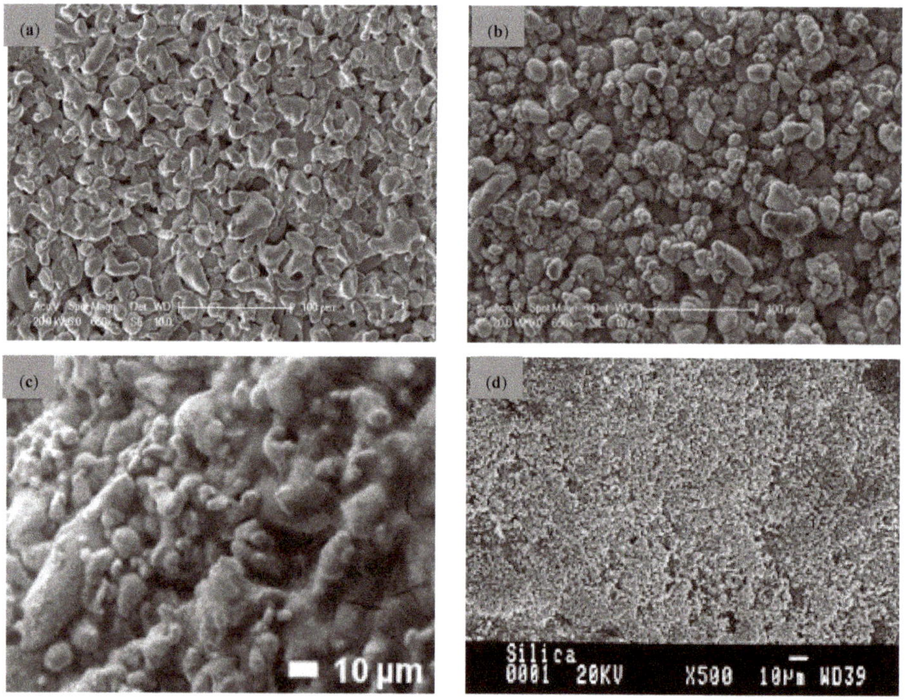

Figure 5. *Cont.*

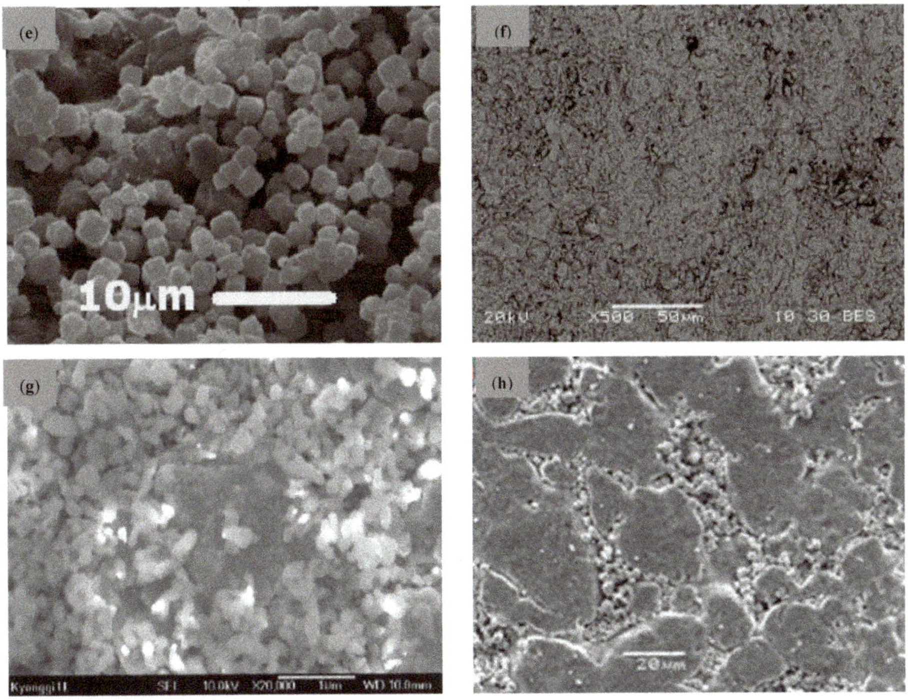

Figure 5. Porous stainless-steel supports before (**a**); and after the incorporation of different materials as intermediate layer: mixed oxides by calcination in air (**b**); alumina (**c**); amorphous silica (**d**); zeolite (**e**); zirconia (**f**); ceria (**g**); and tungsten (**h**). Figure adapted from originals published in [47,51,78,99,132,136,153], with permission from © Elsevier.

Table 2. Inorganic commercial supports for supported Pd-based membrane preparation.

Support	Modification Alternative	Particular Details	Selective Layer	Tselective Layer (m)	Permeation Conditions		Permeation Capacity	H_2 Separation Factor	Ref.
					T (°C)	P (kPa)			
PSS	Chemical treatment	HCl, 5 min.	Pd	20.0	350	100	3.11×10^{-4} (a)	5000	[110]
PSS	Chemical treatment	HCl-HNO_3 mixture	Pd	5.0	450–550	100	3.24×10^{-1}–4.34×10^{-1} (c)	n.a.	[111]
Ni	Chemical treatment	HCl	Pd	0.3	450	100	1.44×10^{-1} (c)	1600	[113]
Al_2O_3	Mechanical treatment	Sandpapers: #320, #500 and #800	Pd	0.5	n.a.	n.a.	n.a.	n.a.	[112]
Ni	Mechanical treatment	Sandpapers: #1200	PdCuNi	12.0	350–500	138–276	1.30×10^{-7}–3.80×10^{-7} (b)	∞	[113]
PSS	Mechanical treatment	Ion shot penning	Pd	6.0	400	100	5.80×10^{-2} (c)	n.a.	[116]
PSS	Permanent Intermediate layer	CeO_2 particles	Pd	13.0	550	200	2.75×10^{-1} (c)	∞	[122]
PSS	Permanent Intermediate layer	CeO_2, sol-gel	PdCu	8.0	450	100	74.00 (a)	2369	[123]
PSS	Permanent Intermediate layer	ZrO_2, sol-gel		10.0	500	100	8.30×10^{-2} (c)	n.a.	[124]
PSS	Permanent Intermediate layer	ZrO_2, sol-gel	PdCu	10.0	480	100	1.10×10^{-7} (b)	∞	[125]
PSS	Permanent Intermediate layer	ZrO_2, sol-gel, vacuum assisted method	PdAu	10.0	400	100	1.10×10^{-3} (a)	>10,000	[93]
PSS	Permanent Intermediate layer	YSZ particles	Pd	27.7	350–450	30–400	4.50×10^{-4} (a)	∞	[70]
PSS	Permanent Intermediate layer	YSZ particles	Pd	13.8	350–450	0–250	4.10×10^{-5}–4.10×10^{-4} (a)	∞	[78]
Hast X	Permanent Intermediate layer	YSZ–Al_2O_3/YSZ	PdAg	4.0–5.0	400–600	100	100.00×10^{-8} (b)	>200,000	[98]
PSS	Permanent Intermediate layer	γ-Al_2O_3, dip-coating	Pd	11.0	n.a.	n.a.	n.a.	n.a.	[130]
PSS	Permanent Intermediate layer	Graded Al_2O_3 particles	Pd	<5.0	500	n.a.	2.94×10^{-3} (a)	1124	[131]
PSS	Permanent Intermediate layer	SiO_2 particles	PdCu	2.0	450	n.a.	8.37×10^{-7} (d)	70,000	[132]
PSS	Permanent Intermediate layer	Silicalite-1, sol-gel and dip-coating	Pd	5.0	350–450	50–250	1.42×10^{-4} (a)	∞	[62]
PSS	Permanent Intermediate layer	Zeolite NaA	Pd	19.0	450	50	1.10×10^{-3} (a)	608	[136]
PSS	Permanent Intermediate layer	Zeolite FAU-type	Pd	1.0	200	100	1.20×10^{-4} (a)	n.a.	[139]
Al_2O_3	Permanent Intermediate layer	Zeolite TS-1	Pd	2.0	350–450	50–500	1.48×10^{-1} (c)	148	[141]
PSS	Permanent Intermediate layer	Fe_2O_3-Cr_2O_3, oxidation in air (T = 600 °C)	Pd	33.0	300	n.a.	2.66×10^{-4} (a)	n.a.	[143]
PSS	Permanent Intermediate layer	Fe_2O_3-Cr_2O_3, oxidation in air (T = 600 °C)	Pd	19.0	n.a.	n.a.	n.a.	n.a.	[144]
PSS	Permanent Intermediate layer	Tungsten particles	PdCu	5.0–20.0	n.a.	n.a.	n.a.	n.a.	[47]
PSS	Temporary intermediate layer	Aluminum hydroxide gel/polymer	Pd	5.0	600	200	3.50×10^{-3} (a)	∞	[150]
Al_2O_3	Permanent Intermediate layer	Graphite-Clay (from 2B pencil)	Pd	5.0	450	100	3.10×10^{-1} (c)	3700	[151]
Al_2O_3	Permanent Intermediate layer	Pd(II)-modified bohamite sol	Pd	1.0	450	n.a.	2.23×10^{-2}–1.07 (c)	20–130	[152]
Al_2O_3	Permanent Intermediate layer	YSZ particles	Pd	5.0	150–500	150–400	0.10–0.60 (c)	n.a.	[50]

Permeation capacity: (a) Permeance ($mol \cdot m^{-2} \cdot s^{-1} \cdot Pa^{-0.5}$), (b) Permeance ($mol \cdot m^{-2} \cdot s^{-1} \cdot Pa^{-1}$) or (c) Permeation flux ($mol \cdot m^{-2} \cdot s^{-1}$), n.a.: non available.

3. Palladium Incorporation by Electroless Plating

3.1. Electroless Plating Standard Method

The term Electroless Plating (ELP) was coined for the first time in the middle forties by Brenner and Riddell to define the metal deposition in the absence of an external source of electric current [154]. The application of ELP technology to the palladium incorporation on porous supports has been widely used to prepare hydrogen selective membranes for years. As it is previously mentioned in the introduction, this technique does not require any expensive equipment and neither high operational costs due to the absence of electrodes and external electricity sources. Moreover ELP is able to create homogeneous films on complex geometries and non-conducting materials [111,155–157], being the option usually preferred over other methods. Here, a general description of the method is presented, including the most relevant advances carried out during the last years in case of using both metallic and ceramic supports.

Essentially, the use of ELP for the preparation of H_2-selective membranes is based on the palladium deposition (or related alloying materials, as it is discussed later) onto a support target surface from an aqueous solution containing the metal precursor. Usually, this precursor is dissolved and stabilized with some ligand in order to form a complex prior to be reduced through a controlled autocatalytic chemical reaction [158,159]. In recent years, most published manuscripts use ammonium hydroxide and ethylenediaminetetraacetic acid to complex the palladium precursor. On the other hand, hydrazine is preferred as reducing agent due to the generation of nitrogen as unique by-product of the chemical reaction, avoiding other prohibited deposits in the film, i.e., phosphorous [157,159,160]. Hydrazine is a powerful reducing agent in both acid and alkaline media. The reduction of higher valent metal ions to lower metal ones or to the zero valent state is possible depending on the reaction conditions [161–163]. Hereunder, the main chemical reactions involved in the process for palladium deposition are:

$$\begin{array}{l} 2Pd(NH_3)_4^{2+} + 4e^- \rightarrow 2Pd^0 + 8NH_3 \ E^0 = 0.95\ V \\ N_2H_4 + 4OH^- \rightarrow N_2 + 4H_2O + 4e^- \ E^0 = 1.12\ V \end{array} \quad (2)$$

being the global reaction of the process as follows:

$$2Pd(NH_3)_4^{2+} + N_2H_4 + 4OH^- \rightarrow 2Pd^0 + 8NH_3 + N_2 + 4H_2O \ E^0 = 2.07\ V \quad (3)$$

In order to achieve a homogeneous Pd deposition, good adherence and reasonable induction times to spontaneously initiate the chemical reactions, the supports need to be seeded with a first nano-sized Pd nuclei before the main plating step [164]. Conventionally, this step has been carried out by repetitive immersions in acidic tin and palladium solutions, also known as sensitization-activation treatment [165]. However, some studies advise problems in membrane stability at high operating temperatures caused by tin residues, which lead to the formation of defects and pinholes in the Pd film, as Paglieri et al. indicated for the first time in the late nineties [166] and other authors endorsed most recently [167]. A detailed study about the correlation between presence of tin residues and membrane stability has been lately published by Wei et al. [167]. Considering these negative effects of classical sensitization-activation treatments, alternative methods avoiding the use of tin solutions have been proposed. Different approaches have been used, such as, the use of activated particles with Pd nuclei for intermediate layers preparation [125,127,146,168,169]; catalysed anodic alumina surfaces to facilitate the Pd electroless plating [170]; the increase of nuclei deposition rate and rupture of conglomerates by applying ultrasounds [171]; the incorporation, decomposition and reduction of a palladium acetate solution in chloroform onto the surface [172]; or, directly, the generation of nano-sized Pd particles by direct reduction of a highly diluted solution with a mixture of ammonia-hydrazine [51,71]. However, up to date it has not been found a better solution and the classical method remains as the top choice for many researchers [28,46,161].

3.2. Recent Developments in Electroless Plating

In last years, great efforts have been carried out to reduce the overall cost of Pd-based membranes manufacturing, mainly focusing in reducing the palladium layer thickness but ensuring absence of defects in the coating [35,37,38,173]. As previously commented, one strategy is based on the preparation of supported membranes, which usually involves the modification of raw supports to facilitate the incorporation of an ultrathin free-defects palladium layer [21,62]. Other strategies are focused to improve the metal deposition process, particularly the electroless plating, to achieve better adherence, homogeneity, greater pores coverage or, in a general way, better stability of the Pd-based H_2-selective layer with minimum thickness. Along following lines, the most relevant developments in this context are presented, particularly focusing on the advances published in recent years.

Thereby, Uemiya et al. [174] increased the metal incorporation rate by immersing the porous support in a solution containing hydrazine prior to each electroless plating step. Other authors tried to improve the palladium incorporation in deep areas on the surface, where deposited metal particles effectively close the pore mouths of the support and complete a fully dense and continuous layer by the bridge mechanism [175]. Zhao et al. [152] and Zhang et al. [176] reported the use of vacuum in the inner size of the supports to achieve uniform microstructure of Pd layer with a lower average thickness to that of conventional electroless plating. Similar results were obtained by other researchers, such as Yeung [177], Souleimanova [178] or Li [179] when osmotic effect is generated between the plating solution and an aqueous sucrose solution.

Pacheco Tanaka et al. [50,52,180] went a step further by preparing supported membranes in which the incorporation of Pd or Pd-based alloys was carried out by vacuum-assisted electroless plating between two zirconia oxide ceramic layers, one of them activated with a previous Pd seed, deposited onto a tubular alumina support. This particular kind of membranes in which the selective layer is placed into a sandwich-type structure was denoted as pore-filled type membranes. The main advantages outlined by the authors includes the ability to operate the membrane below the critical temperature and to maintain a fully mechanical stability, unlike other supported membranes based on a conventional external coating, where fatal damages usually occur. Moreover, the sandwich structure also provides to the selective layer an additional protection against poisoning. Recently, a clear scheme describing in detail the fundamentals of this alternative has been reported by Arratibel et al. [90], as shown in Figure 6 (with permission).

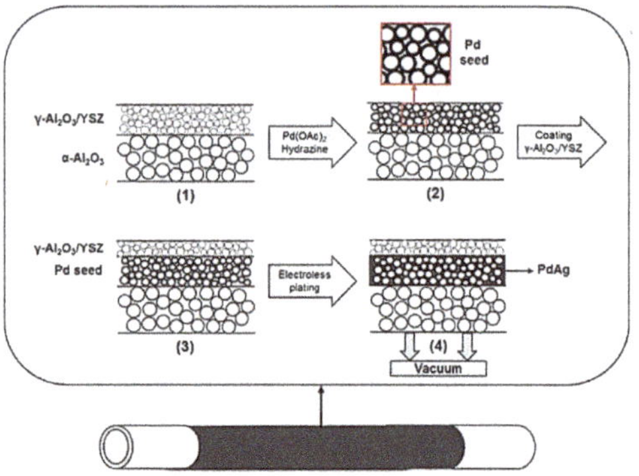

Figure 6. Procedure to prepare pore-filled type membranes [90] with permission from © Elsevier.: (1) Incorporation of a first γ-Al_2O_3/YSZ layer; (2) Pd seed on smaller ceramic particles; (3) incorporation of a top additional γ-Al_2O_3/YSZ layer and (4) incorporation of a Pd-based layer by vacuum-assisted ELP.

On the other hand, different studies are focused on modifying the plating baths composition to improve the final properties of the palladium film. In this context, it has been demonstrated that conventional electroless plating baths containing ethylenediaminetetraacetic acid (EDTA) present good stability at different temperatures, although it results in limited purity of the palladium layer due to the incorporation of carbon deposits from the EDTA complex within the metal particles [97,181,182]. These carbon deposits could diminish the membrane performance by CO_2 formation at some operating conditions. Thus preparation of free-EDTA baths has been also investigated, achieving acceptable palladium deposition yields with good stability of plating baths in absence of these stabilizer [97,181,182].

Other authors have studied the influence of fluid dynamics between support and plating bath. Thus, the rotation of the support during the electroless plating has been found to increase the plating rate and the homogeneity of Pd layer, as reported Chi et al. [183]. Compared with static ELP, the use of rotation of the support during the process reached more uniform and smoother surfaces of Pd membranes, which in turns enhances the stability of the supported system. These authors reported a membrane permeance of 3.0×10^{-3} mol·m^{-2}·s^{-1}·Pa$^{-0.5}$ with ideal hydrogen separation factor upper than 400 for only 5 μm thick (T = 400 °C, P = 4 bar).

Despite the effort in the research to improve both quality and cost-efficiency of Pd membranes, other many studies are focused on diminishing the rejected membranes due to the presence of defects or cracking during the fabrication processes. In this mean, novel alternatives to repair possible defects generated on the Pd surface have been recently reported. Thus, Li et al. [179] use the fundamentals of previously reported osmotic effect to incorporate Pd particles preferentially in defect areas for repairing the Pd layer. Following this procedure, they ensure the complete disappearance of defects and a consequent meaningful increase in the ideal hydrogen separation factor without noticeable reduction of the permeation flux nor thickness growth. With similar fundamentals, point plating has been also proposed by Zeng et al. [184] to repair located defects in supported Pd-based membranes. In this case, the method forces the chemical reaction for palladium reduction around defects by feeding both metal source and hydrazine baths from opposite sides of the supported membrane.

Based on these repairing procedures, other researchers have recently reported the separated supply of Pd source and reducing agent bath to prepare Pd-based membranes directly on rough commercial PSS supports [49,51,71,78–80]. This novel procedure, denoted as Electroless Pore-Plating (ELP-PP), uses the wall of the support itself to maintain separated both Pd source and hydrazine solutions. At these conditions, hydrazine preferentially diffuses through the pores of the support and reacts with the amino-palladium complex near the pore area. Ideally, in case of proper activation of the inner pore surface, this reduction initiates from the internal porosity of the support in a similar way to the sealing method previously [49,51,185] asserted that it is possible to save palladium source and to minimize the number of rejected membranes following this methodology, consequently reducing the overall cost of membrane preparation. This is possible because the contact between reactants turns progressively difficult during the Pd incorporation up to the complete block of pores, moment at which the process stops. The comparison of both conventional ELP and ELP-PP alternatives is shown in Figure 7. This method hinders the increasing of palladium incorporation after blocking the pores, in contrast to the behaviour reached by conventional ELP, resulting in a fully dense film with minimum thickness.

In spite of the preferential incorporation of Pd inside the pores of the support, authors revealed the generation of an external film on both commercial and modified PSS supports caused by the wide variety of pore diameters in these supports [49,51,185]. The hydrazine cannot pass through the smallest pores because they become fully closed by palladium particles in a relative short time, while the reducing agent can diffuse through the widest ones, partially covered, until the outer surface in contact with the palladium bath, where the external layer is formed. In this manner, it is obvious that several parameters affect the ELP-PP process: (i) pore characteristics of support (average pore diameter and porosity), (ii) reducing and metal plating baths formulation and (iii) ratio between membrane length and volume of solutions. In fact, supports with smaller pores, achieved by direct oxidation of commercial supports in air, provide membranes with an apparent thickness around 10 μm, a half of the value reached in case of using unmodified supports (20 μm). However, this apparent thickness, determined by gravimetric analysis, is 2–6 μm greater than the real value obtained from SEM characterization due to the Pd introduction in the pores of the support is not considered for the estimation. All membranes obtained by this ELP-PP alternative exhibited high stability at different simulated and real operating conditions in a WGS membrane reactor with permeances in the range 1–6·10^{-4} mol·m^{-2}·s^{-1}·$Pa^{-0.5}$ and complete H_2 selectivity [51]. Within the recent past, this novel method has been also reported for preparation of supported membranes on ceramic supports with noticeable smaller pores respect to typical PSS supports, confirming the great importance of support characteristics on the plating performance (primarily average pore diameter and pore size distribution). In this way, thinner membranes were achieved even though palladium is still present in both internal pores and external surface [71].

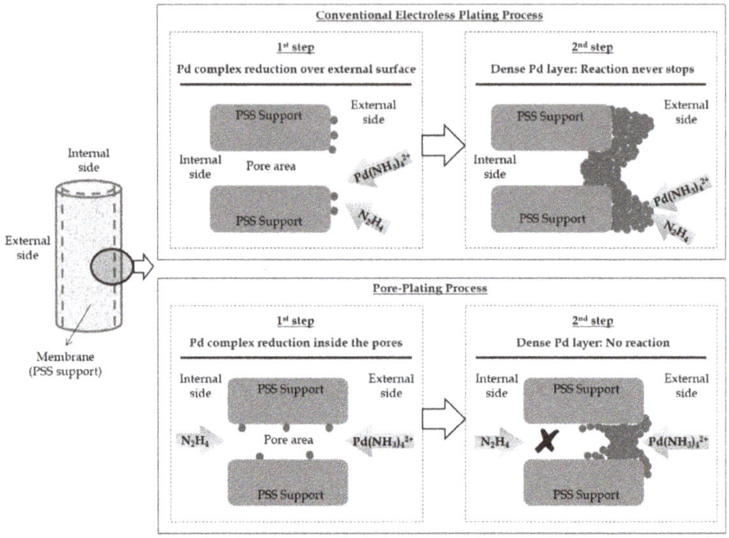

Figure 7. Pd incorporation around pores in both conventional electroless plating (ELP) and pore-plating (ELP-PP) alternatives [185], with permission from © Elsevier.

Finally, some researchers propose the improvement of membranes properties, mainly the increase of permeation rate and thermal stability simultaneously to the presence of defects is decreased, by using a further thermal treatment step (>640 °C) after the palladium plating [42]. Although this alternative is not strictly an improvement of the metal deposition process, it can be used to enhance the previously prepared membrane. In fact, a heat treatment of as-prepared membranes improved the Pd layer microstructure, achieving a densification of the metal film, as it can be seen in Figure 8.

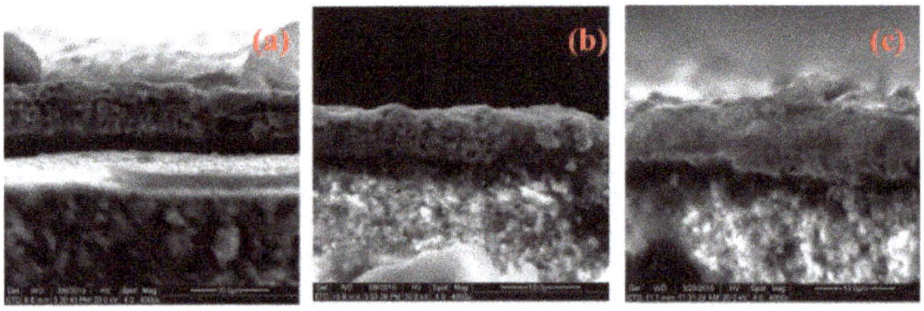

Figure 8. Microstructural modification on Pd films prepared by ELP after different thermal treatments: (**a**) as prepared; (**b**) 168 h at 550 °C; and (**c**) 72 h at 700 °C [42], with permission from © Elsevier.

Like previous information about main alternatives for modification of commercial supports, here we summarize the most relevant advances for Pd incorporation by electroless plating in Table 3. Key improvements and experimental details are summarized beside information about support material, support modifications, thickness of selective layer and permeation properties.

Table 3. Recent improvements on electroless plating to prepare supported Pd-based membranes.

ELP Improvement	Particular Details	Support	Support Modification	Tselective Layer (m)	Permeation Conditions		Permeation Capacity	H_2 Separation Factor	Ref.
					T (°C)	P (kPa)			
Deposition around pores	Vacuum asisted-deposition	Al_2O_3	-	6.0	500	n.a.	8.78×10^{-4} [(a)]	3000	[176]
Deposition around pores	Vacuum asisted-deposition	Al_2O_3	Pd(II)-modified bohamite sol	1.0	450	n.a.	2.23×10^{-2}–1.07 [(b)]	20–130	[152]
Deposition around pores	Osmotic effect with aqueous sucrose solution	Vycor glass	-	1.6	n.a.	n.a.	n.a.	n.a.	[177]
Deposition around pores	Osmotic effect with aqueous sucrose solution	Vycor glass	-	2.5	n.a.	n.a.	n.a.	n.a.	[178]
Protecting selective layer	Pore- filled, vacuum asissted-deposition between two ZrO_2 layers	Al_2O_3	YSZ particles	5.0	150–500	150–400	0.10–0.60 [(b)]	n.a.	[50]
Reduction of carbon deposits	Free-EDTA baths	Al_2O_3	ZrO_2	1.3	365	138	394.61 [(a)]	n.a.	[181]
Reduction of carbon deposits	Free-EDTA baths	PSS	Al_2O_3	5.0	400	100	$3.05 \cdot \times 10^{-3}$ [(a)]	500	[182]
Increase film homogeneity	Support rotation	Al_2O_3	ZrO_2	5.0	350–450	100–400	$3.00 \cdot \times 10^{-3}$ [(a)]	>400	[183]
Membrane repairing	Osmotic effect to close defects without thickness increase	PSS	-	10.0	425–475	68–136	$2.00 \cdot 10^{-4}$ [(b)]	400–1600	[179]
Membrane repairing	Point plating to close defects without thickness increase	α-Al_2O_3	γ-Al_2O_3	n.a.	500	100	7.20×10^{-1}–8.50×10^{-1} [(b)]	n.a.	[184]
Reducing rejected membranes	ELP-PP. Pd-source and reducing agent from opposite sides of support	PSS	Fe_2O_3-Cr_2O_3	11.0–20.0	350–450	100–250	1.00×10^{-4}–6.00×10^{-4} [(a)]	∞	[51]
Pd microstructure	Heat treatment at T > 640 °C	PSS	YSZ	4.9	600	82	2.40×10^{-3} [(a)]	200–2000	[42]

Permeation capacity: [(a)] Permeance ($mol\ m^{-2} \cdot s^{-1} \cdot Pa^{-0.5}$) or [(b)] Permeation flux ($mol \cdot m^{-2} \cdot s^{-1}$). n.a.: non availabl.

4. Pd-Alloy Membranes

Independently of using conventional or improved electroless plating processes, many researchers endorse the preparation of alloys in which palladium is combined with some amounts of other metals in order to improve the permeation behaviour, the thermal and mechanical stability and the poison tolerance of the membrane [56,186–188]. Thus, in this section we present an overview of most frequent Pd-based alloys, detailing the preparation procedures and main reported benefits, as well as recent trends and future perspectives for new formulations with improving properties.

Pure Pd usually suffers the so-called hydrogen embrittlement phenomenon due to the lattice expansion provoked by the α to β phase transition that occurs when the metal is exposed to hydrogen atmosphere at temperatures and pressures below 298 °C and 2 MPa, respectively. This phase transition generates tensile stress, especially in case of tubular geometries, which often leads to cracking the Pd layer and thus, a subsequent loos of hydrogen selectivity of the membrane. This drawback can be avoided by working at operating conditions above the mentioned critical point when membrane is exposed to hydrogen or modifying the Pd phase diagram [189]. The last option can be realized by alloying pure Pd with other metals, i.e., silver [52,106,190–192], copper [99,102], ruthenium [187,193] or gold [194,195]. It is demonstrated that Pd-based alloys with specific concentrations of these metals modify the metal-hydride phase diagram avoiding the mentioned embrittlement phenomena [28].

Other problem that negatively affects to the permeability of dense Pd-based membranes is the irreversible poisoning by chemical contaminants, such as carbon monoxide or sulphur. These molecules are chemisorbed over the metallic layer, being also possible a chemical reaction with hydrogen to form species that block the active sites on the surface and hinder the hydrogen permeation. Some alloys help to avoid this poisoning effect while maintaining an ideal complete hydrogen separation factor [90], even in presence of sulphur compounds that traditionally causes irreversibly poisoning in pure Pd films [28,102,186,196].

4.1. Alloy Preparation

The preparation of efficient Pd-based alloys by electroless plating with an accurate composition is currently one of the most important milestones for industrial membranes implementation. Physical vapour deposition provides multiple possibilities for incorporating different metals to the membrane with a really good control of the alloy composition [61,197–199]. However, this technique has some difficulty to generate defect-free layers on rough surfaces and high investments costs [60]. Thus, at the present date, the metal incorporation by electroless plating is widely adopted [182].

In general, the incorporation of metals by electroless plating for preparation of alloys can be carried out in different ways after a previous activation of the support, as illustrated in Figure 9. First, a unique plating bath containing all alloy constituents, i.e., materials A and B, can be used to deposit simultaneously all of them, being denoted as co-deposition (Figure 9a). In this case, the alloy constituents are randomly distributed in the selective layer with similar composition in both longitudinal and transversal directions. Thus, the following thermal treatment to form the alloy is favoured. However, this option is only possible in case of using metals with analogous properties that can be reduced in similar conditions, i.e., palladium and silver with comparable bath compositions and identical reducing agents. However, kinetics of the reduction process can be different for each component and, consequently, it is not easy to define the bath conditions to achieve a desired alloy composition [28,74,200,201].

Other possibility to prepare Pd-alloy membranes is based on sequential depositions of each constituent, incorporating all required amount of material B onto a previous layer formed by the material A (Figure 9b) or *vice versa* (Figure 9c), denoting both alternatives as consecutive methods. The alternation of different layers formed by each constituent until achieving the desired composition and layer thickness is also viable (Figure 9d,e, alternate methods). In these cases, it is possible to incorporate metals from different plating baths by using the same or different reducing agents on the condition that galvanic displacement does not occur. The kinetics of deposition processes can be easily controlled and the final alloy composition is determined by recurrences of each constituent plating. The most relevant drawback of sequential depositions is the difficulty to achieve a good alloy homogeneity through the whole layer thickness [202].

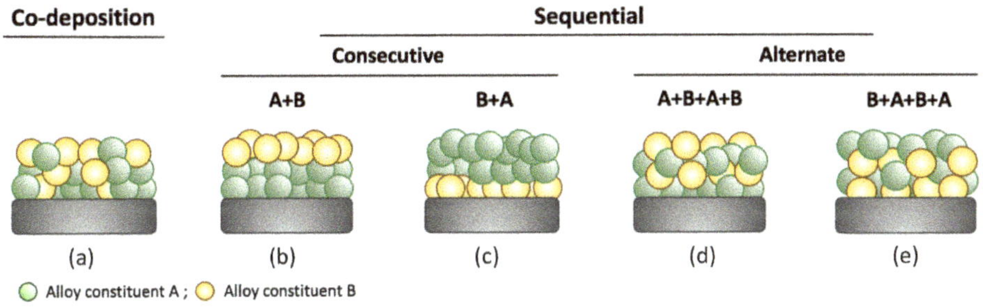

Figure 9. Different possibilities to prepare binary alloys by electroless plating: (**a**) co-deposition; (**b**,**c**) sequential deposition; (**d**,**e**) alternative deposition.

Anyway, a further thermal treatment is always required to achieve the diffusion of atoms within the solid material to form the alloy, independently of using sequential or co-deposited alternatives for the incorporation of metals. This process, also known as annealing, can be carried out under inert atmosphere (usually by using Ar, He or N_2 as inert gas) [99,146,203] or in presence of hydrogen [52,94,187,200,204]. Traditional annealing processes for Pd-based membranes use inert environment at ambient pressure and require quite long times [53]. However, recent developments prefer faster processes in pressurized hydrogen atmosphere. In this case, it is proposed that dissolved hydrogen forms vacancies in the crystal lattice of palladium favouring the mobility of other alloy constituents and, consequently, reducing the time required to obtain the alloy [195]. As previously mentioned, layers prepared by co-deposition need softer thermal treatments (shorter times or lower temperatures) for annealing as compared with layers generated by sequential deposition (either consecutive or alternative) [95,202]. Taking into account that preparation of alloys with accurate control is a decisive challenge for the large-scale application of Pd-based membranes [205], following sections summarize the most relevant advances in this field, distinguishing the preparation of binary and ternary alloys.

4.2. Binary Alloys

Among the large number of feasible alloys from different metal pairs, the Pd-based binary alloys are the most frequently studied and used for hydrogen production. As mentioned before, alloying palladium with other component can avoid the hydrogen embrittlement as well as improve mechanical and chemical properties. In some specific cases, the hydrogen permeability may be even increased, depending on the alloy composition (Figure 10). Some alloys can improve the hydrogen permeability of the membrane only in a narrow composition window, while others also work in a wide range of compositions. Deviations from these target compositions or differences in composition inside the bulk metal may deteriorate noticeably the permeation behaviour respect pure palladium. For instance, this occurs when exceeding 36 wt % or 21 wt % in case of alloying with silver or gold, respectively. For palladium-copper alloys, small deviations from a target $Pd_{60}Cu_{40}$ value reach to a drastic decrease in hydrogen permeability.

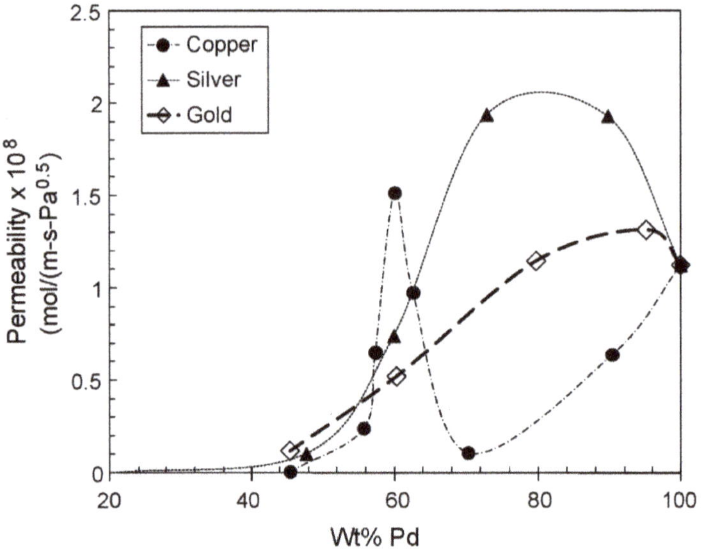

Figure 10. H_2 permeability at 350 °C for different Pd-based alloys containing Ag, Cu and Au [206], with permission from © Elsevier.

4.2.1. PdAg Membranes

One of the first alternatives to prepare binary Pd-based alloys is based on the addition of silver, used since the eighties for separation of hydrogen isotopes [207,208]. It is widely reported that mechanical strength against hydrogen embrittlement is significantly improved after silver incorporation on a palladium layer [201], as well as original hydrogen permeability can be also increased for some particular conditions [209]. As previously shown in Figure 10, the addition of silver on bulk palladium to prepare a binary PdAg alloy increases the membrane permeability in a wide range of compositions, from very low Ag percentages up to around 36 wt % [206]. Particularly, it is proved that using a $Pd_{75}Ag_{25}$ composition the hydrogen permeation reached a maximum and, thus, many researchers around the world have adopted this composition as main target for membrane preparation. Concerning the preparation procedure to obtain this alloy by electroless plating, both co-deposition [95] and sequential deposition [202] alternatives can be widely found in literature.

The research group headed by Yi Hua Ma has reported some studies on the binary PdAg alloy [83,210]. For example, one of the most interesting ones, published by Rajkumar et al. [211], presents the results obtained for incorporating Ag by both electroless and electro-plating techniques, after a first Pd incorporation by electroless plating. These PdAg supported membranes were prepared directly onto a commercial porous Inconel supports of 0.1 μm media grade, obtaining H_2 selective layers under 10 μm thick with complete He retention up to a transmembrane pressure difference of 10^5 Pa. All membranes were annealed during 24 h at 550 °C in H_2 atmosphere. They observed that electro-plated Ag exhibited an optimal penetration in the pores of the support, although a non-uniform growth with dendritic morphology was achieved. In contrast, the use of electroless plating for silver incorporation provides a uniform growth without dendritic morphology and a lower penetration into the pores.

The contributions of the research group headed by Laura Cornaglia, based on the use of PdAg membranes prepared by electroless plating, are also widely reported in the literature [201,209,212–214]. One example is the work published by Bosko et al. [202], in which PdAg supported membranes were prepared by sequential electroless plating on tubular stainless-steel supports with thickness ranged from 20 to 26 μm. The supports were previously modified with both α-Al_2O_3 and γ-Al_2O_3

particles by a vacuum assisted-coating method. All membranes were annealed at 500 °C, exhibiting a hydrogen permeability of 3.1 × 10^{-4} mol·m^{-2}·s^{-1}·$Pa^{-0.5}$ at 450 °C and 100 kPa and a H_2/N_2 ideal selectivity of around 954 at same conditions. They observed that annealing the membranes at higher temperatures created defects which deteriorated the selective layer, obtaining lower selectivity and higher permeability.

Others works based on electroless-plated PdAg membranes to be highlighted are those of Andreas Goldbach and co-workers. Recently, Zeng et al. [53] reported the use of sequential electroless plating to prepare H_2 selective PdAg membranes on Al_2O_3 tubular porous asymmetric supports. They incorporated the palladium prior to the required amount of silver to achieve a final composition of $Pd_{77}Ag_{23}$, after a thermal treatment at 500 °C in the presence of H_2 between each deposition. Additionally, intermediate surface activations with Pd seeds were also carried out. In this way, membranes with a final thickness in the range 2.3–2.5 μm were achieved. Despite the limited thickness of these membranes, authors indicated the need to extend the annealing treatment up to 800 h maintaining a temperature of 500 °C under atmospheric H_2 pressure. The progress of annealing between palladium and silver was monitored by XRD during the entire process as it is shown in Figure 11, to assess the formation of a homogeneous PdAg alloy. These membranes exhibited a H_2/N_2 selectivity between 3770 and 5600, with permeance values ranged from 5.77 to 3.86 × 10^{-8} mol·m^{-1}·s^{-1}·$Pa^{-0.5}$, respectively. Similar to other researchers, they also reported the frequent membrane failure during the fabrication process due to the fragility of ceramic supports [89,95].

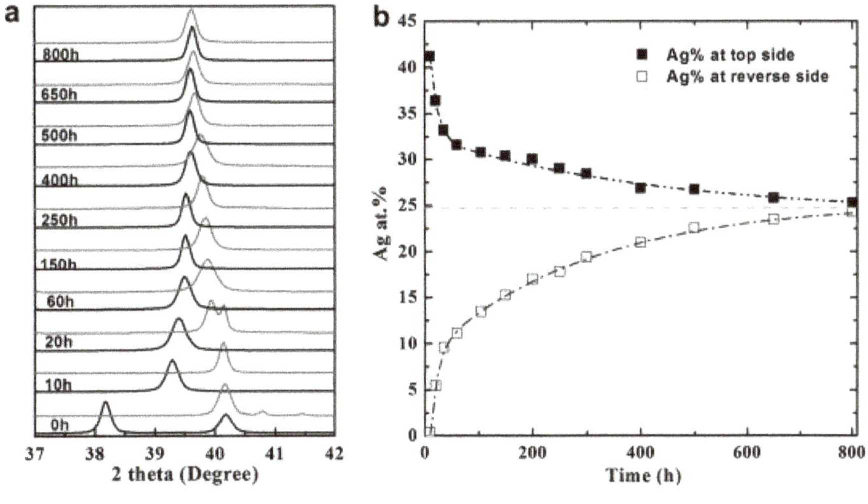

Figure 11. (**a**) 111 XRD reflections from the top (black) and reverse surface (grey) of a PdAg membrane during alloying at 550 °C and (**b**) convergence of the corresponding alloy lattice parameters [53], with permission from © Elsevier.

PdAg membranes prepared by Tecnalia innovation centre are also relevant. For instance, Ekain et al. [95] have recently reported a procedure to prepare thin PdAg membranes (thickness of around 3.2 μm) on tubular ceramic supports by using Pd and Ag simultaneous electroless plating during 210–240 min. The final alloy was achieved after annealing at 550 °C for 2 h, using N_2 carrier gas for both heating and cooling rates but a mixture $10H_2$-$90N_2$ when the annealing temperature has been reached. These membranes exhibit a H_2 permeance of 3.10 × 10^{-6} mol·m^{-2}·s^{-1}·Pa and ideal selectivity in the range 8000–10,000, calculated for a H_2 partial pressure of 1 bar and 400 °C. As other studies, these authors indicate that the use of ceramic supports generates fragile membranes, despite the good quality of the Pd-based selective layer. In this case, authors specifically recommended to avoid exceeding a maximum torque value when coupling the membranes into the reactor, usually made in stainless-steel.

4.2.2. PdCu Membranes

The use of copper as alloying element of Pd-based membranes not only improves mechanical strength against hydrogen embrittlement, also increases slightly the permeation rate as compared with palladium pure membranes while retaining the permeation capacity in gas mixtures containing sulphur compounds [205]. Moreover, copper is quite cheaper than palladium and thus, the percentage reduction of the last one in the selective layer (optimal composition around $Pd_{60}Cu_{40}$) reduces its cost. The preparation of PdCu membranes by electroless plating is usually carried out by sequential incorporation of palladium in first place and the copper, followed by an annealing treatment at high temperature. In this case, stable co-deposition is really difficult due to the different nature of each metal and the order of incorporation is also determined by the galvanic displacement of copper by palladium due to the lower reduction potential of the first one [215].

However, the use of PdCu alloys with face-centred-cubic metal structure (Figure 12) is restricted to a range of composition due to the drastic fall in H_2 permeate when small variations in the 40% content of Cu are produced.

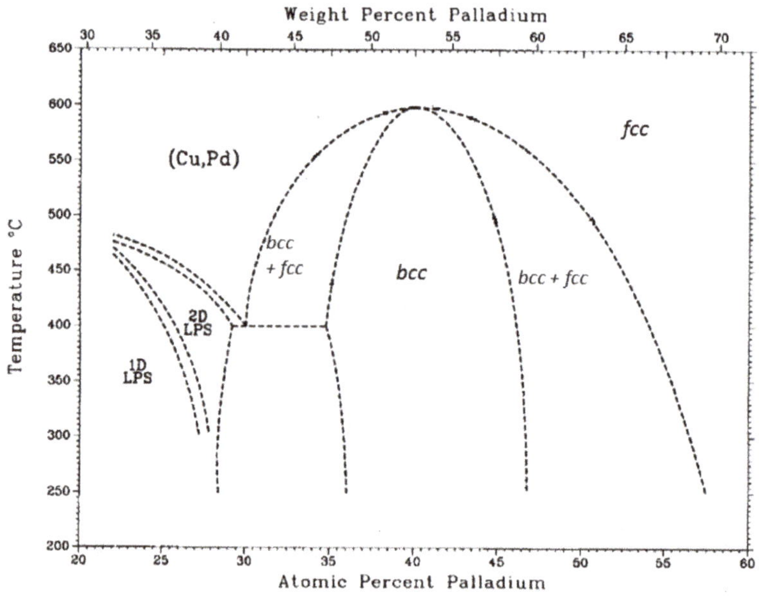

Figure 12. Pd-cu phase diagram [205], with permission from © Elsevier.

Moreover, sulphur inhibition depended strongly on temperature. Zhao et al. [54] developed PdCu membranes on ceramic supports by using sequential deposition of metals via electroless plating and performed tests to analyse the influence of H_2S. They revealed that single gas H_2 permeation rates were significantly reduced after tests performed with 35 ppm H_2S at 400 °C, although they fully restored the original operation capacity of the membrane after flowing H_2 at 500 °C. However, the precedent flow rate obtained for N_2 leaks increase noticeably (thus, decreasing $\alpha_{H2/N2}$ up to 1194), presumably due to sulphide formation in defect sites at lower temperatures.

The research group headed by Douglas J. Way maintains a wide activity on the study of PdCu systems for years [102,181,216,217]. Among all of them, it can be pointed out the preparation of

PdCu alloys on tubular porous supports of symmetric α-alumina published by Roa et al. [84]. In this work, the incorporation of the H_2 selective layer was carried out by sequential electroless plating of both metals, palladium and copper, followed by an annealing treatment at high temperature in flowing H_2. Thus, a PdCu membrane with 11 μm thick was obtained, exhibiting a H_2 permeate flux of 0.8 $mol \cdot m^{-2} \cdot s^{-1}$ at T = 450 °C and ΔP = 345 kPa with a H_2/N_2 ideal separation factor of 1150.

Similar supported membranes based on electroless deposition of a $Pd_{60}Cu_{40}$ alloy layer were prepared by Qiao et al. [123] but using porous stainless-steel modified by the incorporation of a CeO_2 as intermediate layer. In this work, copper was incorporated at room temperature on a first Pd layer prepared at T = 55 °C and a further annealing treatment at 480 °C under hydrogen atmosphere. The alloy layer thickness was around 8 μm, that yielded a hydrogen permeability of 5.92×10^{12} $mol \cdot m^{-2} \cdot s^{-1}$ at T = 450 °C (pressure was maintained at 10^5 Pa). No nitrogen permeate was detected at ΔP = 10^5 Pa and room temperature.

4.2.3. PdAu Membranes

The addition of gold in a palladium layer brings out similar benefits to that of copper in Pd-based alloys, mainly attending to sulphur tolerance [91], although the cost of the resulting membranes is higher due to the higher cost of gold respect to that of copper and the need of a higher content in palladium for these types of alloys, usually upper than 80% [206]. However, the range of compositions in the PdAu that exhibits FCC structure without phase segregation, which ensures greater permeability than pure Pd membranes with diverse alloy formulation, is wider [205]. This fact is very beneficial in the membrane preparation since it is possible to achieve the alloy varying the gold content up to 21%, while maintaining improved permeation and sulphur tolerance. Among these possibilities, a composition target of $Pd_{90}Au_{10}$ seems to offer the best properties in the resulting membrane [206].

Attending to the preparation of these membranes, the combination of Pd electroless plating with a consecutive Au incorporation by galvanic displacement is the procedure widely described in most published works. Thus, Yi Hua Ma and co-workers prepared PdAu membranes that ensure good resistance to sulphur presence in a mixture 54.8 ppm H_2S/H_2 at temperatures ranged from 350 to 500 °C [94]. These membranes, prepared on a PSS support, have a gold content around 8–11 wt % with a total thickness below 15 μm. PdAu membrane experienced a decline in H_2 permeance by ~85% after the H_2S exposure. Despite this, a value of ~65% of the original permeance could be recovered in pure H_2 at the poisoning temperature of 400 °C.

Similar PdAu alloys were also prepared by Tarditi et al. [93], in a ZrO_2-modified porous stainless-steel support. Both metals, palladium and gold, were incorporated by sequential electroless deposition: palladium was first deposited at 50 °C in two steps, while a unique step at 60 °C was used for incorporating the gold. This process was repeated several times up to obtain a N_2 non-permeable membrane at room temperature and ΔP = 10^5 kPa at, obtaining a total thickness of around 10 μm. Finally, the membranes were annealed at 500 °C in H_2 atmosphere to promote the alloy formation. A H_2 permeation flux of 0.14 $mol \cdot s^{-1} \cdot m^{-2}$ was obtained at T = 400 °C and ΔP = 100 kPa, showing a H_2/N_2 ideal selectivity upper than 10,000. It was also possible to recover ~65% H_2 flux, in a $Pd_{91}Au_9$ membrane with an initial permeability of 1.1×10^{-8} $mol \cdot m^{-1} \cdot s^{-1} \cdot Pa^{-0.5}$ at 400 °C after 24 h exposure of 54.8 ppm H_2S in H_2 [201].

4.2.4. Others Binary Alloys

The previously mentioned binary alloys based on palladium, silver, copper and gold are the most common alternatives used by researchers for H_2 selective separation. However, it is also possible to find the combination of palladium with other metals to achieve additional advantages in cost-reduction or permeation behaviour. In this context, the principal limitation to explore new alloys it is the chance to incorporate the metals by electroless plating, being preferred to use the magnetron sputtering [61,199]. However, as we previously detailed, the use of electroless plating for H_2 selective membranes preparation is recommended in terms of cost efficiency [33]. In this context, the use of nickel [148,218] or platinum [85,187] to prepare Pd-based alloys by electroless plating has been explored. Pd-Ni alloy supported membranes show long-term thermal stability at 300 °C under hydrogen permeation. An example of this alloy can be found in the work published Lu et al. in which Pd and Ni were consecutively incorporated by ELP on a capillary α-Al_2O_3 substrate [218]. The PdNi selective layer thickness is ~7 μm and shows a hydrogen permeance of 2.74×10^{-3} mol·s^{-1}·m^{-2}·Pa$^{-0.5}$, close to that obtained for a similar supported membrane of pure palladium. However, the manuscript does not mention the particular alloy composition achieved. In contrast to PdNi alloy, PdRu and PdPt [187] alloys showed long-term thermal stability at high temperatures. Membranes based on a PdRu thin film are prepared by co-deposition with really low contents on ruthenium (<2 wt %) [193], while the PdPt membranes are formed by alternating layers of Pd and Pt [85] with a final platinum load of around 25 wt %. El Hawa et al. prepared these types of alloys on modified porous stainless-steel with yttria-stabilized zirconia [187], reaching an average thickness of 6 μm. The H_2 permeance achieved at 550 °C for membranes based on PdRu and PdPt alloys were 2.1×10^{-3} mol·s^{-1}·m^{-2}·Pa$^{-0.5}$ and 1.39×10^{-3} mol·s^{-1}m^{-2}·Pa$^{-0.5}$, respectively.

4.3. Ternary Alloys

The formulation of ternary alloys has been also considered to combine simultaneously the improvements of each constituent [74]. However, published researches about preparation of multicomponent alloys by electroless plating is still scarce, being initiated only some years ago. First works suggest that particular compositions seems to reach an additional improvement on the membrane properties as compared to binary alloys, in terms of increasing hydrogen permeability and/or chemical resistance [56]. Alloying Pd simultaneously with two or more other metals (i.e., Ag, Cu, or Au) it is possible to improve not only the membrane permeability but also the mechanical and chemical resistances to sulphur poisons at the same time [41,201]. Additionally, the use of cheaper materials (i.e., Ag or Cu) in these formulations reduces the membrane cost [219–222]. On the other hand, copper and gold present higher melting points than silver, although lightly lower permeability of their binary alloys with palladium. Hence adding these metals on PdAg alloys to conform a ternary alloy could increase the thermal stability of the membrane [41,56,223].

Generally, the procedure to prepare these ternary alloys is similar to the previously described for obtaining binary alloys, including co-deposition or sequential electroless plating following by an annealing process at high temperature to finally obtain free-defects, homogeneous and continuous layers [74]. For instance, Tarditi et al. [224] presented the fabrication procedure of a PdAgCu ternary alloy on stainless-steel supports by consecutive deposition of palladium, silver and copper, in this order. They evidenced a hydrogen permeation flux of the ternary alloy about 70% higher than in case of considering a binary PdCu alloyed membrane with similar copper loading and average thickness of the selective layer. The annealing process for these membranes consists of a thermal treatment at 500 °C for 162 h, to obtain the FCC phase of the ternary alloy by XRD after the treatment. Figure 13 shows the evolution of the alloy during the annealing process.

In this work, no uniform distribution of metals was achieved, increasing silver content as going towards top surface in radial direction. Authors explain this effect by the lower surface tension of silver compared with this value in both palladium and copper, obtaining a surface segregation of silver after the annealing treatment [55].

Other researchers add gold to a previously prepared PdAg membrane to increase the resistance against H_2S poison, in a similar way as using PdCu or PdAu binary alloys but with higher permeabilities derived from the presence of silver. For instance, Melendez et al. has lately published the preparation and testing of ternary PdAgAu alloys by incorporation of gold over a previously electroless plated PdAg membrane (using co-deposition of metals) on an asymmetric tubular Al_2O_3 support, with a thickness of 2.71 μm and final composition near to $Pd_{91.7}Ag_{4.8}Au_{3.5}$ [41]. Prior to gold incorporation, the Pd-Ag layer was annealed at 550 °C for 4 h and the process was repeated again after incorporating gold. This membrane exhibited a H_2 permeance of 4.71×10^{-3} mol·s^{-1}·m^{-2}·Pa$^{-0.5}$ at

600 °C. The authors affirm that the H_2/N_2 ideal separation factor achieved with the ternary PdAgAu membrane was maintained relatively high after H_2S exposure, in comparison with other binary PdAg membranes taken as reference. In fact, a $Pd_{96.1}Ag_{3.9}$ membrane suffered a decrease in hydrogen selectivity from 1308 to 18 after exposure to 9 ppm H_2S for 15.25 h. However, the selectivity of a ternary alloy composition with composition like that of previously described, $Pd_{91.5}Ag_{4.7}Au_{3.8}$, experienced a lower decrease in hydrogen selectivity, from 4115 to 800 after 9 ppm of H_2S for 12.5 h. Moreover, ternary alloy membranes recovered original permeation rates in an 85% after sulphur exposure, whereas the hydrogen flux of PdAg membranes maintained below detectable values.

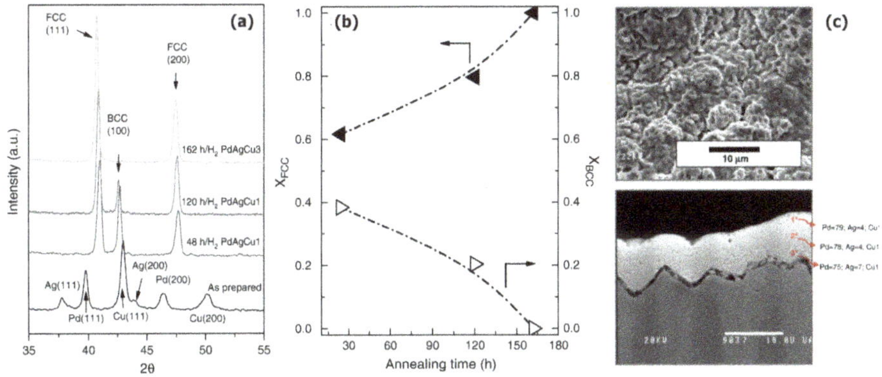

Figure 13. Ternary PdAgCu alloy formation: (**a**) XRD patterns after annealing up to 500 °C in H_2 at different times, (**b**) Microstructure evolution with annealing time and (**c**) SEM images after annealing of both top surface and cross-section [224], with permission from © Elsevier.

Membranes formed by the alloy PdAuCu were also prepared by Tarditi et al. onto PSS disks with sequential electroless deposition of each metal [200]. First, the supports were cleaned and oxidized before incorporating a ZrO_2 intermediate layer by using a vacuum-assisted dip-coating method. Palladium was deposited onto this modified support in two steps of 60 min, following by the incorporation of gold. Later, the membrane was rinsed and dried at 120 °C for 12 h prior to be activated again by conventional sensitization-activation process with $SnCl_2$-$PdCl_2$ solutions to allow for copper deposition on the top layer. Finally, in order to achieve an alloy with homogeneous composition, the membrane was annealed at 500 °C in H_2 atmospheric pressure. After using this synthesis procedure, the best membrane was reached with 14 μm in thickness of the alloy composition $Pd_{69}Au_{17}Cu_{14}$, obtaining a hydrogen permeance of 8.7×10^{-9} mol·s^{-1}·m^{-1}·Pa$^{-0.5}$ at 400 °C and 50 kPa. The permeability value taken as reference in clean conditions were reduced around 55% after exposure to 100 ppm H_2S/H_2 at 400 °C for 24 h. However, authors affirmed that it was possible to recover around the 80% of the original permeation capacity of the membrane after testing with pure H_2 at 400 °C.

Table 4 summarizes the most relevant information about Pd-alloy membranes detailed in this section, including information about metal incorporation, final thickness, alloy composition and annealing conditions, as well as used support and main permeation properties.

Table 4. Recent advances on preparation of Pd-based alloys.

Alloy Type	Alloy Composition	ELP Metal Incorporation	Support	Support Modification	Tselective Layer (m)	Annealing	Permeation Conditions T (°C)	Permeation Conditions P (kPa)	Permeation Capacity	H_2 Separation Factor	Sulfur Tolerance	Ref.
Binary	$Pd_{75}Ag_{25}$	Sequential	Inconel	-	10.0	500 °C, 24 h	250–500	100	-	60–436	-	[211]
Binary	$Pd_{75}Ag_{25}$	Sequential	PSS	α-Al_2O_3/γ-Al_2O_4	20.0–26.0	500 °C	450	100	3.10×10^{-4} (a)	954	-	[202]
Binary	$Pd_{77}Ag_{23}$	Sequential	α-Al_2O_3/γ-Al_2O_3	-	2.3–2.5	500 °C, 800 h in H_2	500	100	1.61–1.57 $\times 10^{-2}$ (a)	3770–5600	-	[53]
Binary	$Pd_{77}Ag_{23}$	Co-deposition	Al_2O_3	-	3.2	500 °C, 2 h in N_2	400	100	3.10×10^{-6} (a)	8000–10,000	-	[95]
Binary	PdAg	Co-deposition	Hast X	YSZ–Al_2O_3	4.0–5.0	n.a.	4–600	100	100.00×10^{-8} (b)	>200,000	-	[192]
Binary	$Pd_{81}Cu_{19}$	Sequential	Al_2O_3	-	5.0	500 °C, 48 h in N_2	400	100	1.20×10^{-3} (a)	1194	Yes (35 ppm)	[54]
Binary	$Pd_{60}Cu_{40}$	Sequential	α-Al_2O_3/γ-Al_2O_3; α-Al_2O_3/ZrO_2	-	11.0	H_2 atmosphere	450	345	0.80 (b)	1150	Yes	[84]
Binary	$Pd_{62}Cu_{38}$	Sequential	PSS	CeO_2	8.0	480 °C, 6 h in H_2	450	100	74.00 (a)	2369	Yes	[123]
Binary	$Pd_{90}Au_{10}$	Sequential, galvanic displacement	PSS	Oxidation in air (700 °C, 12 h)	<15.0	500 °C, 48 h in H_2	3–500	100	9.35×10^{-4} (a)	∞	Yes (54.8 ppm)	[94]
Binary	$Pd_{91}Au_9$	Sequential, galvanic displacement	PSS	ZrO_2	10.0	500 °C in H_2	400	100	1.10×10^{-3} (a)	>10,000	Yes (54.8 ppm)	[93]
Binary	Pd_xNi_y	Sequential	α-Al_2O_3	-	7.0	n.a.	500	20–120	2.74×10^{-3} (a)	640	-	[218]
Binary	$Pd_{98}Ru_2$	Co-deposition	PSS	YSZ	6.0	n.a.	550	n.a.	2.10×10^{-3} (a)	1860	-	[187]
Binary	$P_{75}Pt_{25}$	Co-deposition	PSS	YSZ	6.0	n.a.	550	n.a.	1.39×10^{-4} (a)	1590	-	[187]
Ternary	$Pd_xAg_yCu_z$	Sequential	PSS	Oxidation in air (500 °C, 12 h)	24.0–27.0	500 °C, 162 h	3–450	10–100	1.70–2.10 $\times 10^{-4}$ (a)	300–10,000	n.a.	[224]
Ternary	$Pd_{91.7}Ag_{4.8}Au_{3.5}$	Co-deposition/ Sequential	α-Al_2O_3/γ-Al_2O_3	-	2.7	550 °C, 8 h	600	n.a.	4.71×10^{-3} (a)	n.a.	Yes (9 ppm)	[41]
Ternary	$Pd_{91.5}Ag_{4.7}Au_{3.8}$	Co-deposition/ Sequential	α-Al_2O_3/γ-Al_2O_4	-	2.7	550 °C, 8 h	600	n.a.	2.32×10^{-3} (a)	4115–793	Yes (9 ppm)	[41]
Ternary	$Pd_{69}Au_{17}Cu_{14}$	Sequential	PSS	ZrO_2	14.0	500 °C in H_2	400	50	6.20×10^{-4} (a)	n.a.	Yes (100 ppm)	[200]

Permeation capacity: (a) Permeance ($mol \cdot m^{-2} \cdot s^{-1} \cdot Pa^{-0.5}$) or (b) Permeation flux ($mol \cdot m^{-2} \cdot s^{-1}$). n.a.: non available.

5. Concluding Remarks and Future Perspectives

This concluding section offers a perspective on current research in Pd-based membranes for H_2 production processes, focused on dense supported membranes prepared by electroless plating, as well as future challenges need to be addressed for the real implementation of this technology in the industry. Supported membranes are preferred to reduce the thickness of the hydrogen selective layer and, consequently, the cost of the process. Moreover, the hydrogen permeation values, as well as thermal and mechanical resistances of the supported membranes, are increased comparing to unsupported membranes. Among the wide variety of materials that can be used as support, porous ceramic and 316L stainless-steel are the prevalent ones in recent researches. Particularly, it is expected that metal supports will prevail for industrial applications, above all at high temperatures, in which most devices are made in stainless-steel. Up to now, suitable ceramic-steel fitting for long operation times is not guaranteed. In this context, considering the wide pore size distribution and the high roughness of the metallic stainless-steel supports, it is usually required to modify its external surface with the aim to facilitate the incorporation of a thin selective palladium layer. One of the best alternatives is the incorporation of an intermediate layer, which simultaneously reduces the average pore size and roughness of the support top surface and avoids inter-diffusion problems. The main features to be considered to select the most suitable material used as interlayer are thermal compatibility between each component (support, barrier and selective layer) and membrane cost. Many of these intermediate layers are formed by metal oxides (i.e., CeO_2, ZrO_2 or SiO_2 based materials) incorporated by dip-coating techniques. However, nowadays it has not been reached any prevalent solution and additional research in this issue is still required. Future directions are aimed to explore new materials or the combination of some of the already studied to achieve improved properties for the intermediate layer.

Attending to the selective layer deposition, three main objectives are being followed: (i) reduction of the selective layer thickness (target: permeability increase and membrane cost reduction while ensuring complete absence of defects), (ii) reduction in the number of rejected membranes (target: to control the overall cost of the fabrication process) and (iii) preparation of Pd-based alloys (target: to improve permeation properties in real operation conditions, i.e., thermal cycles or presence of sulphur). To reach these objectives and to make easier the metal incorporation around the pores, several modifications of the conventional electroless plating method have been proposed in the last years, including the use of osmotic effect, vacuum or feeding both metal source and reducing agent from opposite sides of the support (ELP-PP). Finally, the use of binary Pd-based alloys to avoid hydrogen embrittlement of pure palladium or sulphur poisoning is also frequent, mainly adding specific amounts of silver, copper or gold. In the last years, an increasing number of alloy constituents is observed, trying to obtain membranes that gather particular benefits of different metals saving costs, i.e., hydrogen perm-selectivity of bulk palladium, increase of permeation rate provided by silver or sulphur resistance given by copper and gold. Here, it is very important to indicate the current limitation of possible metal constituents that can be incorporated by electroless plating and reproducibility between published results in terms of alloy composition, annealing conditions and hydrogen permeability. Thus, future trends go towards the combination of the most promising modifications for electroless plating with additional exploration of new Pd-based alloys with better properties in real industrial operation conditions (i.e., thermal cycling, presence of sulphur compounds, low hydrogen concentrations, etc.).

Acknowledgments: First, authors want to acknowledge the invitation to prepare the present review focused on the current state of the art in Pd membranes preparation by electroless plating. The financial support achieved through competitive projects ENE-2007-66959 and CTQ2010-21102-C02-01 during last years has been very important to become expertise in this field. We also thank the European Commission for the contract of David Martinez-Diaz through the Young Employment Initiative program. Lastly, we also acknowledge to the publishers of referenced figures for allowing the reproduction here.

Conflicts of Interest: The authors declare no conflict of interest.

References

1. Grunewald, N.; Klasen, S.; Martínez-Zarzoso, I.; Muris, C. The Trade-off between Income Inequality and Carbon Dioxide Emissions. *Ecol. Econ.* **2017**, *142*, 249–256. [CrossRef]
2. Henriques, I.; Sadorsky, P. Investor implications of divesting from fossil fuels. *Glob. Financ. J.* **2017**. [CrossRef]
3. Furlan, C.; Mortarino, C. Forecasting the impact of renewable energies in competition with non-renewable sources. *Renew. Sustain. Energy Rev.* **2018**, *81*, 1879–1886. [CrossRef]
4. Hanley, E.S.; Deane, J.P.; Gallachóir, B.P.Ó. The role of hydrogen in low carbon energy futures—A review of existing perspectives. *Renew. Sustain. Energy Rev.* **2017**. [CrossRef]
5. Muradov, N. Low to near-zero CO_2 production of hydrogen from fossil fuels: Status and perspectives. *Int. J. Hydrog. Energy* **2017**, *42*, 14058–14088. [CrossRef]
6. Abbasi, T.; Abbasi, S.A. "Renewable" hydrogen: Prospects and challenges. *Renew. Sustain. Energy Rev.* **2011**, *15*, 3034–3040. [CrossRef]
7. Vivas, F.J.; de las Heras, A.; Segura, F.; Andújar, J.M. A review of energy management strategies for renewable hybrid energy systems with hydrogen backup. *Renew. Sustain. Energy Rev.* **2018**, *82*, 126–155. [CrossRef]
8. Valente, A.; Iribarren, D.; Dufour, J. Harmonised life-cycle global warming impact of renewable hydrogen. *J. Clean. Prod.* **2017**, *149*, 762–772. [CrossRef]
9. Li, R. Latest progress in hydrogen production from solar water splitting via photocatalysis, photoelectrochemical and photovoltaic-photoelectrochemical solutions. *Chin. J. Catal.* **2017**, *38*, 5–12. [CrossRef]
10. Moniruddin, M.; Ilyassov, B.; Zhao, X.; Smith, E.; Serikov, T.; Ibrayev, N.; Asmatulu, R.; Nuraje, N. Recent progress on perovskite materials in photovoltaic and water splitting applications. *Mater. Today Energy* **2017**. [CrossRef]
11. Rau, F.; Herrmann, A.; Krause, H.; Fino, D.; Trimis, D. Production of hydrogen by autothermal reforming of biogas. *Energy Procedia* **2017**, *120*, 294–301. [CrossRef]
12. Sengodan, S.; Lan, R.; Humphreys, J.; Du, D.; Xu, W.; Wang, H.; Tao, S. Advances in reforming and partial oxidation of hydrocarbons for hydrogen production and fuel cell applications. *Renew. Sustain. Energy Rev.* **2018**, *82*, 761–780. [CrossRef]
13. Li, X.; Li, A.; Lim, C.J.; Grace, J.R. Hydrogen permeation through Pd-based composite membranes: Effects of porous substrate, diffusion barrier and sweep gas. *J. Membr. Sci.* **2016**, *499*, 143–155. [CrossRef]
14. Coutanceau, C.; Baranton, S.; Audichon, T. Chapter 2—Hydrogen Production from Thermal Reforming BT—Hydrogen Electrochemical Production. In *Hydrogen Energy Fuel Cells Primary*; Academic Press: Cambridge, MA, USA, 2018; pp. 7–15.
15. Hossain, M.A.; Jewaratnam, J.; Ganesan, P. Prospect of hydrogen production from oil palm biomass by thermochemical process—A review. *Int. J. Hydrog. Energy* **2016**, *41*, 16637–16655. [CrossRef]
16. Devasahayam, S.; Strezov, V. Thermal Decomposition of Magnesium Carbonate with Biomass and Plastic Wastes for Simultaneous Production of Hydrogen and Carbon Avoidance. *J. Clean. Prod.* **2018**, *174*, 1089–1095. [CrossRef]
17. Tosti, S.; Cavezza, C.; Fabbricino, M.; Pontoni, L.; Palma, V.; Ruocco, C. Production of hydrogen in a Pd-membrane reactor via catalytic reforming of olive mill wastewater. *Chem. Eng. J.* **2015**, *275*, 366–373. [CrossRef]
18. Tian, T.; Li, Q.; He, R.; Tan, Z.; Zhang, Y. Effects of biochemical composition on hydrogen production by biomass gasification. *Int. J. Hydrog. Energy* **2017**, *42*, 19723–19732. [CrossRef]
19. Zahedi, S.; Solera, R.; García-Morales, J.L.; Sales, D. Effect of the addition of glycerol on hydrogen production from industrial municipal solid waste. *Fuel* **2016**, *180*, 343–347. [CrossRef]
20. Murugan, A.; Brown, A.S. Review of purity analysis methods for performing quality assurance of fuel cell hydrogen. *Int. J. Hydrog. Energy* **2015**, *40*, 4219–4233. [CrossRef]
21. Pinacci, P.; Basile, A. 3–Palladium-based composite membranes for hydrogen separation in membrane reactors. In *Handbook of Membrane Reactors*; Elsevier: Amsterdam, The Netherlands, 2013; pp. 149–182.
22. Di Marcoberardino, G.; Binotti, M.; Manzolini, G.; Viviente, J.L.; Arratibel, A.; Roses, L.; Gallucci, F. Achievements of European projects on membrane reactor for hydrogen production. *J. Clean. Prod.* **2017**, *161*, 1442–1450. [CrossRef]
23. Yin, H.; Yip, A.C.K. A Review on the Production and Purification of Biomass-Derived Hydrogen Using Emerging Membrane Technologies. *Catalysts* **2017**, *7*, 297. [CrossRef]

24. Dittmeyer, R.; Boeltken, T.; Piermartini, P.; Selinsek, M.; Loewert, M.; Dallmann, F.; Kreuder, H.; Cholewa, M.; Wunsch, A.; Belimov, M.; et al. Micro and micro membrane reactors for advanced applications in chemical energy conversion. *Curr. Opin. Chem. Eng.* **2017**, *17*, 108–125. [CrossRef]
25. Al-Mufachi, N.A.; Rees, N.V.; Steinberger-Wilkens, R. Hydrogen selective membranes: A review of palladium-based dense metal membranes. *Renew. Sustain. Energy Rev.* **2015**, *47*, 540–551. [CrossRef]
26. Deveau, N.D.; Ma, Y.H.; Datta, R. Beyond Sieverts' law: A comprehensive microkinetic model of hydrogen permeation in dense metal membranes. *J. Membr. Sci.* **2013**, *437*, 298–311. [CrossRef]
27. Adhikari, S.; Fernando, S. Hydrogen Membrane Separation Techniques. *Ind. Eng. Chem. Res.* **2006**, *45*, 875–881. [CrossRef]
28. Rahimpour, M.R.; Samimi, F.; Babapoor, A.; Tohidian, T.; Mohebi, S. Palladium membranes applications in reaction systems for hydrogen separation and purification: A review. *Chem. Eng. Process. Process Intensif.* **2017**, *121*, 24–49. [CrossRef]
29. Deville, H.; Troost, L. Sur la permeabilitè du fer a haute temperature. *Comptes Rendus* **1863**, *57*, 965–967.
30. Deville, H. Note sur le passage des gaz au travers des corps solides homogènes. *Comptes Rendus* **1864**, *59*, 102.
31. Thomas, G. On the absorption and dialytic separation of gases by colloid septa. *Philos. Trans. R. Soc. A* **1866**, *156*, 399–439.
32. Dunbar, Z.W.; Lee, I.C. Effects of elevated temperatures and contaminated hydrogen gas mixtures on novel ultrathin palladium composite membranes. *Int. J. Hydrog. Energy* **2017**, *42*, 29310–29319. [CrossRef]
33. Yun, S.; Oyama, S.T. Correlations in palladium membranes for hydrogen separation: A review. *J. Membr. Sci.* **2011**, *375*, 28–45. [CrossRef]
34. Helmi, A.; Gallucci, F.; van Sint Annaland, M. Resource scarcity in palladium membrane applications for carbon capture in integrated gasification combined cycle units. *Int. J. Hydrog. Energy* **2014**, *39*, 10498–10506. [CrossRef]
35. Jayaraman, V.; Lin, Y.S. Synthesis and hydrogen permeation properties of ultrathin palladium-silver alloy membranes. *J. Membr. Sci.* **1995**, *104*, 251–262. [CrossRef]
36. Yun, S.; Ko, J.H.; Oyama, S.T. Ultrathin palladium membranes prepared by a novel electric field assisted activation. *J. Membr. Sci.* **2011**, *369*, 482–489. [CrossRef]
37. Maneerung, T.; Hidajat, K.; Kawi, S. Ultra-thin (<1 μm) internally-coated Pd–Ag alloy hollow fiber membrane with superior thermal stability and durability for high temperature H_2 separation. *J. Membr. Sci.* **2014**, *452*, 127–142. [CrossRef]
38. Melendez, J.; Fernandez, E.; Gallucci, F.; van Annaland, M.; Arias, P.L.; Tanaka, D.A.P. Preparation and characterization of ceramic supported ultra-thin (~1 μm) Pd-Ag membranes. *J. Membr. Sci.* **2017**, *528*, 12–23. [CrossRef]
39. Feitosa, J.L.C.S.; da Cruz, A.G.B.; Souza, A.C.; Duda, F.P. Stress effects on hydrogen permeation through tubular multilayer membranes: Modeling and simulation. *Int. J. Hydrog. Energy* **2015**, *40*, 17031–17037. [CrossRef]
40. Dittmar, B.; Behrens, A.; Schödel, N.; Rüttinger, M.; Franco, T.; Straczewski, G.; Dittmeyer, R. Methane steam reforming operation and thermal stability of new porous metal supported tubular palladium composite membranes. *Int. J. Hydrog. Energy* **2013**, *38*, 8759–8771. [CrossRef]
41. Melendez, J.; de Nooijer, N.; Coenen, K.; Fernandez, E.; Viviente, J.L.; van Sint Annaland, M.; Arias, P.L.; Tanaka, D.A.P.; Gallucci, F. Effect of Au addition on hydrogen permeation and the resistance to H_2S on Pd-Ag alloy membranes. *J. Membr. Sci.* **2017**, *542*, 329–341. [CrossRef]
42. El Hawa, H.W.A.; Lundin, S.-T.B.; Paglieri, S.N.; Harale, A.; Way, J.D. The influence of heat treatment on the thermal stability of Pd composite membranes. *J. Membr. Sci.* **2015**, *494*, 113–120. [CrossRef]
43. Lundin, S.-T.B.; Patki, N.S.; Fuerst, T.F.; Wolden, C.A.; Way, J.D. Inhibition of hydrogen flux in palladium membranes by pressure–induced restructuring of the membrane surface. *J. Membr. Sci.* **2017**, *535*, 70–78. [CrossRef]
44. Livshits, A.I. The hydrogen transport through the metal alloy membranes with a spatial variation of the alloy composition: Potential diffusion and enhanced permeation. *Int. J. Hydrog. Energy* **2017**, *42*, 13111–13119. [CrossRef]

45. Fernandez, E.; Helmi, A.; Medrano, J.A.; Coenen, K.; Arratibel, A.; Melendez, J.; de Nooijer, N.C.A.; Spallina, V.; Viviente, J.L.; Zuñiga, J.; et al. Palladium based membranes and membrane reactors for hydrogen production and purification: An overview of research activities at Tecnalia and TU/e. *Int. J. Hydrog. Energy* **2017**, *42*, 13763–13776. [CrossRef]
46. Guo, Y.; Wu, H.; Fan, X.; Zhou, L.; Chen, Q. Palladium composite membrane fabricated on rough porous alumina tube without intermediate layer for hydrogen separation. *Int. J. Hydrog. Energy* **2017**, *42*, 9958–9965. [CrossRef]
47. Nayebossadri, S.; Fletcher, S.; Speight, J.D.; Book, D. Hydrogen permeation through porous stainless steel for palladium-based composite porous membranes. *J. Membr. Sci.* **2016**, *515*, 22–28. [CrossRef]
48. Itoh, N.; Suga, E.; Sato, T. Composite palladium membrane prepared by introducing metallic glue and its high durability below the critical temperature. *Sep. Purif. Technol.* **2014**, *121*, 46–53. [CrossRef]
49. Calles, J.A.; Sanz, R.; Alique, D.; Furones, L.; Marín, P.; Ordoñez, S. Influence of the selective layer morphology on the permeation properties for Pd-PSS composite membranes prepared by electroless pore-plating: Experimental and modeling study. *Sep. Purif. Technol.* **2018**, *194*, 10–18. [CrossRef]
50. Tanaka, D.A.P.; Tanco, M.A.L.; Okazaki, J.; Wakui, Y.; Mizukami, F.; Suzuki, T.M. Preparation of "pore-fill" type Pd–YSZ–γ-Al_2O_3 composite membrane supported on α-Al_2O_3 tube for hydrogen separation. *J. Membr. Sci.* **2008**, *320*, 436–441. [CrossRef]
51. Sanz, R.; Calles, J.A.; Alique, D.; Furones, L. New synthesis method of Pd membranes over tubular {PSS} supports via "pore-plating" for hydrogen separation processes. *Int. J. Hydrog. Energy* **2012**, *37*, 18476–18485. [CrossRef]
52. Pacheco, D.A.; Llosa, M.A.; Niwa, S.; Wakui, Y.; Mizukami, F.; Namba, T.; Suzuki, T.M. Preparation of palladium and silver alloy membrane on a porous α-alumina tube via simultaneous electroless plating. *J. Membr. Sci.* **2005**, *247*, 21–27. [CrossRef]
53. Zeng, G.; Goldbach, A.; Shi, L.; Xu, H. On alloying and low-temperature stability of thin, supported PdAg membranes. *Int. J. Hydrog. Energy* **2012**, *37*, 6012–6019. [CrossRef]
54. Zhao, L.; Goldbach, A.; Bao, C.; Xu, H. Sulfur inhibition of PdCu membranes in the presence of external mass flow resistance. *J. Membr. Sci.* **2015**, *496*, 301–309. [CrossRef]
55. Tarditi, A.M.; Braun, F.; Cornaglia, L.M. Novel PdAgCu ternary alloy: Hydrogen permeation and surface properties. *Appl. Surf. Sci.* **2011**, *257*, 6626–6635. [CrossRef]
56. Jia, H.; Wu, P.; Zeng, G.; Salas-Colera, E.; Serrano, A.; Castro, G.R.; Xu, H.; Sun, C.; Goldbach, A. High-temperature stability of Pd alloy membranes containing Cu and Au. *J. Membr. Sci.* **2017**, *544*, 151–160. [CrossRef]
57. Tosti, S. Supported and laminated Pd-based metallic membranes. *Int. J. Hydrog. Energy* **2003**, *28*, 1445–1454. [CrossRef]
58. Zhang, K.; Gade, S.K.; Hatlevik, Ø.; Way, J.D. A sorption rate hypothesis for the increase in H_2 permeability of palladium-silver (Pd–Ag) membranes caused by air oxidation. *Int. J. Hydrog. Energy* **2012**, *37*, 583–593. [CrossRef]
59. Tosti, S.; Bettinali, L.; Violante, V. Rolled thin Pd and Pd–Ag membranes for hydrogen separation and production. *Int. J. Hydrog. Energy* **2000**, *25*, 319–325. [CrossRef]
60. Tosti, S.; Bettinali, L.; Castelli, S.; Sarto, F.; Scaglione, S.; Violante, V. Sputtered, electroless and rolled palladium-ceramic membranes. *J. Membr. Sci.* **2002**, *196*, 241–249. [CrossRef]
61. Peters, T.A.; Kaleta, T.; Stange, M.; Bredesen, R. Development of thin binary and ternary Pd-based alloy membranes for use in hydrogen production. *J. Membr. Sci.* **2011**, *383*, 124–134. [CrossRef]
62. Calles, J.A.; Sanz, R.; Alique, D. Influence of the type of siliceous material used as intermediate layer in the preparation of hydrogen selective palladium composite membranes over a porous stainless steel support. *Int. J. Hydrog. Energy* **2012**, *37*, 6030–6042. [CrossRef]
63. Peters, T.A.; Stange, M.; Bredesen, R. 2-Fabrication of palladium-based membranes by magnetron sputtering. In *Palladium Membrane Technology Hydrogen Production, Carbon Capture and Other Application*; Doukelis, A., Panopoulos, K., Koumanakos, A., Kakaras, E., Eds.; Woodhead Publishing: Sawston, UK, 2015; pp. 25–41.
64. Huang, L.; Chen, C.S.; He, Z.D.; Peng, D.K.; Meng, G.Y. Palladium membranes supported on porous ceramics prepared by chemical vapor deposition. *Thin Solid Films* **1997**, *302*, 98–101. [CrossRef]
65. Jun, C.-S.; Lee, K.-H. Palladium and palladium alloy composite membranes prepared by metal-organic chemical vapor deposition method (cold-wall). *J. Membr. Sci.* **2000**, *176*, 121–130. [CrossRef]

66. Lee, S.M.; Xu, N.; Kim, S.S.; Li, A.; Grace, J.R.; Lim, C.J.; Boyd, T.; Ryi, S.-K.; Susdorf, A.; Schaadt, A. Palladium/ruthenium composite membrane for hydrogen separation from the off-gas of solar cell production via chemical vapor deposition. *J. Membr. Sci.* **2017**, *541*, 1–8. [CrossRef]
67. Chen, S.C.; Tu, G.C.; Hung, C.C.Y.; Huang, C.A.; Rei, M.H. Preparation of palladium membrane by electroplating on AISI 316L porous stainless steel supports and its use for methanol steam reformer. *J. Membr. Sci.* **2008**, *314*, 5–14. [CrossRef]
68. Chen, C.-H.; Huang, Y.-R.; Liu, C.-W.; Wang, K.-W. Preparation and modification of PdAg membranes by electroless and electroplating process for hydrogen separation. *Thin Solid Films* **2016**, *618*, 189–194. [CrossRef]
69. Yoshii, K.; Oshino, Y.; Tachikawa, N.; Toshima, K.; Katayama, Y. Electrodeposition of palladium from palladium(II) acetylacetonate in an amide-type ionic liquid. *Electrochem. Commun.* **2015**, *52*, 21–24. [CrossRef]
70. Sanz, R.; Calles, J.A.; Alique, D.; Furones, L.; Ordóñez, S.; Marín, P.; Corengia, P.; Fernandez, E. Preparation, testing and modelling of a hydrogen selective Pd/YSZ/SS composite membrane. *Int. J. Hydrog. Energy* **2011**, *36*, 15783–15793. [CrossRef]
71. Alique, D.; Imperatore, M.; Sanz, R.; Calles, J.A.; Baschetti, M.G. Hydrogen permeation in composite Pd-membranes prepared by conventional electroless plating and electroless pore-plating alternatives over ceramic and metallic supports. *Int. J. Hydrog. Energy* **2016**, *41*, 19430–19438. [CrossRef]
72. Ryi, S.-K.; Lee, S.-W.; Oh, D.-K.; Seo, B.-S.; Park, J.-W.; Park, J.-S.; Lee, D.-W.; Kim, S.S. Electroless plating of Pd after shielding the bottom of planar porous stainless steel for a highly stable hydrogen selective membrane. *J. Membr. Sci.* **2014**, *467*, 93–99. [CrossRef]
73. Sari, R.; Yaakob, Z.; Ismail, M.; Daud, W.R.W.; Hakim, L. Palladium-alumina composite membrane for hydrogen separator fabricated by combined sol-gel and electroless plating technique. *Ceram. Int.* **2013**, *39*, 3211–3219. [CrossRef]
74. Tarditi, A.M.; Bosko, M.L.; Cornaglia, L.M. Electroless Plating of Pd Binary and Ternary Alloys and Surface Characteristics for Application in Hydrogen Separation A2—Hashmi, MSJ BT. In *Comprehensive Materials Finishing*; Elsevier: Oxford, UK, 2017; pp. 1–24.
75. Zhang, B. Chapter 1—History–From the Discovery of Electroless Plating to the Present BT. In *Amorphous and Nano Alloys Electroless Depositions*; Elsevier: Oxford, UK, 2016; pp. 3–48.
76. Tosti, S.; Fabbricino, M.; Pontoni, L.; Palma, V.; Ruocco, C. Catalytic reforming of olive mill wastewater and methane in a Pd-membrane reactor. *Int. J. Hydrog. Energy* **2016**, *41*, 5465–5474. [CrossRef]
77. Tosti, S.; Adrover, A.; Basile, A.; Camilli, V.; Chiappetta, G.; Violante, V. Characterization of thin wall Pd–Ag rolled membranes. *Int. J. Hydrog. Energy* **2003**, *28*, 105–112. [CrossRef]
78. Calles, J.A.; Sanz, R.; Alique, D.; Furones, L. Thermal stability and effect of typical water gas shift reactant composition on H_2 permeability through a Pd-YSZ-PSS composite membrane. *Int. J. Hydrog. Energy* **2014**, *39*, 1398–1409. [CrossRef]
79. Sanz, R.; Calles, J.A.; Alique, D.; Furones, L. H_2 production via water gas shift in a composite Pd membrane reactor prepared by the pore-plating method. *Int. J. Hydrog. Energy* **2014**, *39*, 4739–4748. [CrossRef]
80. Sanz, R.; Calles, J.A.; Alique, D.; Furones, L.; Ordóñez, S.; Marín, P. Hydrogen production in a Pore-Plated Pd-membrane reactor: Experimental analysis and model validation for the Water Gas Shift reaction. *Int. J. Hydrog. Energy* **2015**, *40*, 3472–3484. [CrossRef]
81. Masuda, H.; Nishio, K.; Baba, N. Preparation of microporous metal membrane using two-step replication of interconnected structure of porous glass. *J. Mater. Sci. Lett.* **1994**, *13*, 338–340. [CrossRef]
82. Cheng, Y.S.; Yeung, K.L. Palladium-silver composite membranes by electroless plating technique. *J. Membr. Sci.* **1999**, *158*, 127–141. [CrossRef]
83. Augustine, A.S.; Mardilovich, I.P.; Kazantzis, N.K.; Ma, Y.H. Durability of PSS-supported Pd-membranes under mixed gas and water-gas shift conditions. *J. Membr. Sci.* **2012**, *415–416*, 213–220. [CrossRef]
84. Roa, F.; Way, J.D.; McCormick, R.L.; Paglieri, S.N. Preparation and characterization of Pd-Cu composite membranes for hydrogen separation. *Chem. Eng. J.* **2003**, *93*, 11–22. [CrossRef]
85. Lewis, A.E.; Kershner, D.C.; Paglieri, S.N.; Slepicka, M.J.; Way, J.D. Pd-Pt/YSZ composite membranes for hydrogen separation from synthetic water-gas shift streams. *J. Membr. Sci.* **2013**, *437*, 257–264. [CrossRef]
86. Kong, S.Y.; Kim, D.H.; Henkensmeier, D.; Kim, H.J.; Ham, H.C.; Han, J.; Yoon, S.P.; Yoon, C.W.; Choi, S.H. Ultrathin layered Pd/PBI–HFA composite membranes for hydrogen separation. *Sep. Purif. Technol.* **2017**, *179*, 486–493. [CrossRef]

87. Kim, D.H.; Kong, S.Y.; Lee, G.; Yoon, C.W.; Ham, H.C.; Han, J.; Song, K.H.; Henkensmeier, D.; Choi, S.H. Effect of PBI-HFA surface treatments on Pd/PBI-HFA composite gas separation membranes. *Int. J. Hydrog. Energy* **2017**, *42*, 22915–22924. [CrossRef]
88. Kumar, R.; Kamakshi; Kumar, M.; Awasthi, K. Selective deposition of Pd nanoparticles in porous PET membrane for hydrogen separation. *Int. J. Hydrog. Energy* **2017**, *42*, 15203–15210. [CrossRef]
89. Alique, D. Processing and Characterization of Coating and Thin Film Materials. In *Advanced Ceramic and Metallic Coating and Thin Film Materials Energy and Environmental Applications*; Zhang, J., Jung, Y., Eds.; Springer: Cham, Switzerland, 2018.
90. Plazaola, A.A.; Tanaka, D.P.; Van Sint Annaland, M.; Gallucci, F. Recent Advances in Pd-Based Membranes for Membrane Reactors. *Molecules* **2017**, *22*, 51. [CrossRef] [PubMed]
91. Li, H.; Caravella, A.; Xu, H.Y. Recent progress in Pd-based composite membranes. *J. Mater. Chem. A* **2016**, *4*, 14069–14094. [CrossRef]
92. Mardilovich, I.P.; Engwall, E.; Ma, Y.H. Dependence of hydrogen flux on the pore size and plating surface topology of asymmetric Pd-porous stainless steel membranes. *Desalination* **2002**, *144*, 85–89. [CrossRef]
93. Tarditi, A.; Gerboni, C.; Cornaglia, L. PdAu membranes supported on top of vacuum-assisted ZrO_2-modified porous stainless steel substrates. *J. Membr. Sci.* **2013**, *428*, 1–10. [CrossRef]
94. Chen, C.; Ma, Y.H. The effect of H_2S on the performance of Pd and Pd/Au composite membrane. *J. Membr. Sci.* **2010**, *362*, 535–544. [CrossRef]
95. Fernandez, E.; Helmi, A.; Coenen, K.; Melendez, J.; Luis, J.; Alfredo, D.; Tanaka, P.; van Sint, M.; Gallucci, F. Development of thin Pd-Ag supported membranes for fluidized bed membrane reactors including WGS related gases. *Int. J. Hydrog. Energy* **2014**, *40*, 3506–3519. [CrossRef]
96. Shackelford, J.F.; William, A. *Materials Science Engineering Hand Book*, 3rd ed.; CRC Press: Boca Raton, FL, USA, 2001.
97. Ryi, S.K.; Xu, N.; Li, A.; Lim, C.J.; Grace, J.R. Electroless Pd membrane deposition on alumina modified porous Hastelloy substrate with EDTA-free bath. *Int. J. Hydrog. Energy* **2010**, *35*, 2328–2335. [CrossRef]
98. Fernandez, E.; Medrano, J.A.; Melendez, J.; Parco, M.; Viviente, J.L.; van Sint Annaland, M.; Gallucci, F.; Tanaka, D.A.P. Preparation and characterization of metallic supported thin Pd-Ag membranes for hydrogen separation. *Chem. Eng. J.* **2016**, *305*, 182–190. [CrossRef]
99. Ryi, S.K.; Ahn, H.S.; Park, J.S.; Kim, D.W. Pd-Cu alloy membrane deposited on CeO_2 modified porous nickel support for hydrogen separation. *Int. J. Hydrog. Energy* **2014**, *39*, 4698–4703. [CrossRef]
100. Zhang, D.; Zhou, S.; Fan, Y.; Xu, N.; He, Y. Preparation of dense Pd composite membranes on porous Ti–Al alloy supports by electroless plating. *J. Membr. Sci.* **2012**, *387–388*, 24–29. [CrossRef]
101. Kim, S.S.; Xu, N.; Li, A.; Grace, J.R.; Lim, C.J.; Ryi, S.K. Development of a new porous metal support based on nickel and its application for Pd based composite membranes. *Int. J. Hydrog. Energy* **2015**, *40*, 3520–3527. [CrossRef]
102. Kulprathipanja, A.; Alptekin, G.O.; Falconer, J.L.; Way, J.D. Pd and Pd-Cu membranes: Inhibition of H_2 permeation by H_2S. *J. Membr. Sci.* **2005**, *254*, 49–62. [CrossRef]
103. Okazaki, J.; Tanaka, D.A.P.; Tanco, M.A.L.; Wakui, Y.; Ikeda, T.; Mizukami, F.; Suzuki, T.M. Preparation and Hydrogen Permeation Properties of Thin Pd-Au Alloy Membranes Supported on Porous α-Alumina Tube. *Mater. Trans.* **2008**, *49*, 449–452. [CrossRef]
104. Wang, W.P.; Thomas, S.; Zhang, X.L.; Pan, X.L.; Yang, W.S.; Xiong, G.X. H_2/N_2 gaseous mixture separation in dense Pd/α-Al_2O_3 hollow fiber membranes: Experimental and simulation studies. *Sep. Purif. Technol.* **2006**, *52*, 177–185. [CrossRef]
105. Nair, B.K.R.; Choi, J.; Harold, M.P. Electroless plating and permeation features of Pd and Pd/Ag hollow fiber composite membranes. *J. Membr. Sci.* **2007**, *288*, 67–84. [CrossRef]
106. Incelli, M.; Santucci, A.; Tosti, S.; Sansovini, M.; Carlini, M. Heavy water decontamination tests through a Pd-Ag membrane reactor: Water Gas Shift and Isotopic Swamping performances. *Fusion Eng. Des.* **2016**, *124*, 692–695. [CrossRef]
107. Yu, C.; Xu, H. An efficient palladium membrane reactor to increase the yield of styrene in ethylbenzene dehydrogenation. *Sep. Purif. Technol.* **2011**, *78*, 249–252. [CrossRef]
108. Brunetti, A.; Caravella, A.; Fernandez, E.; Tanaka, D.A.P.; Gallucci, F.; Drioli, E.; Curcio, E.; Viviente, J.L.; Barbieri, G. Syngas upgrading in a membrane reactor with thin Pd-alloy supported membrane. *Int. J. Hydrog. Energy* **2015**, *40*, 10883–10893. [CrossRef]

109. Rahman, M.A.; García-García, F.R.; Li, K. Development of a catalytic hollow fibre membrane microreactor as a microreformer unit for automotive application. *J. Membr. Sci.* **2012**, *390–391*, 68–75. [CrossRef]
110. Mardilovich, P.P.; She, Y.; Ma, Y.H.; Rei, M.-H. Defect-free palladium membranes on porous stainless-steel support. *AIChE J.* **1998**, *44*, 310–322. [CrossRef]
111. Li, A.; Grace, J.R.; Lim, C.J. Preparation of thin Pd-based composite membrane on planar metallic substrate: Part II. Preparation of membranes by electroless plating and characterization. *J. Membr. Sci.* **2007**, *306*, 159–165. [CrossRef]
112. Jayaraman, V.; Lin, Y.S.; Pakala, M.; Lin, R.Y. Fabrication of ultrathin metallic membranes on ceramic supports by sputter deposition. *J. Membr. Sci.* **1995**, *99*, 89–100. [CrossRef]
113. Ryi, S.-K.; Park, J.-S.; Kim, S.-H.; Kim, D.-W.; Cho, K.-I. Formation of a defect-free Pd–Cu–Ni ternary alloy membrane on a polished porous nickel support (PNS). *J. Membr. Sci.* **2008**, *318*, 346–354. [CrossRef]
114. Pinacci, P.; Drago, F. Influence of the support on permeation of palladium composite membranes in presence of sweep gas. *Catal. Today* **2012**, *193*, 186–193. [CrossRef]
115. Ryi, S.-K.; Park, J.-S.; Hwang, K.-R.; Lee, C.-B.; Lee, S.-W. Repair of Pd-based composite membrane by polishing treatment. *Int. J. Hydrog. Energy* **2011**, *36*, 13776–13780. [CrossRef]
116. Jemaa, N.; Shu, J.; Kaliaguine, S.; Grandjean, B.P.A. Thin Palladium Film Formation on Shot Peening Modified Porous Stainless Steel Substrates. *Ind. Eng. Chem. Res.* **1996**, *35*, 973–977. [CrossRef]
117. Huang, Y.; Dittmeyer, R. Preparation of thin palladium membranes on a porous support with rough surface. *J. Membr. Sci.* **2007**, *302*, 160–170. [CrossRef]
118. Huang, Y.; Li, X.; Fan, Y.; Xu, N. Palladium-based composite membranes: Principle, preparation and characterization. *Prog. Chem.* **2006**, *18*, 230–238.
119. Collins, J.P.; Way, J.D. Catalytic decomposition of ammonia in a membrane reactor. *J. Membr. Sci.* **1994**, *96*, 259–274. [CrossRef]
120. Huang, Y.; Shu, S.; Lu, Z.; Fan, Y. Characterization of the adhesion of thin palladium membranes supported on tubular porous ceramics. *Thin Solid Films* **2007**, *515*, 5233–5240. [CrossRef]
121. Coronas, J.; Santamaría, J. Catalytic reactors based on porous ceramic membranes. *Catal. Today* **1999**, *51*, 377–389. [CrossRef]
122. Tong, J.; Matsumura, Y.; Suda, H.; Haraya, K. Thin and dense Pd/CeO_2/MPSS composite membrane for hydrogen separation and steam reforming of methane. *Sep. Purif. Technol.* **2005**, *46*, 1–10. [CrossRef]
123. Qiao, A.; Zhang, K.; Tian, Y.; Xie, L.; Luo, H.; Lin, Y.S.; Li, Y. Hydrogen separation through palladium-copper membranes on porous stainless steel with sol-gel derived ceria as diffusion barrier. *Fuel* **2010**, *89*, 1274–1279. [CrossRef]
124. Wang, D.; Tong, J.; Xu, H.; Matsumura, Y. Preparation of palladium membrane over porous stainless steel tube modified with zirconium oxide. *Catal. Today* **2004**, *93–95*, 689–693. [CrossRef]
125. Gao, H.; Lin, J.Y.S.; Li, Y.; Zhang, B. Electroless plating synthesis, characterization and permeation properties of Pd–Cu membranes supported on ZrO_2 modified porous stainless steel. *J. Membr. Sci.* **2005**, *265*, 142–152. [CrossRef]
126. Lee, C.-B.; Lee, S.-W.; Park, J.-S.; Ryi, S.-K.; Lee, D.-W.; Hwang, K.-R.; Kim, S.-H. Ceramics used as intermetallic diffusion barriers in Pd-based composite membranes sputtered on porous nickel supports. *J. Alloys Compd.* **2013**, *578*, 425–430. [CrossRef]
127. Zhang, K.; Gao, H.; Rui, Z.; Liu, P.; Li, Y.; Lin, Y.S. High-Temperature Stability of Palladium Membranes on Porous Metal Supports with Different Intermediate Layers. *Ind. Eng. Chem. Res.* **2009**, *48*, 1880–1886. [CrossRef]
128. Yepes, D.; Cornaglia, L.M.; Irusta, S.; Lombardo, E.A. Different oxides used as diffusion barriers in composite hydrogen permeable membranes. *J. Membr. Sci.* **2006**, *274*, 92–101. [CrossRef]
129. Li, A.; Grace, J.R.; Lim, C.J. Preparation of thin Pd-based composite membrane on planar metallic substrate: Part I: Pre-treatment of porous stainless steel substrate. *J. Membr. Sci.* **2007**, *298*, 175–181. [CrossRef]
130. Broglia, M.; Pinacci, P.; Radaelli, M.; Bottino, A.; Capannelli, G.; Comite, A.; Vanacore, G.; Zani, M. Synthesis and characterization of Pd membranes on alumina-modified porous stainless steel supports. *Desalination* **2009**, *245*, 508–515. [CrossRef]
131. Chi, Y.-H.; Yen, P.-S.; Jeng, M.-S.; Ko, S.-T.; Lee, T.-C. Preparation of thin Pd membrane on porous stainless steel tubes modified by a two-step method. *Int. J. Hydrog. Energy* **2010**, *35*, 6303–6310. [CrossRef]

132. Nam, S.-E.; Lee, K.-H. Hydrogen separation by Pd alloy composite membranes: Introduction of diffusion barrier. *J. Membr. Sci.* **2001**, *192*, 177–185. [CrossRef]
133. Van Gestel, T.; Hauler, F.; Bram, M.; Meulenberg, W.A.; Buchkremer, H.P.; van Gestel, T.; Hauler, F.; Bram, M.; Meulenberg, W.A.; Buchkremer, H.P. Synthesis and characterization of hydrogen-selective sol–gel SiO_2 membranes supported on ceramic and stainless steel supports. *Sep. Purif. Technol.* **2014**, *121*, 20–29. [CrossRef]
134. Kanezashi, M.; Fuchigami, D.; Yoshioka, T.; Tsuru, T. Control of Pd dispersion in sol–gel-derived amorphous silica membranes for hydrogen separation at high temperatures. *J. Membr. Sci.* **2013**, *439*, 78–86. [CrossRef]
135. Zheng, L.; Li, H.; Xu, T.; Bao, F.; Xu, H. Defect size analysis approach combined with silicate gel/ceramic particles for defect repair of Pd composite membranes. *Int. J. Hydrog. Energy* **2016**, *41*, 18522–18532. [CrossRef]
136. Bosko, M.L.; Ojeda, F.; Lombardo, E.A.; Cornaglia, L.M. NaA zeolite as an effective diffusion barrier in composite Pd/PSS membranes. *J. Membr. Sci.* **2009**, *331*, 57–65. [CrossRef]
137. Mobarake, M.D.; Jafari, P.; Irani, M. Preparation of Pd-based membranes on Pd/TiO_2 modified NaX/PSS substrate for hydrogen separation: Design and optimization. *Microporous Mesoporous Mater.* **2016**, *226*, 369–377. [CrossRef]
138. Yu, J.; Qi, C.; Zhang, J.; Bao, C.; Xu, H. Synthesis of a zeolite membrane as a protective layer on a metallic Pd composite membrane for hydrogen purification. *J. Mater. Chem. A* **2015**, *3*, 5000–5006. [CrossRef]
139. Sato, K.; Natsui, M.; Hasegawa, Y. Preparation of Double Layer Membrane Combined with Palladium Metal and FAU Zeolite for Catalytic Membrane Reactor. *Mater. Trans.* **2015**, *56*, 473–478. [CrossRef]
140. Wang, X.; Tan, X.; Meng, B.; Zhang, X.; Liang, Q.; Pan, H.; Liu, S. TS-1 zeolite as an effective diffusion barrier for highly stable Pd membrane supported on macroporous α-Al_2O_3 tube. *RSC Adv.* **2013**, *3*, 4821–4834. [CrossRef]
141. Abate, S.; Díaz, U.; Prieto, A.; Gentiluomo, S.; Palomino, M.; Perathoner, S.; Corma, A.; Centi, G. Influence of Zeolite Protective Overlayer on the Performances of Pd Thin Film Membrane on Tubular Asymmetric Alumina Supports. *Ind. Eng. Chem. Res.* **2016**, *55*, 4948–4959. [CrossRef]
142. Ma, Y.H.; Mardilovich, P.P.; She, Y. Hydrogen Gas-Extraction Module and Method of Fabrication. U.S. Patent 6152987 A, 28 November 2000.
143. Guazzone, F.; Engwall, E.E.; Ma, Y.H. Effects of surface activity, defects and mass transfer on hydrogen permeance and n-value in composite palladium-porous stainless steel membranes. *Catal. Today* **2006**, *118*, 24–31. [CrossRef]
144. Mateos-Pedrero, C.; Soria, M.A.; Rodríguez-Ramos, I.; Guerrero-Ruiz, A. Modifications of porous stainless steel previous to the synthesis of Pd membranes. *Stud. Surf. Sci. Catal.* **2010**, *175*, 779–783. [CrossRef]
145. Nam, S.-E.; Lee, K.-H. Preparation and Characterization of Palladium Alloy Composite Membranes with a Diffusion Barrier for Hydrogen Separation. *Ind. Eng. Chem. Res.* **2005**, *44*, 100–105. [CrossRef]
146. Ayturk, M.E.; Mardilovich, I.P.; Engwall, E.E.; Ma, Y.H. Synthesis of composite Pd-porous stainless steel (PSS) membranes with a Pd/Ag intermetallic diffusion barrier. *J. Membr. Sci.* **2006**, *285*, 385–394. [CrossRef]
147. Lee, J.-H.; Han, J.-Y.; Kim, K.-M.; Ryi, S.-K.; Kim, D.-W. Development of homogeneous Pd–Ag alloy membrane formed on porous stainless steel by multi-layered films and Ag-upfilling heat treatment. *J. Membr. Sci.* **2015**, *492*, 242–248. [CrossRef]
148. Pujari, M.; Agarwal, A.; Uppaluri, R.; Verma, A. Role of electroless nickel diffusion barrier on the combinatorial plating characteristics of dense Pd/Ni/PSS composite membranes. *Appl. Surf. Sci.* **2014**, *305*, 658–664. [CrossRef]
149. Tong, J.; Suda, H.; Haraya, K.; Matsumura, Y. A novel method for the preparation of thin dense Pd membrane on macroporous stainless steel tube filter. *J. Memb. Sci.* **2005**, *260*, 10–18. [CrossRef]
150. Tong, J.; Su, L.; Haraya, K.; Suda, H. Thin Pd membrane on α-Al_2O_3 hollow fiber substrate without any interlayer by electroless plating combined with embedding Pd catalyst in polymer template. *J. Membr. Sci.* **2008**, *310*, 93–101. [CrossRef]
151. Hu, X.; Chen, W.; Huang, Y. Fabrication of Pd/ceramic membranes for hydrogen separation based on low-cost macroporous ceramics with pencil coating. *Int. J. Hydrog. Energy* **2010**, *35*, 7803–7808. [CrossRef]
152. Zhao, H.-B.; Pflanz, K.; Gu, J.-H.; Li, A.-W.; Stroh, N.; Brunner, H.; Xiong, G.-X. Preparation of palladium composite membranes by modified electroless plating procedure. *J. Membr. Sci.* **1998**, *142*, 147–157. [CrossRef]
153. Bottino, A.; Broglia, M.; Capannelli, G.; Comite, A.; Pinacci, P.; Scrignari, M.; Azzurri, F. Sol-gel synthesis of thin alumina layers on porous stainless steel supports for high temperature palladium membranes. *Int. J. Hydrog. Energy* **2014**, *39*, 4717–4724. [CrossRef]

154. Brenner, A.; Riddell, G.E. Nickel plating on steel by chemical reduction. *J. Res. Natl. Bur. Stand.* **1946**, *37*, 31–34. [CrossRef]
155. Zornoza, B.; Casado, C.; Navajas, A. Chapter 11—Advances in Hydrogen Separation and Purification with Membrane Technology. In *Renewable Hydrogen Technologies*; Gandía, L.M., Arzamendi, G., Diéguez, P.M., Eds.; Elsevier: Amsterdam, The Netherlands, 2013; pp. 245–268.
156. De Falco, M.; Iaquaniello, G.; Palo, E.; Cucchiella, B.; Palma, V.; Ciambelli, P. 11—Palladium-based membranes for hydrogen separation: Preparation, economic analysis and coupling with a water gas shift reactor. In *Handbook of Membrane Reactors*; Basile, A., Ed.; Woodhead Publishing: Sawston, UK, 2013; pp. 456–486.
157. Den Exter, M.J. 3—The use of electroless plating as a deposition technology in the fabrication of palladium-based membranes. In *Palladium Membrane Technology Hydrogen Production, Carbon Capture and Other Application*; Doukelis, A., Panopoulos, K., Koumanakos, A., Kakaras, E., Eds.; Woodhead Publishing: Sawston, UK, 2015; pp. 43–67.
158. Basile, A.; Tong, J.; Millet, P. 2—Inorganic membrane reactors for hydrogen production: An overview with particular emphasis on dense metallic membrane materials. In *Handbook of Membrane Reactors*; Basile, A., Ed.; Woodhead Publishing: Sawston, UK, 2013; pp. 42–148.
159. Cheng, Y.S.; Yeung, K.L. Effects of electroless plating chemistry on the synthesis of palladium membranes. *J. Membr. Sci.* **2001**, *182*, 195–203. [CrossRef]
160. Dogan, M.; Kilicarslan, S. Effects of process parameters on the synthesis of palladium membranes. *Nucl. Instrum. Methods Phys. Res. Sect. B* **2008**, *266*, 3458–3466. [CrossRef]
161. Yeung, K.L.; Christiansen, S.C.; Varma, A. Palladium composite membranes by electroless plating technique: Relationships between plating kinetics, film microstructure and membrane performance. *J. Membr. Sci.* **1999**, *159*, 107–122. [CrossRef]
162. Djokić, S.S. Fundamentals of Electroless Deposition BT. In *Reference Module in Chemistry, Molecular Sciences and Chemical Engineering*; Elsevier: Oxford, UK, 2016.
163. Mallory, G.O.; Hajdu, J.B. *Electroless Plating: Fundamentals and Applications*; American Electroplaters and Surface Finishers Society: Orlando, FL, USA, 1990.
164. Shu, J.; Grandjean, B.P.A.; Ghali, E.; Kaliaguine, S. Simultaneous deposition of Pd and Ag on porous stainless steel by electroless plating. *J. Membr. Sci.* **1993**, *77*, 181–195. [CrossRef]
165. Rothenberger, K.S.; Cugini, A.V.; Howard, B.H.; Killmeyer, R.P.; Ciocco, M.V.; Morreale, B.D.; Enick, R.M.; Bustamante, F.; Mardilovich, I.P.; Ma, Y.H. High pressure hydrogen permeance of porous stainless steel coated with a thin palladium film via electroless plating. *J. Membr. Sci.* **2004**, *244*, 55–68. [CrossRef]
166. Paglieri, S.N.; Way, J.D. Innovations in palladium membrane research. *Sep. Purif. Methods* **2002**, *31*, 1–169. [CrossRef]
167. Wei, L.; Yu, J.; Hu, X.; Wang, R.; Huang, Y. Effects of Sn residue on the high temperature stability of the H_2-permeable palladium membranes prepared by electroless plating on Al_2O_3 substrate after $SnCl_2$–$PdCl_2$ process: A case study. *Chin. J. Chem. Eng.* **2016**, *24*, 1154–1160. [CrossRef]
168. Chi, Y.-H.; Uan, J.-Y.; Lin, M.-C.; Lin, Y.-L.; Huang, J.-H. Preparation of a novel Pd/layered double hydroxide composite membrane for hydrogen filtration and characterization by thermal cycling. *Int. J. Hydrog. Energy* **2013**, *38*, 13734–13741. [CrossRef]
169. Guo, Y.; Jin, Y.; Wu, H.; Zhou, L.; Chen, Q.; Zhang, X.; Li, X. Preparation of palladium membrane on Pd/silicalite-1 zeolite particles modified macroporous alumina substrate for hydrogen separation. *Int. J. Hydrog. Energy* **2014**, *39*, 21044–21052. [CrossRef]
170. Seshimo, M.; Ozawa, M.; Sone, M.; Sakurai, M.; Kameyama, H. Fabrication of a novel Pd/γ-alumina graded membrane by electroless plating on nanoporous γ-alumina. *J. Membr. Sci.* **2008**, *324*, 181–187. [CrossRef]
171. Touyeras, F.; Hihn, J.Y.; Delalande, S.; Viennet, R.; Doche, M.L. Ultrasound influence on the activation step before electroless coating. *Ultrason. Sonochem.* **2003**, *10*, 363–368. [CrossRef]
172. Paglieri, S.N.; Foo, K.Y.; Way, J.D.; Collins, J.P.; Harper-Nixon, D.L. A New Preparation Technique for Pd/Alumina Membranes with Enhanced High-Temperature Stability. *Ind. Eng. Chem. Res.* **1999**, *38*, 1925–1936. [CrossRef]
173. Zhu, B.; Tang, C.H.; Xu, H.Y.; Su, D.S.; Zhang, J.; Li, H. Surface activation inspires high performance of ultra-thin Pd membrane for hydrogen separation. *J. Membr. Sci.* **2017**, *526*, 138–146. [CrossRef]

174. Uemiya, S.; Sato, N.; Ando, H.; Kikuchi, E. The water gas shift reaction assisted by a palladium membrane reactor. *Ind. Eng. Chem. Res.* **1991**, *30*, 585–589. [CrossRef]
175. Shi, Z.; Wu, S.; Szpunar, J.A.; Roshd, M. An observation of palladium membrane formation on a porous stainless steel substrate by electroless deposition. *J. Membr. Sci.* **2006**, *280*, 705–711. [CrossRef]
176. Zhang, X.; Xiong, G.; Yang, W. A modified electroless plating technique for thin dense palladium composite membranes with enhanced stability. *J. Membr. Sci.* **2008**, *314*, 226–237. [CrossRef]
177. Yeung, K.L.; Sebastian, J.M.; Varma, A. Novel preparation of Pd/Vycor composite membranes. *Catal. Today* **1995**, *25*, 231–236. [CrossRef]
178. Souleimanova, R.S.; Mukasyan, A.S.; Varma, A. Effects of osmosis on microstructure of Pd-composite membranes synthesized by electroless plating technique. *J. Membr. Sci.* **2000**, *166*, 249–257. [CrossRef]
179. Li, A.; Liang, W.; Hughes, R. Characterisation and permeation of palladium/stainless steel composite membranes. *J. Membr. Sci.* **1998**, *149*, 259–268. [CrossRef]
180. Tanaka, D.A.P.; Tanco, M.A.L.; Nagase, T.; Okazaki, J.; Wakui, Y.; Mizukami, F.; Suzuki, T.M. Fabrication of hydrogen-permeable composite membranes packed with palladium nanoparticles. *Adv. Mater.* **2006**, *18*, 630–632. [CrossRef]
181. Thoen, P.M.; Roa, F.; Way, J.D. High flux palladium–copper composite membranes for hydrogen separations. *Desalination* **2006**, *193*, 224–229. [CrossRef]
182. Gade, S.K.; Thoen, P.M.; Way, J.D. Unsupported palladium alloy foil membranes fabricated by electroless plating. *J. Membr. Sci.* **2008**, *316*, 112–118. [CrossRef]
183. Chi, Y.-H.; Lin, J.-J.; Lin, Y.-L.; Yang, C.-C.; Huang, J.-H. Influence of the rotation rate of porous stainless steel tubes on electroless palladium deposition. *J. Membr. Sci.* **2015**, *475*, 259–265. [CrossRef]
184. Zeng, G.; Goldbach, A.; Xu, H. Defect sealing in Pd membranes via point plating. *J. Membr. Sci.* **2009**, *328*, 6–10. [CrossRef]
185. Sanz, R.; Calles, J.A.; Ordóñez, S.; Marín, P.; Alique, D.; Furones, L. Modelling and simulation of permeation behaviour on Pd/PSS composite membranes prepared by "pore-plating" method. *J. Membr. Sci.* **2013**, *446*, 410–421. [CrossRef]
186. Hatlevik, Ø.; Gade, S.K.; Keeling, M.K.; Thoen, P.M.; Davidson, A.P.; Way, J.D. Palladium and palladium alloy membranes for hydrogen separation and production: History, fabrication strategies and current performance. *Sep. Purif. Technol.* **2010**, *73*, 59–64. [CrossRef]
187. Abu El Hawa, H.W.; Paglieri, S.N.; Morris, C.C.; Harale, A.; Way, J.D. Identification of thermally stable Pd-alloy composite membranes for high temperature applications. *J. Membr. Sci.* **2014**, *466*, 151–160. [CrossRef]
188. Conde, J.J.; Maroño, M.; Sánchez-Hervás, J.M. Pd-Based Membranes for Hydrogen Separation: Review of Alloying Elements and Their Influence on Membrane Properties. *Sep. Purif. Rev.* **2017**, *46*, 152–177. [CrossRef]
189. Shu, J.; Grandjean, B.P.A.; van Neste, A. Catalytic Palladium-Based Membrane Reactors: A Review. *Can. J. Chem. Eng.* **1991**. [CrossRef]
190. Lin, W.H.; Chang, H.F. Characterizations of Pd-Ag membrane prepared by sequential electroless deposition. *Surf. Coat. Technol.* **2005**, *194*, 157–166. [CrossRef]
191. Santucci, A.; Borgognoni, F.; Vadrucci, M.; Tosti, S. Testing of dense Pd-Ag tubes: Effect of pressure and membrane thickness on the hydrogen permeability. *J. Membr. Sci.* **2013**, *444*, 378–383. [CrossRef]
192. Medrano, J.A.; Fernandez, E.; Melendez, J.; Parco, M.; Tanaka, D.A.P.; van Sint Annaland, M.; Gallucci, F. Pd-based metallic supported membranes: High-temperature stability and fluidized bed reactor testing. *Int. J. Hydrog. Energy* **2016**, *41*, 8706–8718. [CrossRef]
193. Abu El Hawa, H.W.; Paglieri, S.N.; Morris, C.C.; Harale, A.; Way, J.D. Application of a Pd-Ru composite membrane to hydrogen production in a high temperature membrane reactor. *Sep. Purif. Technol.* **2015**, *147*, 388–397. [CrossRef]
194. Abu El Hawa, H.W.; Lundin, S.-T.B.; Patki, N.S.; Way, J.D. Steam methane reforming in a PdAu membrane reactor: Long-term assessment. *Int. J. Hydrog. Energy* **2016**, *41*, 10193–10201. [CrossRef]
195. Patki, N.S.; Lundin, S.T.; Way, J.D. Rapid annealing of sequentially plated Pd-Au composite membranes using high pressure hydrogen. *J. Membr. Sci.* **2016**, *513*, 197–205. [CrossRef]
196. Morreale, B.D.; Ciocco, M.V.; Howard, B.H.; Killmeyer, R.P.; Cugini, A.V.; Enick, R.M. Effect of hydrogen-sulfide on the hydrogen permeance of palladium–copper alloys at elevated temperatures. *J. Membr. Sci.* **2004**, *241*, 219–224. [CrossRef]

197. Kurokawa, H.; Yakabe, H.; Yasuda, I.; Peters, T.; Bredesen, R. Inhibition effect of CO on hydrogen permeability of Pd-Ag membrane applied in a microchannel module conFiguration. *Int. J. Hydrog. Energy* **2014**, *39*, 17201–17209. [CrossRef]
198. Peters, T.A.; Liron, O.; Tschentscher, R.; Sheintuch, M.; Bredesen, R. Investigation of Pd-based membranes in propane dehydrogenation (PDH) processes. *Chem. Eng. J.* **2016**, *305*, 191–200. [CrossRef]
199. Peters, T.A.; Kaleta, T.; Stange, M.; Bredesen, R. Hydrogen transport through a selection of thin Pd-alloy membranes: Membrane stability, H_2S inhibition and flux recovery in hydrogen and simulated WGS mixtures. *Catal. Today* **2012**, *193*, 8–19. [CrossRef]
200. Tarditi, A.M.; Imhoff, C.; Braun, F.; Miller, J.B.; Gellman, A.J.; Cornaglia, L. PdCuAu ternary alloy membranes: Hydrogen permeation properties in the presence of H_2S. *J. Membr. Sci.* **2015**, *479*, 246–255. [CrossRef]
201. Braun, F.; Tarditi, A.M.; Miller, J.B.; Cornaglia, L.M. Pd-based binary and ternary alloy membranes: Morphological and perm-selective characterization in the presence of H_2S. *J. Membr. Sci.* **2014**, *450*, 299–307. [CrossRef]
202. Bosko, M.L.; Miller, J.B.; Lombardo, E.A.; Gellman, A.J.; Cornaglia, L.M. Surface characterization of Pd-Ag composite membranes after annealing at various temperatures. *J. Membr. Sci.* **2011**, *369*, 267–276. [CrossRef]
203. Gao, H.; Lin, Y.S.; Li, Y.; Zhang, B. Chemical Stability and Its Improvement of Palladium-Based Metallic Membranes. *Ind. Eng. Chem. Res.* **2004**, *43*, 6920–6930. [CrossRef]
204. Okazaki, J.; Ikeda, T.; Tanaka, D.A.P.; Sato, K.; Suzuki, T.M.; Mizukami, F. An investigation of thermal stability of thin palladium-silver alloy membranes for high temperature hydrogen separation. *J. Membr. Sci.* **2011**, *366*, 212–219. [CrossRef]
205. Zhang, K.; Way, J.D. Palladium-copper membranes for hydrogen separation. *Sep. Purif. Technol.* **2017**, *186*, 39–44. [CrossRef]
206. Gade, S.K.; Payzant, E.A.; Park, H.J.; Thoen, P.M.; Way, J.D. The effects of fabrication and annealing on the structure and hydrogen permeation of Pd-Au binary alloy membranes. *J. Membr. Sci.* **2009**, *340*, 227–233. [CrossRef]
207. Bond, R.A.; Evans, J.; Harris, I.R.; Ross, D.K. Hydrogen isotope separation using palladium alloy membranes. In Proceedings of the 12th Symposium on Fusion Technology, Juelich, Germany, 13–17 September 1982; Elsevier: Amsterdam, The Netherlands, 1983; pp. 537–542.
208. Evans, J.; Harris, I.R.; Ross, D.K. A proposed method of hydrogen isotope separation using palladium alloy membranes. *J. Less Common Met.* **1983**, *89*, 407–414. [CrossRef]
209. Pérez, P.; Cornaglia, C.A.; Mendes, A.; Madeira, L.M.; Tosti, S. Surface effects and CO/CO_2 influence in the H_2 permeation through a Pd–Ag membrane: A comprehensive model. *Int. J. Hydrog. Energy* **2015**, *40*, 6566–6572. [CrossRef]
210. Augustine, A.S.; Ma, Y.H.; Kazantzis, N.K. High pressure palladium membrane reactor for the high temperature water-gas shift reaction. *Int. J. Hydrog. Energy* **2011**, *36*, 5350–5360. [CrossRef]
211. Bhandari, R.; Ma, Y.H. Pd-Ag membrane synthesis: The electroless and electro-plating conditions and their effect on the deposits morphology. *J. Membr. Sci.* **2009**, *334*, 50–63. [CrossRef]
212. Bosko, M.L.; Yepes, D.; Irusta, S.; Eloy, P.; Ruiz, P.; Lombardo, E.A.; Cornaglia, L.M. Characterization of Pd-Ag membranes after exposure to hydrogen flux at high temperatures. *J. Membr. Sci.* **2007**, *306*, 56–65. [CrossRef]
213. Bosko, M.L.; Múnera, J.F.; Lombardo, E.A.; Cornaglia, L.M. Dry reforming of methane in membrane reactors using Pd and Pd-Ag composite membranes on a NaA zeolite modified porous stainless steel support. *J. Membr. Sci.* **2010**, *364*, 17–26. [CrossRef]
214. Bosko, M.L.; Lombardo, E.A.; Cornaglia, L.M. The effect of electroless plating time on the morphology, alloy formation and H_2 transport properties of Pd-Ag composite membranes. *Int. J. Hydrog. Energy* **2011**, *36*, 4068–4078. [CrossRef]
215. Huerta, E.O. Corrosión y degradación de materiales, n.d. *Comput. Fraud Secur.* **2013**, *2000*, 3.
216. Roa, F.; Block, M.J.; Way, J.D. The influence of alloy composition on the H_2 flux of composite Pd-Cu membranes. *Desalination* **2002**, *147*, 411–416. [CrossRef]
217. Roa, F.; Way, J.D. The effect of air exposure on palladium-copper composite membranes. *Appl. Surf. Sci.* **2005**, *240*, 85–104. [CrossRef]

218. Lu, H.; Zhu, L.; Wang, W.; Yang, W.; Tong, J. Pd and Pd-Ni alloy composite membranes fabricated by electroless plating method on capillary α-Al_2O_3 substrates. *Int. J. Hydrog. Energy* **2015**, *40*, 3548–3556. [CrossRef]
219. Reboredo, J.C.; Ugolini, A. Quantile causality between gold commodity and gold stock prices. *Resour. Policy* **2017**, *53*, 56–63. [CrossRef]
220. Zhu, H.; Peng, C.; You, W. Quantile behaviour of cointegration between silver and gold prices. *Financ. Res. Lett.* **2016**, *19*, 119–125. [CrossRef]
221. Liu, C.; Hu, Z.; Li, Y.; Liu, S. Forecasting copper prices by decision tree learning. *Resour. Policy* **2017**, *52*, 427–434. [CrossRef]
222. Balcilar, M.; Hammoudeh, S.; Asaba, N.-A.F. A regime-dependent assessment of the information transmission dynamics between oil prices, precious metal prices and exchange rates. *Int. Rev. Econ. Financ.* **2015**, *40*, 72–89. [CrossRef]
223. Sumrunronnasak, S.; Tantayanon, S.; Kiatgamolchai, S. Influence of layer compositions and annealing conditions on complete formation of ternary PdAgCu alloys prepared by sequential electroless and electroplating methods. *Mater. Chem. Phys.* **2017**, *185*, 98–103. [CrossRef]
224. Tarditi, A.M.; Cornaglia, L.M. Novel PdAgCu ternary alloy as promising materials for hydrogen separation membranes: Synthesis and characterization. *Surf. Sci.* **2011**, *605*, 62–71. [CrossRef]

MDPI
St. Alban-Anlage 66
4052 Basel
Switzerland
Tel. +41 61 683 77 34
Fax +41 61 302 89 18
www.mdpi.com

Membranes Editorial Office
E-mail: membranes@mdpi.com
www.mdpi.com/journal/membranes

www.ingramcontent.com/pod-product-compliance
Lightning Source LLC
Chambersburg PA
CBHW041214220326
41597CB00033BA/5898